Oracle
性能优化与诊断案例精选

盖国强 李轶楠 主编
杨廷琨 熊军 李真旭 杨俊 黄宸宁 等编著

人民邮电出版社
北京

图书在版编目（CIP）数据

Oracle性能优化与诊断案例精选 / 盖国强，李轶楠主编 ; 杨廷琨等编著. -- 北京 : 人民邮电出版社，2016.11
ISBN 978-7-115-43827-0

Ⅰ. ①O… Ⅱ. ①盖… ②李… ③杨… Ⅲ. ①关系数据库系统 Ⅳ. ①TP311.138

中国版本图书馆CIP数据核字(2016)第244053号

内 容 提 要

Oracle 数据库是关系型数据库领域最重要的产品之一，在市场上赢得了广大用户和技术爱好者的信赖。在使用数据库产品的过程中，如何通过优化提升性能，通过诊断分析解决问题，一直是这个领域最重要的议题。随着 Oracle 12c 版本的深入使用和云技术的蓬勃发展，关于 12c 的版本特性也备受关注。

本书汇聚了 Oracle 数据库领域的一批技术专家，通过成长历程分享、最佳技术经验讲解、诊断和优化案例分析，将其独特的经验和视角分享给广大读者。作者希望通过这些各具特色的最佳实践分享，让读者找到对自己有益的学习方法和诊断优化思路。

本书适用于对 Oracle 数据库技术有一定的了解，希望深入学习的数据库从业人员，尤其是希望深入研究 Oracle 数据库的管理人员。

◆ 主　编　盖国强　李轶楠
　　编　著　杨廷琨　熊　军　李真旭　杨　俊　黄宸宁　等
　　责任编辑　杨海玲
　　责任印制　焦志炜

◆ 人民邮电出版社出版发行　北京市丰台区成寿寺路11号
　　邮编　100164　电子邮件　315@ptpress.com.cn
　　网址　http://www.ptpress.com.cn
　　北京艺辉印刷有限公司印刷

◆ 开本：800×1000　1/16
　　印张：23.5
　　字数：502 千字　　　　　　　2016年11月第1版
　　印数：1-2 000 册　　　　　　2016年11月北京第1次印刷

定价：79.00 元

读者服务热线：(010)81055410　印装质量热线：(010)81055316
反盗版热线：(010)81055315
广告经营许可证：京东工商广字第 8052 号

作者简介

盖国强　云和恩墨创始人

Oracle ACE 总监，中国地区首位 Oracle ACE，曾获评"2006年中国首届杰出数据库工程师"奖；拥有超过15年的数据库实施和顾问咨询经验，长于数据库诊断、性能调整与 SQL 优化等，尤其擅长 Oracle 内核技术；中国地区最著名的 Oracle 技术推广者之一，已编著出版 13 本 Oracle 专业书籍，其专著《深入解析 Oracle》《循序渐进 Oracle》等受到 Oracle 技术爱好者的广泛好评；2010 年，他与 Oracle ACE 总监张乐奕先生共同创立 ACOUG(中国 Oracle 用户组)，持续推动 Oracle 技术圈的地面活动与技术交流；2011 年，创建了云和恩墨，致力于为中国数据库用户提供专业的数据库服务。

个人博客 http://www.eygle.com。

张乐奕　云和恩墨副总经理

Oracle ACE 总监，ACOUG 联合创始人，云和恩墨的联合创始人，致力于通过不断的技术探索，帮助中国用户理解和接触新技术，推广数据库技术应用；曾先后任职于 UT 斯达康、电讯盈科、甲骨文等知名企业，担任 DBA 及技术顾问工作，具备丰富的行业经验与技术积累，对于数据库技术具有深刻的理解；热切关注 Oracle 技术和其他相关技术，对于 Oracle 数据库 RAC 以及高可用解决方案具有丰富的实践经验；长于数据库故障诊断，数据库性能调优。作为社区和网络的活跃者，在公开演讲和出版方面，多有建树。

作者简介

杨廷琨　云和恩墨 CTO

Oracle ACE 总监，ITPUB Oracle 数据库管理版版主，人称"杨长老"，数年如一日坚持进行 Oracle 技术研究与写作，号称"Oracle 的百科全书"。到目前为止，他已经在自己的博客上发表了超过 3000 篇文章。2004 年参与编写《Oracle 数据库性能优化》一书，2007 年被 Oracle 公司授予 ACE 称号，2010 年与盖国强共同主编出版了《Oracle DBA 手记》一书。

个人博客 http://www.yangtingkun.net。

侯圣文　恩墨学院院长

北京大学理学硕士，Oracle ACE 总监，OCM 和 ACCUG 创始人，ITPUB 版主。曾任职于海关总署数据中心，负责运维国家级海量数据库；后任国际航空运输协会（IATA）高级数据架构师，负责国际化高可用海量数据库系统架构设计、开发实施及运维。

为多家大中型企业提供过 Oracle、MySQL、Hadoop 大数据及云计算相关课程培训，培训经验丰富，讲课富有激情和感染力，善于理论联系实践，以通俗易懂、诙谐幽默的语言讲解枯燥深奥的数据库理论，并凭借丰富的实践经验，教导学员学以致用、融会贯通，使学员受教于课堂之上，受益于工作之中。作为 OCM 认证金牌讲师，目前已培养 OCM 大师 600 余位，培训 DBA 近万人。

个人博客 http://www.secooler.me。

李轶楠　Oracle ACE、高级培训专家

Oracle ACE，高级培训专家，ITPUB 论坛版主，从事 Oracle 培训及技术服务行业 12 年，擅长基于 Oracle 数据库的应用需求分析、架构设计、数据建模、代码开发、数据库管理、灾难恢复和性能优化等。目前培训企业超过 500 家，培养的学员已经在很多大型行业用户成为主要企业核心人才（分布在银行、证券、保险、移动、电信、税务、海关和各大领头企业）。作为负责人为众多行业客户提供 Oracle 技术服务，致力于为客户设计部署稳定的系统架构，解决各种数据库疑难故障，优化数据库应用系统性能，保证客户数据库系统的正常运行。

2005 年 6 月，担任主要作者出版了《Oracle 数据库性能优化》一书，2008 年 7 月，被 Oracle 公司授予 Oracle ACE 称号。

作者简介

熊军　云和恩墨西区总经理

Oracle ACE 总监，ACOUG 核心专家，著名数据库紧急恢复软件 OracleODU 作者，曾任 ITPUB 论坛 Oracle 专题深入讨论版版主。

一直从事 Oracle 数据库专业技术服务工作，有超过 10 年的 Oracle 数据库技术顾问经验。所服务的对象包括电信、金融保险、政府机关以及制造业等多个行业的客户。在 Oracle 数据库领域具有深厚的理论基础、丰富的实践经验、有大量省级电信行业系统性能优化成功案例，对于 IT 系统的架构也有深入理解。

独立开发了 Oracle 数据库紧急恢复软件 ODU，能够在数据库损坏并且没有备份的情况下直接解析数据块进行数据恢复。ODU 帮助大量的客户挽救了重要数据，累计数据量超过 100 TB。

个人博客 http://www.laoxiong.net。

李真旭　云和恩墨西北区技术总监

Oracle ACE，Oracle Young Expert，网名 Roger，拥有超过 7 年的 Oracle 运维管理使用经验，参与过众多移动、电信、联通和银行等大型数据库交付项目，具有丰富的运维管理经验，对 Oracle 数据库管理运行机制、锁机制和优化机制等具有深入理解，擅长 Oracle 数据库的性能调优、故障排除以及异常恢复。

个人博客：http://www.killdb.com。

Joel Perez　云和恩墨首席技术专家

Oracle ACE 总监，Maximum Availability OCM，获 11g 和 12C 的 OCM 证书，在超过 50 个国家的 Oracle 研讨会上发表演讲，2003 年获得 OTN 专家的荣誉称号，也是全球第一批 ACE 称号获得者，他十数年潜心学习钻研 Oracle 技术，致力于数据库高可用、灾难恢复、升级迁移和数据复制等方向设计和实现解决方案。

怀晓明　云和恩墨性能优化专家

ITPUB 社区版主。兴趣广泛，视野广阔，目前专注于 SQL 审核与优化工作。是一个细心敏锐的 Troubleshooter，善于使用以搜索为主要手段去处理解决各类问题。曾获得第一届 ITPUB 最佳建议奖，并在多个大型 IT 企业多年的工作历练中，积累了丰富的系统架构设计经验，擅长数据库和 Web 的设计和开发，精于故障诊断与处理,具有丰富的省部级电子政务行业工作经验及项目管理经验。合著作品有《剑破冰山——Oracle 开发艺术》《Oracle DBA 手记 2》。

作者简介

黄宸宁　云和恩墨高级技术专家

目前服务于云和恩墨西区，担任 Oracle 技术高级工程师职务，拥有长达 10 年的 Oracle 技术服务工作经验，多年在通信运营商和电力行业担任技术服务顾问。

杨俊　云和恩墨西区技术专家

6 年以上 IT 工作经历，涵盖软件开发、应用运维和中间件实施维护，擅长数据库运维管理、集成部署和数据迁移，先后服务公安、保险和银行等行业，具有丰富的数据库运维管理经验，熟悉数据库容灾建设、高可用架构和高性能 zData 实施。

个人博客：http://www.ora-dbca.com。

刘旭　云和恩墨性能优化专家

2005 年开始接触 Oracle 数据库，2008 年获得 10g OCP 证书。曾任职于某大型智能交通企业，开发过基于 Oracle 数据库的智能交通管理系统；后任职于全球领先的通信行业 IT 解决方案和服务提供商，提供电信级数据库的规划与运维支持。目前就职于云和恩墨性能管理部门，长期服务于某大型移动运营厂商的 SQL 审核项目，具有丰富的 SQL 审核售前和售后交付经验。

孙雪　云和恩墨市场专员

Oracle 11g OCP，2016 年毕业于广东外语外贸大学。曾获中国大学生计算机设计大赛三等奖。热爱生活，乐观善良。广泛涉猎各类书籍，钟情文字的世界！

序

当今世界信息技术日新月异，以新一代移动通信、下一代互联网以及物联网为代表的新一轮信息技术革命，催生出了很多新技术、新产品和新应用，同时也促进了电子商务、工业互联网和互联网金融等新兴产业的蓬勃发展。信息技术如同一把双刃剑，一方面给传统产业造成了巨大的冲击，另一方面也给了传统产业一个再次焕发新活力的机会。以此为契机，传统产业与信息产业互联互通、相互融合，并且不断涌现出新的市场需求。数量庞大的用户和更广泛的产业领域连接，使得企业的数据量正以一种令人难以想象的速度在扩张，全球产生的数据量正以大约每两年翻一番的速度增长。

数据作为日渐重要的新型资产，其拥有者需要有与数据体量相匹配的处理能力来应对数据创造、采集、管理和存储等过程中会出现的种种技术和应用方面的问题，数据巨大的发展空间吸引着大批企业投身并服务于这一市场之中。戴尔秉承"在中国，为中国"理念，通过提供优质的 IT 方案为中国的"大数据"建设贡献一份力量。数据库系统作为承载各种类型数据最主要的存储管理平台，已经在 IT 领域中广泛应用了 30 多年。随着 IT 应用的持续深入，数据库应用已经进入了相对成熟的时代。作为 IT 最为核心的应用，数据库也早已不是可以孤立存在的系统。它与服务器系统、存储系统和网络系统等硬件紧密相连，它与操作系统、中间件系统和应用编码等软件紧密相连，它也与用户访问模式、用户使用频度和数据承载压力等紧密相连。数据库具有的多重关联性及系统本身的复杂性，决定了企业要高效地应对瞬息万变的市场，同时也说明了围绕着数据库系统建设软件和硬件一体化的平台有多么重要。

戴尔作为行业中最大的设备和方案的供应商，在与各个行业客户交流沟通的过程中深刻地感受到了一点，在企业的核心数据库系统建设过程中，由于受到了采购成本、运维成本、系统升级风险、技术复杂度和性能优化等一系列因素的影响，越来越多的企业选择 Unix Migration，Unix Migration 成为企业核心数据库系统平台向前发展不可逆转的趋势。2015 年上半年，戴尔发布了最新的 4 路服务器 PowerEdge R930，也正是 R930 在技术架构上的不断自我创新，功能上的丰富增强和性能上的极大优化，使得 R930 在推出一年半的时间内得到了客户的广泛认可。目前，R930 服务器在中国为上万企业客户的核心数据库系统提供高效、可靠的 IT 支撑平台，这些客户覆盖了国内主要的行业和领域，R930 也成为了戴尔公司数据库解决方案中最重要的组件之一。

云和恩墨汇集了国内数据库领域的众多顶级技术专家，围绕着客户的核心数据架构提供专业的咨询、建设、管理、整合、优化和保护等服务，并且成为了该领域的佼佼者。云和恩墨在以"数据驱动，成就客户未来"为宗旨服务客户的同时，也在时刻关注着客户的业务变化以及行业发展所带来的各种最新的技术。伴随着闪存技术和高速网络技术的不断更新和广泛使用，用户对数据库性能的要求越来越高。通过单一的优化应用系统，抑或是仅仅提升硬件的性能越来越难满足当下客户的需求，软硬件一体化整合交付的方案成为了越来越多的客户购买数据库基础平台的首选。戴尔与云和恩墨一起使用云和恩墨自主研发的 zData Light（软件定义存储组件）。它把服务器、闪存和高速网络有机融合为一体，为客户提供标准开放架构的高性能数据库一体化解决方案。

我与本书的主编——盖国强相识于黄山的一次技术峰会。盖国强是国内 Oracle 技术粉丝熟知的顶级技术专家之一，也是云和恩墨的创始人。这是一本汇集了多位顶级技术专家知识、经验和点评的专业书籍。我相信无论是数据库技术爱好者还是我们的广大用户，都能通过本书受益匪浅。读者不仅可以从理论概念层面、更能从实际应用层面上更好地理解和把握数据库的技术机理，还可以循序渐进地探索其内部的技术细节。我相信本书能够让技术爱好者沉浸其中，感同身受，有所收获。

曹志平

戴尔全球副总裁

2016 年 9 月 29 日

前言

数据驱动，成就未来

最近两年来，很多朋友经常会问我，接下来会不会继续写书，会写一本什么样的书。

其实我也一直在思考，什么样的作品能够以最小的篇幅带来超越时间的价值，尽可能地帮助那些准备进入和刚刚进入这个领域的广大技术人员。

本书缘起

2015 年年底，我在成都和老熊聊天的时候，忽然有了一个想法，如果我能够将云和恩墨的专家团队聚集起来，让每个人都把自己最宝贵的经验方法、经典案例呈现出来，那累计超过 100 年的从业经验一定可以帮助很多人更深层次地了解数据库技术。

于是，我们动手将这个想法付诸实施。如今，云和恩墨的技术团队有 6 位 Oracle ACE 总监，2 位 Oracle ACE，还有 SQL 大赛的冠军选手、SQL 专家以及众多的 OCM 专家，集众智而成城，这些经验、感悟铸成的技术之犁一定可以对大家有所帮助。而且，不同作者的风格、方法、经验既迥然相异，又往往不谋而合，这样的融合既可以互相映衬，又可以彼此补充。若读者可以找到切合自己经验的内容并以此为鉴，找到学习案例方法从而完善、成就自我，如此则为本书之幸、之意义所在。

本书内容

很高兴这个愿望最终在 2016 年得以实现。本书分为四篇。

第一篇成长之路，收录 3 章内容。第 1 章是我的职业生涯回顾，最初写在 Oracle 38 周年庆祝之时，是对我技术生涯的一点总结。第 2 章来自恩墨学院院长侯圣文，从 DBA 到 OCM，再到金牌讲师和 Oracle ACED，他的学习和奋斗历程值得借鉴。第 3 章来自南美的第一位 Oracle ACED、云和恩墨的外籍专家 Joel，介绍了他如何接触并走上 Oracle 的技术道路。如果认真阅读这一篇，你一定可以从我们的学习历程中，找到很多共同的方法、特质，希望这些对你有所帮助。

第二篇知识基础，收录 4 章内容。这一篇是和基础知识相关的技术文章。第 4 章是我的一个总结。诊断数据库的方法是让我一直受益的最宝贵的经验，我对过去的积累进行了梳理和总结，把我认为最宝贵的经验分享给读者。第 5 章是张乐奕（Kamus）的总结。其中收录的案例体现了 Kamus 随时动手获取真知的实战派风格，这些信手拈来的方法来源于 Kamus 的不断学习、长期积累。第 6 章是年轻专家杨俊的实战总结。今天的很多企业都在从小型机走向 X86 架构，XTTS 技术帮助我们的大量客户成功实现了这一跨平台迁移。这一章希望可以帮助那些走在 U2L 道路上的朋友们。第 7 章来自我和杨廷琨的文章合集。我们都认为，对最简单技术的理解往往正是最考验一个人的技术成熟度。虽然有可能小到参数设置，但是这些往往会在实践中扑朔迷离，一点小的改动都有可能造成不可想象的后果。我将我和老杨关于这个主题的内容汇聚删节，作为这一章的核心。本章中提到的多个案例都是来源于客户生产实践，尤其是在 12c 多租户中的问题。DBA 的工作无小事，这一章呈现的就是这样一种态度。

第三篇 SQL 之美，收录 3 章内容。本篇主要包含与 SQL 知识、案例相关的内容。第 8 章是杨廷琨的学习与分享。老杨总结了他的学习过程、方法和对行业的观点，然后修订了他经典的关于 Oracle NULL 的解析，这是老杨认为他自己最重要的分享之一，希望能够给大家以启发。第 9 章是刘旭的实战案例总结。刘旭是云和恩墨为中国移动服务的 SQL 专家，这则案例来自于客户现场，是一个 Cache buffers chains 的 Latch 诊断案例。虽然它只是优化过程中的一个小小缩影，但是读者可以从中体会到，在高并发的客户现场，如何谨慎、严谨地处理实际问题，其最终解决也体现了理解业务的重要。第 10 章是怀晓明的 SQL 生涯笔记。他是 ITPUB 论坛几届 SQL 大赛的评委。他的技术生涯和 SQL 一直紧密相连，他对于 SQL 的理解一直深刻且独特。这一章既包含了他的知识分享，也包含了他的学习历程和建议，希望可以对读者有所帮助。

第四篇诊断分析，收录 4 章内容。本篇着重实战的诊断案例。第 11 章来自云和恩墨西区技术专家黄宸宁，本章通过对一个常见问题的层层分析，展示了其出人意料的因由。第 12 章是熊军的 DBA 手记，既总结了其学习经验又提炼了最新的案例精华。老熊是中国西部唯一的一位 Oracle ACED，解题思路清晰，直指本源，每一篇文章都十分精彩。第 13 章，是李真旭的反思与总结。作为中国地区 Oracle 数据库技术领域最年轻的一位 Oracle ACE，真旭走过的路曲折但不轻松，或许他的经验可以让大家少一些弯路，多一些捷径。第 14 章来自李轶楠的分享。作为国内最早期的 Oracle ACE 之一，李轶楠经历了从售后到售前，从培训到市场，从技术到管理等全方位的岗位考验。勤奋与汗水、思考与经验，一定会带给你不一样的感受。

技术之外

2016 年是云和恩墨公司成立的 5 周年，这本书的面世也可以看作云和恩墨奉献给大家的一个礼物。5 年之路，行之不易，然而分享之心，从未改变。

2016 年 9 月 9 日，我们在北京举行了一场名为"数据云端"的新产品发布会，展示了我们对于行业的思考、行动。在此我摘录了现场演讲中的一些片段，兼做行业的展望与读者分享，本书的内容也正是对这些理念的阐释。

数聚云端

数聚和云端，这两个词可以用来标识我们想和大家分享的行业的发展和演变。

第一，数聚。在过去的 5 年服务过程中，我们注意到很多客户在自发地走向数据整合和集中的历程。大家可能知道，在传统的 IT 建设中，很多用户建设了大量孤立分散的数据系统。在如今云和大数据的时代，这些用户开始自发地将这些数据进行整合、集中和汇聚。很多用户都经历过这样的阶段：

- 物理集中，将物理的 IT 资源、服务器、硬件统一起来，在统一的数据中心管理；
- 数据集中，进一步通过数据的整合和集中来构建整合的数据环境，从而降低运维成本。这在今天达成另外一种目标，降低数据库软件的 License 成本，使得企业走向合规正版化；
- 应用集中，有了整合的数据中心和集中的数据，很多企业开始进行应用重构，将应用推向了 SaaS 化、服务化。

这是我们看到的很多国内的传统企业自然而然发生的一个演进，即数据的整合和集中。

第二，云端。时代的变化是越来越多的应用开始走向云化和云端。在应用和业务走向云端的过程中，不可避免会改变我们所从事的运维领域的格局。我用一组数据和大家分享一下我们的看法。

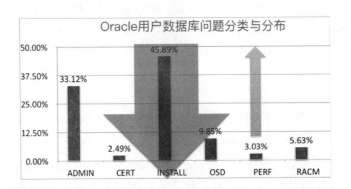

这个数据是来自于最典型的 Oracle 的数据库产品——用户在使用中所遭遇到的问题场景。我们看到在传统的数据应用环境中，绝大多数用户是在 Oracle 的安装和初始化过程中面临大量的问题，这些问题在未来的云场景中会大幅缩减甚至不复存在。而原来非常小众的性能问题，大约只占 3%的比例，我们认为在未来、在云上这些问题会更加突出，它们会成为用户关注的核心。

所以，我们认为优化能够一直为用户带来核心价值，而不像传统的基础的这部分服务。

重在优化

对用户最有价值的优化服务，实际上也是云和恩墨最重要的服务亮点。我们持续通过数据库的性能优化，来帮助客户缩减成本、改善用户体验。在优化方向上，我们坚信通过种种的优化手段，我们一定能够帮助用户带来高达几十倍的性能提升和大幅的用户体验改进。

和大家分享两个案例。第一个是北京电信，客户的 CRM 系统一度面临性能瓶颈，CPU 平均使用率高达 60%，高峰期 CPU 资源基本耗尽。云和恩墨的专家团队通过最有价值的优化服务，经过两个阶段的调节和优化，将 CPU 平均使用率降低到了 30%。这样的优化工作切实帮助用户缩减了硬件投入，减少了硬件升级更新的频度，真正帮助用户解决了问题。

通过专业的数据库优化服务,缩减了用户的硬件投资,改善了响应时间,提升了系统稳定性。

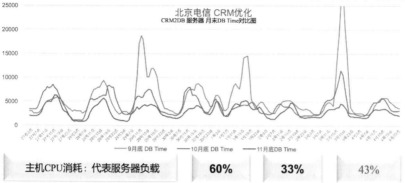

在另外一个客户——大地保险的核心场景下,我们通过最初的优化服务解决了困扰用户的 CPU 100%的高负载问题。这个问题严重影响用户业务,即使硬件升级也未能奏效,我们的优化帮助用户走出了这个困境,将 CPU 的使用率降低到了 60%以下,这是优化为客户带来的价值。但是如果仅仅通过优化帮助客户解决一时的困境还不是我们的全部价值所在,我们希望通过持续的解决方案提供,帮助用户走在技术架构的前沿上,在保证稳定高效的同时保持数据架构的领先性。在客户的真实场景里,我们通过长达 4~5 年的数据汇总,详细分析用户系统的使用瓶颈、使用问题,找出用户数据架构中存在的痛点,然后为用户提供整体的架构咨询和数据架构改进意见,并最终和用户一起确定数据架构的演进方向和推进步骤。

大地保险:从优化到整体架构咨询

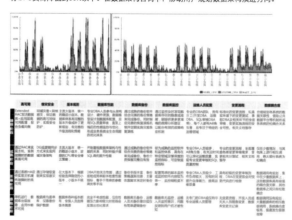
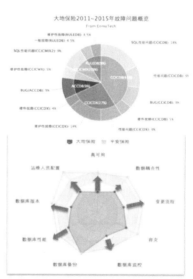

我们一直希望通过基础的服务如性能优化,来帮助用户构建稳定、高效的数据环境,进而通过数据架构咨询和规划服务,帮助用户设计整体数据架构的演进方向,让用户在

稳定、安全的前提下朝着正确的方向实现架构演进。

数据运维现状与未来

在过去云和恩墨 5 年的公司生涯中，也包括我个人在数据库行业 15 年的职业生涯中，我们经历和见证了行业的发展过程，同时也看到了这个行业中存在的一些不足。

当用户在使用数据环境和数据系统时，日常运行过程中大约会分为 4 个阶段：

- 日常监控，了解系统的运行状况，发现和警示问题；
- 定期巡检，了解系统的健康状况，发现和整理问题隐患；
- 维护变更，根据日常的业务需求进行按需变更和维护；
- 优化提升，当遇到问题时，进行应急的故障处理和优化抢险。

这是正常的数据运维过程中所经历的必要环节，在这些环节中我们注意到，运维的现状存在很多不足和亟待改进之处。这些地方包括：

- 在日常的监控过程中，众多的产品和软件存在纷繁复杂、监易控难的局面。很多软件提供大量的检测维度帮助用户监控系统，但是用户往往难以抓住重点，软件能够实现帮助用户监视的目的，但是往往难以控制，当用户发现问题后需要通过各种工具手段和脚本来处理和消除问题。
- 定期巡检过程中的问题更加突出。绝大多数用户获取的巡检报告往往流于形式，甚至很多报告是通过复制粘贴得来的，这样的巡检往往以点代面，质量参差。传统巡检的质量完全取决于个人，如果用户接受的是有三年工作经验工程师的体检报告，可能认为系统运行正常，但是在我们这样的 15 年的老司机眼里，你可能会看到这个系统漏洞百出。
- 维护变更中，我们更多看到的是人力叠加，缺乏规范，缺乏产品化监控。甚至导致了在日常维护期故障频发。我们见到太多的用户因为释放空间误操作删除了一个数据文件甚至一个数据库，在变更环节导致故障。
- 优化提升，是指在系统运行的过程中出现问题，我们要紧急诊断或是通过优化消除一些故障。这些属于事后救火的范畴，往往停留在头痛医头、脚痛医脚的境界，暂时解决了系统问题，但是对于系统中存在的隐患，我们仍然是一无所知。

这就是我们看到的在数据运维领域存在的一些问题。那么，我们期望这个行业的未来是什么样的呢？

过去十多年间数据库健康检查几乎一成不变，数据库的应急优化也周而复始。我们面对大量的全表扫描、隐式转换等问题，其优化方法数十年间几乎是一成不变的，这样的情况需要向自动化、云化和智能化的未来演进。自动化产品有助于改善运维现状，加强规范性，减少人为的故障。通过云化的平台，去应对今天用户云化的业务场景和应用环境。最重要的是智能化，通过对人的智力的总结、积累形成智能化产品，通过产品帮助更广大的用户，走出循环往复的低层次循环，提升整个行业的应用能力。

云和恩墨的产品探索

自动化、云化和智能化是运维的未来，但是运维现状和我们期望的未来之间仍然有着巨大的差距。面对这些问题，云和恩墨做出了一些思考、尝试和创新。

第一，针对监控和运维，云和恩墨推出一款监控产品 ZONE，我更愿意称其为监控运维一体化平台。

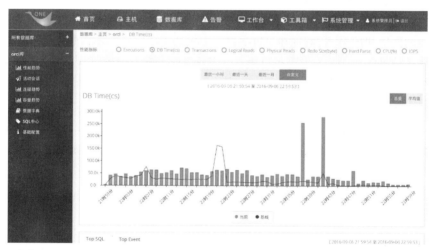

市场上绝大多数的监控产品只完成了监视的作用，达不到控制的目的，通过 ZONE 我们希望将用户的日常监控和运行维护融合起来，实现监控和运维结合的一体化。通过监控发现的问题同样可以通过监控平台提供的功能进行处理。比如工具箱，我们通过工具箱去整合、汇聚 DBA 最常用的问题处理脚本工具，在这个平台内处理，同时实现审计功能。还可以通过工作台实现变更管理、配置管理，加强变更的规范化审核。我们期望通过这样一个软件平台，解决部分用户所面临的监与控分离的问题。

第二，针对定期巡检，数据库的定期巡检类似于大家每年要去医院进行的健康体检。我觉得功能非常相似，大家希望通过健康体检发现身体中存在的健康隐患，并且通过一定的治疗去消除这些隐患。但是在传统的服务过程中，这样的一个过程被模式化了，往往千篇一律，不能够真正帮助用户找出存在的问题。如今云和恩墨新开发了一款产品，是集合了众多专家智慧的云化的智能巡检平台。

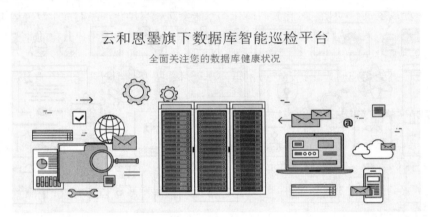

对于云和恩墨来说最值得自豪的就是专家团队和员工，将专家的经验和智慧集中起来，固化到这个平台中，就可以为更多的用户服务。传统的健康体检中面对的往往是刚毕业的医生和护士，我们现在可以提供给大家的是一个经验丰富的保健专家、主治医师，整个系统可以在云端为大家提供服务。当你需要了解系统健康状况的时候，将采样数据上传到系统中，立即调用后台所有专家的建议和经验，帮你生成最详尽的健康检查报告，并且给出诊断建议。

我们希望通过这样一个产品，能够改变这个行业中十多年来一成不变的服务体系，解决那些以点带面，复制粘贴带来的无效服务。我们认为所有这些对用户核心资产不负责任的做法，可以通过这样一个软件产品在云端做出改变。这个产品的名字也体现了我们的决心，叫做 Bethune，我们希望具备白求恩医生一样的精神、品质，帮助更多的人消除病痛，所以整个服务是完全免费的，以 SaaS 的方式提供。

Bethune 不仅提供数据库智能巡检服务，对于用户来说最重要的价值还在于该平台会针对用户数据库体检数据，为用户推送有价值的提醒信息。例如，在用户密码即将到期的前三天，通知用户修改即将过期的用户密码；定期向用户推送 Oracle 每个季度发布的 PSU 补丁分析；不定期向用户推送数据库存在和潜在的 Bug 分析。一键采集、一键上传，剩下一切专业的分析和及时提醒都可以交给 Bethune。Bethune 致力于让每个工程师都成为运维专家。

第三，针对事后优化，我们更主张事前防范，云和恩墨的 Z3 - SQL 审核工具就着眼于事前。在我们的视角里，Z3 也是开发运维一体化平台。SQL 审核是一件什么样的事情?我们说传统的用户在系统运行的过程中出现了性能问题,如性能故障、业务中断等,这时候用户往往需要找到专家来救火,抢救式的优化,暂时的头痛医头。可是一旦系统遇到了这样的问题,用户业务已经遭受了损失。我们希望把这件事情推进到开发测试环节,帮用户找出系统中隐藏的性能瓶颈和隐患,防患于未然地消除它。这就是我们想做的事。

整个的 Z3 SQL 审核平台结合了专家团队的经验与智慧。我们已经做到的就是将案例中总结的经验变成算法,然后将算法内置到软件中,通过软件帮助用户筛选开发测试环境或者是线上环境,找出核心 SQL 代码中的性能问题和隐患,然后提出修改建议,最终通过工作流指引用户完成整个优化过程,从而达到防患于未然的目的。Z3 SQL 审核平台这一产品,其理念和今天最热门的 DevOps 非常吻合,我们认为 SQL 审核就是 DevOps 理念在数据库领域的最佳落地点。

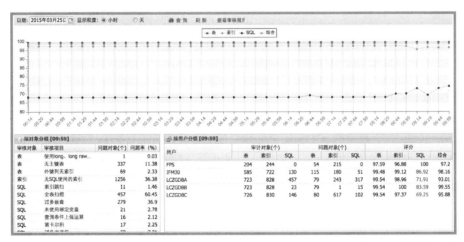

云和恩墨,今天我们坚定认为应当将用户场景中 SQL 层次的问题抽取出来,由一个独立的环节、最专业的群体完成,通过审核、优化把问题消灭掉,保证线上系统的稳定,这就是我们所提出的 SQL 审核理念,也是 Z3 产品实现的开发运维一体化平台。

软硬结合整体优化

当然,除了软件层面的优化,我们也可以通过硬件层面的优化来改善系统性能,尤其是在今天,很多用户在大力推进去小型机化。云和恩墨基于对于数据库的理解,结合 Oracle RAC 进行了深度优化整合和调校,推出了分布式存储数据库优化解决方案——zData。

zData 是云和恩墨自主研发的软硬件整合一体机产品,通过高带宽低时延的 InfiniBand 互联,将高性能 X86 作为计算节点和以闪存为核心 X86 作为存储节点整合起来,形成软硬件紧密耦合的高性能分布式存储架构、同时兼顾高可用及数据容灾的整体解决方案，zData 还具备资源池化、动态扩展能力，配合多租户功能的数据库软件可以实现多租户、高可用性、高扩展性的云数据库服务，一站式交付，一键式业务部署。zData 的一体化大大简化运维，降低了对 IT 管理人员技能要求，可以大幅节省客户采购与管理成本。

zData 既可以通过软硬件整合一体机产品的形式交付给用户，也可以通过我们核心的分布式存储软件 zData Light Storage 将用户的 X86 服务器整合成一体化的数据服务平台，在典型的案例中通过三台 X86 服务器作为存储，能为用户带来超过 150 万的 IOPS 处理能力，这样的架构可以改善传统的 IT 环境，同时带来巨大的性能改进和提升。zData 的解决方案，在过去已经和戴尔合作为用户创造了卓越的价值呈现。

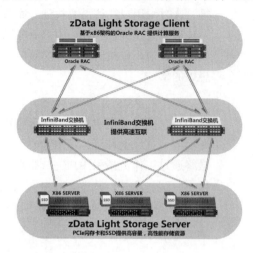

zData 事实上已经在很多用户的真实生产环境中取得了成功，尤其是在运营商的环境中，zData 已经帮助很多用户实现了去 IE 的最佳实践。下图是我们在北京电信实现的 ODS 系统去小型机化，ODS 系统是客户经分系统的核心，数据量超过 10TB。zData 一体化环境替换了用户的小型机架构，在保证了高性能、低成本的情况下，彻底革新了用户的 IT 基础设施。

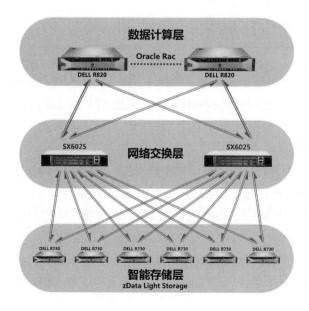

依托对数据的深入理解，今天不管是云和恩墨软件层面的优化服务，还是基于分布式存储的硬件方面的优化，都能够给客户带来卓越的价值实现。

展望未来

最后，回到我们所看到的未来，我们认为在未来的数据服务领域会坚定的走向自动化、云化和智能化。

自动化在运维的基础领域会成为一个常态，通过自动化的产品逐步替代人繁重的工作，通过自动化产品去提升规范化、流程化和安全化，改善原有运维中存在的问题，这会成为基础设施常态化的场景。

在云化的方向上，我们相信随着用户业务场景的云化，服务领域也必然通过云化的平台，为用户提供随时可用的解决方案。更重要的是，我们希望通过云化服务，为更广大的中小企业甚至个人用户提供技术支持，Bethune 就是为了这样一个目标而努力的。

在智能化方向上，不仅仅是前面提到的，通过整合和利用现有专家运维的经验实现运维产品的智能化，我们希望在发展的过程中，能够实现通过机器学习和智能算法，帮助用户发现数据中蕴藏的价值，帮实现数据和数据运营，这是我们看重的更广阔、长久的未来。

鸣谢

本书的成书首先要感谢云和恩墨的专家团队和作者们，大家在紧张的工作之余笔耕不辍，才有了这样的呈现。分享精神一直是云和恩墨的技术基因。特别感谢孙雪同学，她帮助我收集和整理了本书的大部分内容，并且翻译了 Joel 的英文稿件以及第 4 章的附录部分，她的协助是本书得以及时呈现的重要助力；感谢张海军先生为第 7 章提供了测试范例，他最早向我提示参数文件 COPY 参数的增强。

感谢 ITPUB 论坛，我们一直的精神家园。本书的一些专家经验总结，最早来自 ITPUB

的记者访谈，我们成长的印记，一直留在 ITPUB 的成长历程中。

  本书还要特别感谢戴尔公司全球副总裁曹志平先生，在得知云和恩墨和戴尔一直保持良好合作之后，他慨然提笔为本书作序，展望了数据与存储行业的未来，并对云和恩墨和戴尔的合作表示了深切的期待。

  最后还要感谢人民邮电出版社的杨海玲、武晓燕老师。她们在近乎不可能的时间里完成了本书的编辑、审稿和出版工作，使得本书能够在 2016 年呈现在读者面前，感谢你们的信任和支持。

盖国强 2016 年 9 月 30 日于北京

目录

第一篇 成长之路

第 1 章 三十八载，Oracle 伴我同行 2
1.1 缘起边陲，恰同学风华正茂 2
1.2 京师磨练，转眼已历十二载 4
1.3 笔耕不辍，年少曾怀作家梦 6
1.4 三重境界，见山见水见真我 9
1.5 云和恩墨，数据服务起征途 12
1.6 理想实践，开发运维一体化 13
1.7 快乐生活，此心安处是吾乡 16

第 2 章 回首向来萧瑟处，也无风雨也无晴 18
2.1 人生若只如初见 18
2.2 日久生情见真心 19
2.3 衣带渐宽终不悔 19
2.4 天下谁人不识君 20
2.5 回首向来萧瑟处，也无风雨也无晴 21

第 3 章 正确抉择，丰富人生 23
3.1 初闻 Oracle 23
3.2 选择适合自己的路 24
    3.2.1 选择合适的资料书 24
    3.2.2 融入圈子 24
    3.2.3 关于大学的课程 25

3.3 树立目标 26
3.3.1 参与论坛分享 26
3.3.2 写作 26
3.3.3 参加技术研讨会 27
3.4 关于认证 27

第二篇 知识基础

第 4 章 Oracle 数据库的跟踪和分析方法 30
4.1 SQL_TRACE 及 10046 事件 31
4.2 SQL_TRACE 说明 31
4.2.1 DBMS_SYSTEM 跟踪案例 32
4.2.2 系统递归调用的跟踪 36
4.3 10046 与等待事件 39
4.3.1 通过跟踪理解数据库的初始化 42
4.3.2 远程支持之 10046 事件 50
4.3.3 通过 10046 事件跟踪解决未知问题 52
4.3.4 通过 10046 解决数据库 RAC 集群案例 56
4.4 Oracle 跟踪总结 62
4.5 参考资料：数据类型比较规则 62
4.5.1 数值类型 62
4.5.2 日期类型 62
4.5.3 字符类型 62
4.5.4 对象类型 65
4.5.5 数组和嵌套表类型 65

第 5 章 兴趣、思考与实践 72
5.1 Kamus 谈技术、学习与分享 72
5.1.1 Oracle 的适用场景 72
5.1.2 为什么 Oracle 广受欢迎 73
5.1.3 如何实现灾备 73
5.1.4 数据库发展对 DBA 工作的影响 73
5.1.5 Oracle 的学习方法 74
5.1.6 如何成为 ACE 74
5.2 以 12c Identity 类型示范自我探索式学习方法 75
5.3 Dump Block 是否会引起 Block 读入内存 84
5.4 Dump Block 是否会引起脏数据写入磁盘 85

5.5 如何验证 ASM 的块头备份块的位置 ·· 88
5.6 如何利用文件句柄恢复误删除的文件 ·· 90
5.7 从一道面试题看分析问题的思路 ·· 93
    5.7.1 检查被阻塞会话的等待事件 ··· 93
    5.7.2 查找 blocker ··· 94
    5.7.3 乙方 DBA 需谨慎 ·· 94
    5.7.4 清除 blocker ··· 94
    5.7.5 深入一步 ··· 95
5.8 涓涓细流终聚海 ··· 96

第 6 章 使用 XTTS 技术进行 U2L 跨平台数据迁移 ·········· 97

6.1 XTTS 概述 ··· 97
6.2 XTTS 技术迁移应用场景 ··· 99
    6.2.1 应用场景一：全国"去 IOE"战略实施 ······································· 99
    6.2.2 应用场景二："云平台"数据中心建设 ······································· 99
    6.2.3 应用场景三：老旧环境淘汰改造 ··· 100
    6.2.4 应用场景四：数据库分布式存储重构 ······································· 100
    6.2.5 应用场景五：其他应用场景 ·· 101
6.3 XTTS 迁移步骤 ··· 101
6.4 XTTS 迁移方式 ··· 101
    6.4.1 方式一：dbms_file_transfer ··· 102
    6.4.2 方式二：RMAN Backup ·· 103
    6.4.3 方式三：手工 XTTS 迁移 ·· 104
6.5 XTTS 前置条件检查 ··· 106
6.6 XTTS 最佳实践方案论证 ··· 109
    6.6.1 技术方案概况 ··· 109
    6.6.2 技术方案实施步骤 ··· 110
    6.6.3 技术方案模型 ··· 110
    6.6.4 方案可行性说明 ··· 110
    6.6.5 方案优缺点论述 ··· 111
    6.6.6 技术方案论证结论 ··· 111
6.7 XTTS RMAN Backup 步骤 ·· 111
6.8 XTTS 实战案例分享 ··· 113
    6.8.1 案例现状介绍 ··· 113
    6.8.2 系统现状评估 ··· 114
    6.8.3 迁移需求分析 ··· 114

  6.8.4 迁移方案选型 ··· 115
  6.8.5 迁移的具体实施 ··· 116
6.9 XTTS 风险预估 ·· 122
6.10 XTTS 总结 ··· 123

第 7 章 Oracle 的参数和参数文件 ··· 124
7.1 参数的分类 ··· 124
  7.1.1 推导参数 ··· 125
  7.1.2 操作系统依赖参数 ·· 125
  7.1.3 可变参数 ··· 126
  7.1.4 不推荐参数 ·· 126
  7.1.5 废弃参数 ··· 126
7.2 参数文件管理和使用 ·· 128
  7.2.1 参数文件的创建 ··· 128
  7.2.2 12c create spfile 的警示 ··· 130
7.3 12c 参数与参数文件新特性 ·· 134
  7.3.1 参数表的引入 ··· 135
  7.3.2 参数表在 PDB 启动中的作用 ·· 138
7.4 参数修改及重置 ·· 139
  7.4.1 解决参数文件的修改错误 ·· 144
  7.4.2 通过 event 事件来跟踪对参数文件的修改 ·································· 145
7.5 参数的查询 ··· 145
  7.5.1 参数查询的基本方式 ·· 146
  7.5.2 参数值的可选项 ··· 148
7.6 不同查询方法之间的区别 ·· 149
  7.6.1 V$PARAMETER 和 V$PARAMETER2 的区别 ···························· 149
  7.6.2 V$PARAMETER 和 V$SYSTEM_PARAMETER 的区别 ················· 150
  7.6.3 GV$SPPARAMETER 和 V$SPPARAMETER 的区别 ····················· 152
7.7 RAC 下参数的维护 ·· 154
  7.7.1 RAC 下共享 spfile ··· 154
  7.7.2 使用 ASM 存储参数文件 ··· 155
  7.7.3 谨慎修改 RAC 参数 ·· 156
  7.7.4 RAC 环境下初始化参数的查询方法 ·· 156
7.8 参数文件备份 ··· 158
7.9 参数文件恢复 ··· 160

第三篇　SQL 之美

第 8 章　学习与分享 · 164
8.1　对数据库开发和运维的认识 · 164
8.2　行业发展给 DBA 带来的挑战 · 165
8.3　个人学习经验分享 · 165
8.4　Oracle 中的 NULL 剖析 · 166
8.4.1　NULL 的基础概念和由来 · 167
8.4.2　NULL 的布尔运算的特点 · 168
8.4.3　NULL 的默认数据类型 · 173
8.4.4　空字符串''与 NULL 的关系 · 176
8.4.5　NULL 和索引 · 179
8.4.6　NULL 的其他方面特点 · 183

第 9 章　诊断 Cache buffers chains 案例一则 · 185
9.1　详细诊断过程 · 185
9.2　总结 · 194

第 10 章　戒骄戒躁、细致入微 · 195
10.1　我的职业生涯 · 195
10.2　运维的现状及发展 · 196
10.3　如何提高数据库的开发水平 · 196
10.4　DBA 面临的挑战 · 197
10.5　数据库优化的思考 · 197
10.6　提问的智慧 · 199
10.7　细致入微方显价值——通过真实案例认识 SQL 审核 · 200
10.7.1　案例一　仅仅是 NULL 的问题 · 200
10.7.2　案例二　想不到的优化方式 · 202
10.8　号段选取应用的 SQL 技巧 · 205
10.8.1　问题的提出 · 205
10.8.2　相关基础知识 · 206
10.8.3　解决问题 · 208
10.8.4　小结 · 221
10.9　connect by 的作用与技巧 · 221
10.9.1　connect by 是什么 · 222
10.9.2　connect by 可以做什么 · 222

第四篇　诊断分析

第 11 章　抽丝剥茧——一次特殊的 ORA-04030 故障处理　238
- 11.1　聚集数据的信息采集和分析　238
- 11.2　聚焦疑点的跟踪测试与验证　242
- 11.3　解析原理的问题总结与建议　245

第 12 章　不积跬步，无以至千里　247
- 12.1　技术生涯有感　247
- 12.2　自我定位及规划　248
- 12.3　对数据库运维工作的认识　249
- 12.4　学习理念分享　249
- 12.5　RAC 数据库频繁 hang 问题诊断案例　249
  - 12.5.1　案例现象及概要　250
  - 12.5.2　故障详细分析　251
  - 12.5.3　案例总结　260
- 12.6　Exadata 环境下 SQL 性能问题诊断案例　262
  - 12.6.1　AWR 报告　262
  - 12.6.2　生成 SQL 报告　263
  - 12.6.3　检查历史数据　266
  - 12.6.4　判断问题产生的流程　268
  - 12.6.5　查询历史数据　268
  - 12.6.6　并列执行的序列过程　268
  - 12.6.7　检查 call tack　271
  - 12.6.8　检查并行会话　271
- 12.7　关于 RAC 数据库 load alance 案例分析　276
- 12.8　总结　285

第 13 章　反思与总结：轻松从菜鸟到专家　286
- 13.1　一波三折：释放内存导致数据库崩溃的案例　287
- 13.2　层层深入：DRM 引发 RAC 的故障分析　292
- 13.3　始于垒土：应用无法连接数据库问题分析　300
- 13.4　变与不变：应用 SQL 突然变慢优化分析　308
- 13.5　实践真知：INSERT 入库慢的案例分析　314
- 13.6　按图索骥：Expdp 遭遇 ORA-07445 的背后　319
- 13.7　城门失火：Goldengate 引发的数据库故障　323

第 14 章　勤奋与汗水 329

- 14.1　我的职业生涯与思考 329
- 14.2　如何看待企业运维 329
- 14.3　对性能问题的认识 331
- 14.4　学习方法 332
- 14.5　所有奇异的故障都有一个最简单的本质 332
- 14.6　案例一：意料之外的 RAC 宕机祸首——子游标 333
    - 14.6.1　信息采集，准确定位问题 333
    - 14.6.2　层层分析，揪出罪魁祸首 336
    - 14.6.3　对症下药，排除数据故障 341
    - 14.6.4　深入总结，一次故障长久经验 342
- 14.7　案例二：异常诡异的 SQL 性能分析 342
    - 14.7.1　信息收集 342
    - 14.7.2　新特性分析 344
- 14.8　总结 345

参考文献 346

>>> 第一篇

成 长 之 路

- 第 1 章　三十八载，Oracle 伴我同行
- 第 2 章　回首向来萧瑟处，也无风雨也无晴
- 第 3 章　正确抉择，丰富人生

第 1 章 三十八载，Oracle 伴我同行

——记我的职业成长之路（盖国强）

> **题记**
>
> 2015 年是 Oracle 公司 38 周年，2015 年我 38 岁。在 Oracle 庆祝 38 岁生日之际，谨以此文作为回顾，记录我的 Oracle 技术之路。同时也希望可以给走在技术道路上的朋友们以借鉴。

1.1 缘起边陲，恰同学风华正茂

2000 年大学毕业时，我在第一份工作中第一次接触到 Oracle 数据库。那时我作为一个程序员，参与了一个大型企业 ERP 系统的开发进程。也就是从那时开始，我由网络配置一步一步开始深入 Oracle 数据库的内部。

很幸运，在我作为程序员的职业生涯中，我的第一位师傅把云波先生，不断给我信任和鼓励，也不断促进我学习。在我能够顺利完成开发工作之后，我获得的第一份额外工作是管理公司的域服务器和邮件服务器，从那时开始我深入学习了和 Windows 相关的技术，并且在 ITPUB 论坛上成为了微软技术版主。再然后，师傅说，"跟我一起研究一下 UNIX 和 Oracle 吧"。就这样，我走上了 DBA 之路，自那时起也在 ITPUB 论坛担任了 Oracle 数据库管理版版主，这个职务目前仍然挂在我的头上，已然是近 15 年的时间。

我工作中的第一位师傅是职业生涯中对我影响最深的人。他最让我印象深刻的是，始终能够通过已有的知识积累，经由思考对未知问题做出解答，而且几乎不走弯路。感谢我的第一家公司，那些卓越的项目和优秀的人才让我得以从一个毛头小伙走进了信息技术的深邃世界，并

且在如梭的岁月中从未后悔最初的选择。

接触了 Oracle 数据库之后，同时感受到了互联网的魅力，从那时起就开始在论坛上疯狂的活跃起来，提问和回答，从此一发不可收。在很长一段时间内，我是论坛上发帖最多的那个人。我几乎为每个复杂的问题进行测试，尽我所能的回答网友提出的问题，有的甚至会耗时数日，然而乐此不疲。

第一次听到 Oracle 的认证，是从一位同事那里。他煞有介事地说，获得这个认证，就拿到了一张金字招牌，好工作唾手可得。

我想当时我的眼里一定有光芒闪烁，从那开始，我定下了一个目标：一定要通过这个认证。我确实经历了艰苦的学习过程。自学，从每一本能够找到的官方教材入手，夜以继日。那是大学毕业之后，参加的第一次考试，而且是全英文的上机答题，虽然我已经不太记得考场的情形，但是考试前一天夜晚的紧张和忐忑还历历在目。

最终，经过漫长的 3 个月，完成了 5 门课程的考试，我终于拿到了 Oracle 8i 的 OCP 认证证书（如图 1-1 所示）。那是在 2002 年的昆明。

图 1-1

在当时，每考过一门课程，会首先收到一张成绩单（我保存至今，大约也可以算得上古董了）（如图 1-2 所示），而且要凑齐 5 张才能召唤证书啊！几张薄薄的纸却是沉甸甸的，那是我拿到的第一个 IT 相关的认证。

在后来总结这段工作时，我给一些朋友的建议是：如果你手上已经有了一份工作，那你需要做的是——做好它，哪怕那不是你喜欢的！你必须证明给别人看你有做好一件事情的能力，别人才会给你下一个机会！我正是在不断努力工作的进程中，不断获得学习新事物的机会，并最终找到自己喜爱的道路。

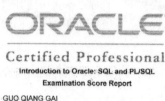

图 1-2

而且，对于任何一门技艺的学习，积累都尤为重要。这个积累的过程需要时间，只有去除浮躁，认真学习，不断积累，寻找机遇，才能够更好的把握自己的职业生涯。

1.2 京师磨练，转眼已历十二载

2003 年 4 月 1 日，我离开原来学习生活了 7 年的城市来到北京，开始寻找新的机会、新的起点。

结束一份熟悉多年的工作，离开一个生活多年的城市，走向一个陌生的城市陌生的街道，这并不是一件容易的事。很多朋友问起我当时的抉择，回想起来感觉重要的有两点：一是感觉遇到了瓶颈；二是做好了知识的积累。然后再有一些契机和触发条件，就很容易做出选择了，当然最重要的，那时候我还年轻。

在北京寻找工作的过程中，我有两个选择的方向。一是作为一个程序员继续做 ERP 软件的开发工作；二是作为数据库管理员（DBA——Database Administrator），寻找一个数据库管理的工程师职位。

机缘巧合，ITPUB 上的一位朋友为我介绍了一份 DBA 的工作（虽然那只是我众多面试的其中之一）。在非典前夕的兵荒马乱之中，这家公司的快速响应让我在 4 月 17 日正式上岗。而在非典之后，还有一些公司陆续通知我去复试或者考虑入职。

在这之后，ITPUB 上几个素未谋面的好友陆续来到北京，叶梁（ITPUB 上的 Coolyl）从广州回到北京，冯春培（ITPUB 上的 biti_rainy）从珠海回到北京，他们和我住到了一起，那所租来的房子成为了几年间我们在北京的重要据点。我们家的技术实力空前壮大，我用四个字来形容那个阶段：黄金时代。那个时候，很多时间在讨论与学习中度过，我们经常彼此提出值得研究的 Oracle 技术内容探讨研究。那一段时间，进步飞快。那个时候 ORA-600、Kamus 都是家里的常客（他们现在都是我创业上的合作伙伴），大家经常在一起打牌、讨论，甚至注册马甲上网吵架，那一个阶段结下的朋友和友谊将永不磨灭。

在北京的最初几年，工作之余，我几乎将所有的业余时间投入到 ITPUB 的培训课程中去。在 ITPUB 的创始人黄志洪（tigerfish）先生的带领下，以培训来支持论坛的持续发展，后来我也因此成为论坛的超级版主从而进入到核心团队中。在我们最初成长的岁月里，论坛始终在我们身边，变成了我们永远无法割舍的精神家园。虽然在 2006 年论坛最终被出售，但是那些一同成长的老朋友始终聚在那里未曾离去。

今天回过头去看，我想，**能找到一些志同道合的人，互相促进、共同奋斗**，是再幸运不过的事情了。

2004 年，我第一次去上海，参加了 Oracle OpenWorld 大会，结识了来自全国数据库领域的朋友们（如图 1-3 所示）。这张照片中的朋友，在数据库应用领域都取得了非常好的个人发展，很多人当时是第一次相聚。Oracle OpenWorld 一直是非常吸引人的技术盛会，在这样一个生态圈茁壮成长。幸运的是，在今天的微信时代，这些朋友又能够辗转相遇，时光流转，大家都在一起前行。

在北京的第一家公司是一个快速成长的企业。我和北京共历了非典，也和这个公司共历了快速成长的整个过程。这个公司给了我宽松的工作环境、良好的同事氛围和足够的成长空间，使得我能够全心地投入到工作和学习中去。很多同事十分优秀，在他们中间，我感受到了工作的乐趣与成长的快乐，从他们身上学到了很多优秀的品质。工作的锻炼、同事的影响和自我的学习，让我快速地成长起来，并且最终在职业上成熟起来。

图 1-3

在职业上成熟，获得充分的自信，能够清晰、冷静、严谨的思考，对于一个技术人员来说尤其重要。我庆幸的是，在这家公司获得了这些。

随着技术上的进步和发展，我的职位开始有所变化。在公司工作的 5 年中，我的职位从工程师到部门经理，再然后是部门总监。职位上的变化让我开始接触新的内容，那就是如何领导和带领更多的同事为了一个共同的目标而努力。作为一个领头人，**你要学习如何为他人着想、从他人的角度看问题，如何带动大家共同进步与发展**。在这个职位上的思考与经历同样让我受益匪浅。

我并不能够清晰地回忆起，何时建立起从他人的角度看问题的思想，但是这一习惯对我至为重要。我一直以来的习惯是，从不轻易对一件事情下判断，哪怕于别人来说可能是一件理所当然的事情。有时候某个人所做的看起来似乎不可理喻的事情，了解之后总有其可原的情理。理解、宽容、不要以恶意去揣度别人，是我一直遵循的人生法则。

在我的职业旅程中，可以说 Oracle 数据库技术帮助我开启了一段新的职业生涯，而社区则帮助我找到了朋友。在今天我相信仍然如此，社区和社群仍然是我们获取知识和结识朋友的重要途径。同声相应，同气相求，所谓相遇，实为必然。

1.3 笔耕不辍，年少曾怀作家梦

在北京工作的这些年，除了做好自己的本职工作之外，我还不断学习，根据自己的实践与积累，写作、编辑、翻译出版了一系列 Oracle 数据库方面的技术书籍。写作的最初想法很简单，那就是把自己积累的知识与经验分享出来，并且可以和朋友们一起为社区与网络生涯留下一点记忆。就这样一路走下来到了今天，自己也在坚持之中受益匪浅。

我年轻时曾经有过一个梦想，那就是成为一个作家，现在这个梦想在技术领域得以局部实现，也算是"失之东隅，收之桑榆"吧。以下这些作品，如图 1-4 所示，或合著，或翻译，或独撰，这期间收获最大的其实是我自己。而《Oracle DBA 手记》系列图书的合著者，今天多数都成为了云和恩墨的合伙人，一起继续奋斗在另一个方向，这其中包括"Oracle 百科全书"杨廷琨和"中国西部唯一的 Oracle ACED"老熊（熊军）。

图 1-4

2004 年 4 月 13 日，也就是我到北京后 1 年多，我在网络上开启了自己个人的博客站点，注册的域名就是 Eygle.com，如图 1-5 所示。在随后的日子里，我基本坚持每天在网站上发表一篇或技术、或生活的个人文章，去记录自己成长的点滴，帮助别人也是帮助自己。

在博客时代，10 几年坚持下来，我的网站上已经累积了数千篇技术和生活感悟文章，这些内容对我来说是无比宝贵的财富。通过网站，我还结识了很多的朋友。最高兴的是，很多文章能够帮助别人。朋友们经常发邮件来和我探讨技术内容或者对某个有帮助的技术文章表示感谢，有很多朋友来自中国台湾、中国香港甚至是国外的很多地方，这些都成为鼓励我坚持下去的动力。

现在经常有网友问我，这么多内容是如何积累起来的，如何构建一个个人站点？其实除了技术之外，只有两个字：**坚持**！如果你能够坚持数年如一日的做一件事，那么最后的成绩一定会让你自己也吃惊的。

图 1-5

我们每个人在学习和成长的过程中,都做过无数的思考和学习。很多时候,我们都只是将这些经验和过程记录在自己的头脑中,时过境迁就可能模糊、遗忘,而如果记录总结出来,不仅可以帮助其他人,还可以对自己做个记录,当然这要有所付出,可是我坚信,**有付出就一定会有收获**。

所以我曾经在《Oracle 数据库性能优化》一书的序言中写到:

兴趣 + 勤奋 + 坚持 + 方法 ≈ 成功

很遗憾我不能给以上公式画上"=",但是无关紧要,只要具备了以上因素,我想我们每个人都会离成功不远了。

在从事一件工作或事业时,能够坚持不懈是多么重要啊!

2006 年 8 月,我和很多朋友一起参加了"中国首届杰出数据库工程师评选"活动,并且获评为"十大杰出数据库工程师"之一,这是外界对我做出的一个非常积极的肯定(图 1-6 是北京大学教授——唐世渭老师为我颁奖的照片)。

2010 年,我和张乐奕(Kamus)一起创立了 ACOUG(All China

图 1-6

Oracle User Group），进一步的推动地面活动和技术交流。

这个阶段我可以作出的总结是：**积累知识，分享经验，收获快乐！**写作的过程是艰苦的，然而分享的收获会超出你的想象。能够帮助别人，分享有价值的经验实在是一件快乐的事情。我计划将这个工作一直坚持下去。

由于个人对于技术的执着和热爱，这么多年来，不管在怎样的工作岗位上，我从来没有停止过对于技术的研究与探索。刚开始在北京做 DBA 的工作时，经常为一个个技术问题废寝忘食。记得有一次在公司思考一个问题未果，吃饭时一直思索，思路顿开始，立即丢下饭不吃，跑回去做实验来推理验证。有时候会持续很多年关注和跟踪某个技术问题，直到某一天豁然开朗，融会贯通。

我相信在学习的过程中，**每个人都会在不同的阶段遇到自己的瓶颈，然而必须在山重水复之后才能有技进乎道的感觉，我相信所有的技艺在最后的层面上都会如此，而只有具备毅力与坚持者方能抵达。**

有一年我去兰州大学做技术交流，兰州大学的一位李老师对我说，最近看我网站上提到的学习方法等内容，感觉到一个字：虚！我当时跟他开玩笑说，我还有更虚无的 8 个字可以送给你，那就是：运用之妙，存乎一心。

这是玩笑，也不是玩笑。有时候对 Oracle 进行了深入的研究与探索之后，接下来如何运用这些知识去解决问题，实际上是非常灵活的。很多时候简单的常规方法经过巧妙运用之后就可以非常神奇，发挥出你意想不到的作用。所以，我们应该花力气去做的仍然是积累、深入和思考，然后才能在遇到问题时举重若轻、运用自如。

这些年在技术方面不断的努力带来的一个额外收获就是 Oracle 公司官方的认同。在 2007 年 3 月，我被 Oracle 公司授予 Oracle ACE 称号，如图 1-7 所示，是国内第一个获此称号的人；在 2008 年 2 月，被 Oracle 公司授予 Oracle ACE Director（ACE 总监）称号。这是 Oracle 公司对 Oracle 公司之外的人所能授予的最高荣誉称号。到 2016 年 8 月，国内目前仅有 10 人保有该称号，其中 Oracle 数据库方向 8 人，MySQL 数据库方向 2 人。我认为在这个技术方向上奋斗的朋友，都可以将此作为一个奋斗目标。

图 1-7

所有的这些积累，都是后来我尝试创业必不可少的重要条件。

1.4 三重境界，见山见水见真我

最近在我们的"云和恩墨微信大讲堂"中，仍然有很多朋友时常向我咨询学习 Oracle 的方法。提到学习之中的艰辛和困惑,我就将自己最有感触的一些经验、观察和总结分享给大家。

最经常被提及的一个问题是,应该如何学习Oracle,怎样才能快速提高?很多人在学习的过程中经常感觉艰辛,甚或阶段性地停滞不前。我想这个旅程的体验不仅仅和Oracle学习相关,和任何一项技术的学习,都有相关。

其实学习任何东西都是一样,没有太多的捷径可走,必须打好了坚实的基础,才有可能在进一步学习中得到快速提高。王国维在他的《人间词话》中曾经概括了为学的三种境界,我在这里借用一下。

古今之成大事业、大学问者,罔不经过三种之境界。

"昨夜西风凋碧树。独上高楼,望尽天涯路。"此第一境界也。

"衣带渐宽终不悔,为伊消得人憔悴。"此第二境界也。

"众里寻他千百度,蓦然回首,那人却在灯火阑珊处。"此第三境界也。

学习Oracle,这也是你必须经历的三种境界。

第一层境界是说,学习的路是漫长的,你必须做好充分的思想准备,如果半途而废还不如不要开始。

这里,注意一个"尽"字,在开始学习的过程中,你必须充分阅读Oracle的基础文档:概念手册、管理手册和备份恢复手册等(这些你都可以在 http://docs.oracle.com 上找到);OCP认证的教材也值得仔细阅读,那些教材撰写得非常详尽和精彩。打好基础之后你才具备了进一步提升的能力,万丈高楼都是由地而起。

第二层境界是说,尽管经历挫折、打击、灰心、沮丧,也都要坚持不放弃。具备了基础知识之后,你可以对自己感兴趣或者工作中遇到的问题进行深入地思考,由浅入深从来都不是轻而易举,甚至很多时候你会感到自己停滞不前,但是不要动摇,学习及理解上的突破也需要时间。

第三层境界是说,经历了那么多努力以后,你会发现,那苦苦思考的问题,那百思不得其解的算法原理,原来答案就在手边。你的思路豁然开朗,宛如拨云见月。这个时候,学习对你来说,不再是个难题,也许是一种享受,是一门艺术。

所以如果你想问我如何速成,那我是没有答案的。"不经一番寒彻骨,哪得梅花扑鼻香。"当然这三种境界在实际中也许是交叉的,在不断的学习中,不断有蓦然回首的收获。

我引用一下杨廷琨在一次访谈中的经验终结,他认为"持之以恒"是关键。

"谈论Oracle技术学习的文章非常多,方法真的不是最重要的,持之以恒不间断的学习才是成功的关键。而除了潜心研究外,多关注新的技术发展和趋势十分关键。低头做事,抬头看路,了解最新的技术发展和行业的趋势可以避免走弯路，对于更好的理解技术的演进很有帮助。"

杨长老罗列了 4 种学习路径，包括他自己的，我稍加总结。

● 杨廷琨从阅读 Oracle 官方文档起步，先看 Concept，再看 Administrator，然后是 Backup、Performance Tunning、RAC、Data Guard、Upgrade、Utilities、Network 等等，通读所有重要官方文档。

● 盖国强多次提过他的学习方法是由点及面，抓住每个技术点，不断的深入下去，最终把整个体系的脉络理清楚。

● 崔华的学习方法是通读 Metalink 文档，他在演讲的时候介绍他每天要看几个小时的 Metalink 文档，每天都会经历多次的页面超时。

● 张乐奕更喜欢关注国外顶级专家的 BLOG 和 Mail List，这样可以快速的获取到业内专家的最新研究成果。

这 4 种学习方法，我概括成两类：Full Scan 和 Index Scan。

● 杨长老有"Oracle 百科全书"的美誉，他看文档是全表扫描，遍历；崔华钻研技术也是如醉如痴、废寝忘食，他读 MOS 是经年累月持之以恒的，也属于全扫。

● 我和张乐奕的方法有点像索引扫描，我推荐由一个根节点下钻，然后你可能发现几个分支，一堆叶节点，通过这样的过程由点及面，形成体系。

方法可以借鉴，但是最终还是要找到适合自己的路径去学习前进。我认为以上探讨的经验和思路，适用于所有领域的学习之中，希望对大家有所帮助。

我自己在学习的过程中，经常是采用"由点及面法"，可以和大家分享。由点及面是指当遇到一个问题后，一定是深入下去，穷究根本，这样你会发现，一个简单的问题也必定会带起一大片的知识点，如果你能对很多问题进行深入思考和研究，那么在深处，你会发现，这些面逐渐接合，慢慢的延伸到 Oracle 的所有层面，逐渐你就能融会贯通。这时候，你会主动去尝试全面学习 Oracle，扫除你的知识盲点，学习已经成为一种需要。

由实践触发的学习才最有针对性，才更能让你深入的理解书本上的知识，正所谓："纸上得来终觉浅，绝知此事要躬行"。实践的经验于我们是至为宝贵的。如果说有，那么这，就是我的捷径。想想自己，经常是"每有所获，便欣然忘食"，兴趣才是我们最好的老师。

作为一个数据库管理人员，你需要做的就是能够根据自己的知识以及经验在各种复杂情况下做出快速正确的判断。当问题出现时，你需要知道应该使用怎样的手段发现问题的根本；找到问题之后，你需要运用你的知识找到解决问题的方法。

这当然并不容易，举重若轻还是举轻若重，取决于你具备怎样的基础以及经验积累。

要是你觉得这一切过于复杂了，那我还有一句简单的话送给大家："不积跬步，无以至千里"。学习正是在逐渐积累的过程中提高。

1.5 云和恩墨，数据服务起征途

时至今日，IT 这个行业仍然是最为吸引毕业生的一个重要行业。记得多年前榕树下的一位朋友"落花如雨"说过一句话：喜欢这个行业，因为这个行业里汇聚了这个时代最聪明的人才与最快速增长的财富。

就因为这两点，众多的年轻人前仆后继的开始涌入这个圈子。那么然后，出路又在何方呢？一直以来大家都认为，程序员或者 IT 领域是年轻人的天下，因为这里有变换迅速的技术和产品，而机遇和压力一直是呈正比增加的。

我也开始探索作为技术人的出路，云和恩墨就是这样一个开始，如图 1-8 所示。

很幸运，我在职业生涯的前 10 年做好了充分的积累，打下了扎实的技术根底，拥有一定的网络影响力，从而具备了开始一项事业最

图 1-8

初的条件；更加幸运的是，我结识了一群值得信赖的伙伴，他们有的擅长管理，有的精通技术，怀着共同的梦想，在前进的途中，逐渐汇聚到云和恩墨公司的大旗之下，为了共同的理想而奋斗。虽然未来充满风险和未知，但是没有什么比挑战自我的极限更值得尝试的了。今天，云和恩墨已经成长为一个超过 200 人的团队，拥有 6 位 Oracle ACED，2 位 Oracle ACE，国内 SQL 大赛冠军，30 多位 OCM 认证大师，同时具备来自互联网公司的 MySQL 专家，打造出国内最卓越的技术服务团队，这其中艰辛与甘甜并尝。

走一条不可预期的路，对自己充满挑战，也充满乐趣，这体验不管成功还是失败，都将是全新的。

云和恩墨是一家依托互联网新型人脉建立起来的公司，也依托互联网进行传播和客户发展，我一直希望这个全新的公司能够带有一些网络的、Web 2.0 的气息，能够跟进时代与潮流，更好的为用户提供服务。"专注、专业、灵动"是我们的目标，专注于数据，专业以人才，灵动于服务，由此实现我们"数据驱动，成就未来"的目标。我们深信，唯有成就客户，才能成就自我，云和恩墨始终将对客户的承诺放在第一位。

在云和恩墨全体同仁的努力之下，公司取得了快速的发展，更获评为 2015 年中关村高成长企业 100 强，如图 1-9 所示，这是我的选择以及恩墨人全体努力所开创出来的一条道路。

当然不管走哪一条路，这里面都有一句潜台词就是：你必须面临严峻的竞争，取得快速的成长！对于企业和个人都是如此。

张爱玲说过，成名要趁早。做技术的也是如此，成长越早越好，越快越好。对于个体，在经历了足够的积累和成长之后，在尽快到达天花板并且超越之后，你会发现前方供你选择的道路会更多、更宽广。而对于企业，也必须具备快速迭代推进目标的能力。有视野、有执行，才能最终成就事业。

这个快速变化的时代给我们的压迫感随时都在，时间与时机总是稍纵即逝，所以进入 IT 这个领域，注定我们要不断跋涉，不能停息。

说了那么多，其实有一个核心的思想：**积累非常重要**！不管在哪一个行业，做什么工作，如果你能够利用自己以前的学习、工作经验，发挥自己积累的技能，那么做事情就会事半功倍；而如果你试图进入一个全新的领域，那么一定要做好充足的功课才行。

总结一下这些年走过的路，零零碎碎有一些话可以和大家分享。

图 1-9

（1）**勤奋、坚持**，这两点非常重要，当然如果能够找到自己的兴趣，作为职业，用正确的方法，走正确的路，那么取得成绩是早晚的事情，我经常写给读者的座右铭就是：天道酬勤。

（2）**在看不清方向的时候，低下头来把手中的工作做好。**

（3）向他人学习，向聪明人学习，借鉴成功者、同行者的经验非常重要。

（4）敞开心胸，平淡看得失。

（5）在正确的时间做正确的事，比如结婚、生子。

（6）行动有时候比思想更重要。

这些话，永不落伍。

1.6　理想实践，开发运维一体化

在数据行业那么久，我们总希望能够通过自己的努力，将好的想法落地，渐渐地改变行业中的不合理之处，让这个技术世界变得美丽一点点。

那么这个行业里有什么迫切需要改变的？

作为资深的 DBA 你可能会发现，我们 10 年前处理的问题和今天没有什么不同。针对数据库的运维巡检日复一日，SQL 优化应对全表扫描或是隐式转换，转眼就耗费了经年的时光。所以我们有一个理想，不要让 DBA 重复在这些无休止的工作上，或者至少能够做得更有价值，也力争能够改变用户在使用数据库的过程中，屡见不鲜的事后救火。

所以我们第一个在国内提出了"SQL 审核""智能巡检"等理念，希望真正能够通过自动化运维、工具化约束，去改善 SQL 开发质量、发现和凸显问题，从而防患于未然，提升系统稳定性，改善数据库运维的现状。我们相信通过规范化、标准化、智能化，才能够不断推动业界向前。

早在 2011 年，我们基于对于业界的思考，就开始开发了一款 SQL 审核产品，称为 z3，如图 1-10 所示。它可以审核开发测试阶段的 SQL，发现问题，提出建议，希望由此将运维 DBA 和开发结合起来。我们从未想过，这居然就是今天最热门的 DevOps 所讨论的范畴。

图 1-10

通过不断地呼吁和倡导，今天我们非常欣喜地看到国内很多企业都开始去开发这方面的工具，去推行 SQL 审核的理念。

那么什么是 DevOps 呢？维基百科的定义如下所示。

DevOps（Development 和 Operations 的组合词）是一种重视"软件开发人员（Dev）"和"IT 运维技术人员（Ops）"之间沟通合作的文化、运动或惯例。透过自动化"软件交付"和"架构变更"的流程，来使得构建、测试、发布软件能够更加地快捷、频繁和可靠。

从这个定义可以看出 DevOps 实际上是一种文化上的改变。开发和运维通过更多的沟通达成更可靠的系统输出，从而为企业的共同目标而加注动力。在 2015 年 Gartner 的技术成熟度曲线上，DevOps 正处于巅峰。

而根据多年的行业经验，我们认为 **DevOps 在 Oracle 数据库的最佳实践应该就是 SQL 审核**。江苏移动技术专家戴建东的一段感触之言为我们提供了来自实践的依据，他明确提到：

"其实在生产中，绝大多数 Oracle 的业务系统出现问题都是 SQL 导致的。但是大多 DBA，尤其是偏运维的 DBA 对 SQL 并不擅长，这些 DBA 承担着数据库运维和维护稳定性的职责，而他们对这些问题可能又无能为力。原本 SQL 的质量应该是开发层负责的问题，但目前的现状是，开发人员管不了，运维人员不擅长。所以当系统出现问题的时候，就需要专业人员"救火"，而事发或事后救火往往是业务已经遭受了损失。"

SQL 审核的理念就是，将这些"开发人员管不了，运维人员不擅长"的核心 SQL 问题抽取出来，作为 DevOps 的范畴。通过来自运维的经验，指导和辅助开发完成高性能的 SQL 改写，并且不断通过自动的 SQL 审核工具和专家的修改建议相结合，推进开发质量的提升，改善系统的稳定性，将性能事故消弭于无形。这也正是 DevOps 的理想所在。

对于开发团队来说，持续的进行 SQL 培训我认为非常重要，开发的 SQL 能力提升了，对于 DBA 只有好处，数据库的稳定性自然会得到提升。DBA 也有职责去和开发沟通，对他们进行面向运维高性能培训。在 Oracle DevOps 时代，DBA 要勇于承担责任，去推进变化。而且在 DBA 的学习过程中，就是要不断深入去了解各个层面的知识，才能不断进步、融会贯通，找到如鱼得水、游刃有余的感觉。也才能从工作中找到自信和乐趣，进而培养和巩固兴趣，在完善自我的同时帮助他人，提升团队。

在今天的云时代，各个领域都在发生变化，DBA 的领域同样面临挑战，表达一下我的观点。

（1）DBA 从后端走向前端才能更充分的体现其技术价值。

（2）应用向着预防问题方向演进永远比事后救火更重要。

所以慢慢很多企业开始在开发环节，以开发 DBA 来进行把关，以 SQL 审核优化来控制质量。我建议 DBA 们关注一下这个方向和变化。在现实中，解决单个问题往往是简单的，但是我们应该思考如何去防范一类问题，让更多的人免于重复落入类似的故障。

从经验到规范，从规范到规则，这是 DBA 工作更高价值的体现。当我们能够将经验固化成 SQL、算法或者程序之后，才能帮助到更多的人。我想，只要我们每个人在自己熟悉的领域都能够努力一点点，就能够一起将我们所从事的行业变得美好一点点，从而也会使得我们的世界变得美好一点点。

云和恩墨在云的时代，正在致力于以团队智慧和经验，衍生产品，以产品服务更多的客户和 DBA 们，进而推进行业的进步。

1.7 快乐生活，此心安处是吾乡

在本文的最后，我还想说几句的是，除了工作之外，不要忘记了生活，没有什么比生活更重要的，家是世界上最重要的地方。

想一想你匆忙的脚步是否已经很久没有为一览风景而停留？想一想你是否已经很久没有陪家人与朋友出游谈天？要记住我们是为了生活而工作，而不是为了工作而生活。在 IT 圈子的朋友们尤其如此，高强度的工作，大量的加班，黑白颠倒，这一切绝不是生活的目标。

在我的一本书的结尾，我写过如下一段话，与大家分享。

2008 年的 9 月 21 日~9 月 25 日，应 Oracle 公司的邀请，我到旧金山参加了 Oracle 2008 Open World 大会。你能猜到大会上最打动我的一句话是什么？

不是 Oracle 发布的 Exadata Programmable Storage Server 也不是 HP Oracle Database Machine（Oracle 软件公司划时代发布的两款硬件产品），而是 Larry 在 Keynote 上发表的演讲时讲到的一段话，他说：在过去七八年间，我的主要工作是去赢得美洲杯（American's Cup），然后才是在 Oracle 的工作。

不管 Larry 想传达的意思是什么，我的理解是，能够快乐地做自己喜爱的事情，才是人生最值得追求的，而工作不过是生活的另一面。

工作是永无穷尽的，而生活则是有限的，快乐的生活比什么都要重要。

作为芸芸众生中的普通一员，在为理想与未来奋斗之余，让我们用更多一点的时间去经历更加快乐的生活吧！

Oracle 公司自 1977 年始，已经度过了 38 年的漫长历史。时至今日，Oracle 公司仍然是充满创新和斗志昂扬的一家公司。我的职业生涯一直有 Oracle 相伴，感谢 Oracle。

根据 2016 年 7 月 28 日的消息，为了实现云上 SaaS 的市场目标，Oracle 公司出价 93 亿美元，收购 NetSuite 公司。NetSuite 是一家提供云上 ERP 服务的公司。在过去的几年间，Larry Ellison 致力于将 Oracle 变革为一家云公司，整个公司正在全力以赴地将所有的产品和服务转移到云上来。Oracle 已经从一家以数据库为主的软件公司，华丽转身成为一家能够在 IaaS、PaaS、SaaS 上提供全面云服务的云公司。

我们相信 Oracle 在云时代必然能够再次"冲上云霄"，赢得未来。

第 2 章　回首向来萧瑟处，也无风雨也无晴

——我的十年 Oracle DBA 奋斗路（侯圣文）

> **题记**
>
> 迄今为止，我觉得这辈子最幸运的两件事，一件是遇见了我太太，另一件就是结识了 Oracle。没有早一步也没有晚一步，刚巧赶上了，在最适合谈恋爱的年纪谈了一场没有分手的恋爱，在最适合干事业的年纪做了一份不曾放弃的事业。

2.1　人生若只如初见

其实，搞事业和谈恋爱是一样的，一半靠"缘"，一半凭"分"，"缘"看天意，"分"靠努力。11 年前，我还是个初出茅庐的计算机专业毕业生，属于干一行爱一行的年纪。我背着书包只身来到北京，本想谋求一个网络工程师的职位，结果面试没通过。不过"失之东隅，收之桑榆"，后来机缘巧合进了海关，做起了与数据打交道的工作。刚开始只是做一些简单的增删改查，没事儿写写 SQL。真正关注 Oracle 是因为公司的几个 DBA 相继离职了，一跳槽都是一万多元的收入，这对于当时只赚三千元的我来说无疑是颇具诱惑的，每天晚上掰着手指头算人家一个月一万多元要怎么才能花完。那个时候数据库技术相当火热，公司里的 DB2 和 Oracle 工程师颇受重视，收入可观，都是妹纸们争抢的对象。于是，我毫不犹豫地选择了 Oracle 数据库作为自己的职业发展方向，我相信学好 Oracle 自有颜如玉，学好 Oracle 自有黄金屋。

2.2 日久生情见真心

为了进入公司的 DBA 组,我需要一个敲门砖,那就是拿到 OCP 认证。在这个世界上有很多无瑕美玉落入泥淖之中,被世人当做了烂石头。OCP 课程实际上就是这样一块美玉。我始终认为 OCP 是 Oracle 开发的最完善最成体系的一套课程。现在的很多人都觉得 OCP 认证无用。当然如果你把它当做一张纸,那这张纸确实没多少分量,但是如果这张纸背后凝结了你的心血,你能够悉心研习整个 OCP 课程,把每个考题分析透,动手做每个实验,那这张纸就是无价之宝。

很长一段时间,我以考取 OCP 为目标,把整个 OCP 课程体系做了系统梳理,每道题都做了仔细的解析,不懂的地方就看书或在网上查。这样有目的又有章法的自学效率极高,我感觉短短一个月学到的东西超过了整个大学四年的收获。这也是我一直以来推崇认证的原因,认证可以给我们的学习确立一个明确的目标,有目标的学习才能事半功倍。

在这样的学习过程中,我渐渐发现 Oracle 的整个体系是非常完善的,这让我对 Oracle 数据库产生了更大的兴趣。这是很自然的现象。如果你跟一个姑娘交往,越接触越发现她的好,一定会喜爱有加;反之,如果发现她败絮其中,自然会厌弃离开。Oracle 无疑是金玉其外,锦绣其中的"好姑娘",我越钻研越觉得它的体系完善,设计精巧,所以渐渐迷上了这门技术。就这样我以满分拿到了 OCP 认证,这对于当时还是小菜鸟的我来说是个不小的鼓舞,我也凭借这个认证以及认证背后的实力顺利升级成了公司的 Oracle DBA。

与程序员的"团伙作案"不同,DBA 行业充满了个人英雄主义色彩。在焦头烂额的故障现场,一个出色的 DBA 往往扮演了救世主的角色,丢失的数据、中断的业务都可能在一瞬间因为你的操作起死回生,几天几夜没合眼的客户也会因为你的出现而得到解脱。这也是数据库技术和 DBA 职业最吸引我的地方。刚当 DBA 那会儿我感觉整个人都处于亢奋状态,海关数据中心虽然待遇一般但绝对是学习的好地方,有大把大把的钞票扔出去买最牛 X 的设备,这里俨然成了我的实验室,于是我每天钻进一年四季都凉嗖嗖的机房干得热火朝天。除了工作的内容,大部分时间我都在鼓捣自己的东西,没事儿搞出个故障来看自己能不能排除。这大概相当于军事演习,虽然没上战场,但也不能荒废了看家的武艺。

2.3 衣带渐宽终不悔

成为 Oracle DBA 是职业选择的问题,如何在这条路上走得更远更成功就是职业发展的问题了。同样拜入少林门下,有人成了不世出的扫地僧,有人则是做一天和尚撞一天钟。

作家柳青曾说:"人生道路虽然漫长,但紧要处常常只有几步,特别是当人年轻的时候"。与 ITPUB 结缘,无疑是我人生中至关重要的一步,因为它为我开启了一扇门、点亮了一盏灯。

DBA 是个实践性极强的工种，涉及问题多，作为一名初级 DBA，我是没有师傅领进门，修行完全靠个人，所以当时上网查资料就成了我解决问题的一个重要途径，然而在浩如烟海的网文中却很难找到真正实用的资料，很多文章不知所云，有的则人云亦云，稍好一些的也感觉言而未尽。因此在 2007 年，Secooler 同学做出了一个重要决定，我要在 ITPUB 上开辟一个个人技术博客，既为自己释疑，也为他人解惑。

然而世事总是知易行难，有太多人选定了方向却没有走下去的毅力和勇气。如果没有当时的坚持，我的博客和我的理想也早已淹没在时间无涯的荒野里。那么当时是何种力量支撑着我继续走下去呢？我想应该是那些读我博文并给予我热情鼓励的网友们，以及那些与我志同道合的同路人。不是所有的成功者都有伟大的梦想。当我站在起点望不到终点时，支撑我跑下去的不是金光闪闪的奖杯，而是沿途的一句句赞叹、一声声问候。

慢慢的在 ITPUB 上撰写技术博文成了我生活的一部分，工作的压力、家人的不解都未曾动摇过我的决心。无数次挑灯夜战，多少个不眠之夜，终于有更多的人发现了我的努力，博客的访问量日增，我成了 ITPUB 的版主之一，这是我在起点时未曾想到的，却是我到终点时最为自豪的回报。

2.4 天下谁人不识君

Eygle 曾经无数次讲起我在一个月黑风高的夜晚打电话询问如何成为 ACE 的故事（图 2-1 是 Oracle 网站上的 ACED 页面），这故事传遍了大江南北，鼓舞着所有从事 Oracle 工作的红男绿女。

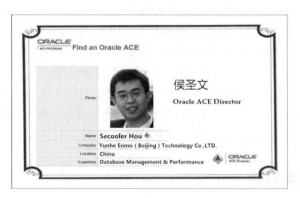

图 2-1

在我看来，成为 ACE，进而成为 ACE 总监绝无捷径。只要坚持研究、乐于分享，总会有人发现你的努力、认可你的成绩。

首先，技术实力是根本。任何成为 ACE 的人都有过刻苦钻研技术、不眠不休的经历。大家的博客和文章都不是掺水的，都是自己实践、研究的结晶。没有几年默默无闻、坐穿板凳的学习和研究，怎么会有高深的技术功力呢。

其次，能写善言是条件。 能写的您就出书、写博客，善言的您就演讲、做讲座，总之要扬己所长将自己所学、所做、所思、所感分享给大家。ACE 鼓励的就是一种分享的精神和态度，ACE 肯定的就是大家对 Oracle 技术的支持与传播。

第三，融入圈子觅知音。 进入任何一个领域，迅速融入这个领域的精英圈子很重要，圈子会给你更多资源，前辈也会给你很好的建议。还是菜鸟的时候我经常去参加 ACOUG 的活动，在那里我认识了圈内的很多知名专家，最开始只是单纯地学习，后来渐渐期望成为其中一员，并最终凤愿得偿。后来发起并创建了 OCMU（中国 OCM 联盟），依然本着"分享技术，提升价值"的理念，每月一次线下的技术分享交流活动，让更多的技术爱好者们有分享交流的平台。

第四，精益求精重细节。 很多人都在坚持写博客，但有的人博客错字连篇，我是很为之可惜的，这样的博客让人看了不会加分，反而会给博主减分。虽然是技术博客，但我写的每一篇文章除了保证技术正确之外，在文字、标点甚至格式上都尽量做到准确无误。切记优秀是一种习惯，敷衍也是一种习惯，决定成败的往往都是细节。

第五，持之以恒是关键。 很多人都会写博客，但是很少有人能坚持每天一篇原创博客而且一写就是以年计数的。曾经有两年左右的时间，我几乎每天都写博客，这件事渐渐成了习惯，就像刷牙一样，每天晚上睡觉前第一件事刷牙，第二件事发文章。

做一个决定只要一秒钟，付诸行动只需一刻钟，但是要得到一个成功的结果，却需要一年、10 年，甚至一辈子。几年如一日的博客写作既奠定了扎实的技术功底又提升了自己在圈内的认可度和知名度。

机会总会垂青有准备的人，在积累了足够的实力和人脉之后，成为 ACE 仿佛是水到渠成的事情。

2.5 回首向来萧瑟处，也无风雨也无晴

从 ACE 到 ACE 总监我用了两年的时间。这两年里我的 Calendar 几乎满配。六七次大型会议演讲，十几次小规模技术交流活动，天南海北的校园分享活动和企业分享活动，还有每周雷打不动的 OCM 课程。最辛苦的时候，一周跑两个城市，风餐露宿。一个月 20 多天，不是在上课，就是在做活动，剩下的时间是在奔赴活动现场的路上。终于这样的奔波辛劳为我赢得了来自 Oracle 的最高奖赏——我得到 ACE 总监的称号，这是来自 Oracle 的最高荣誉，是我寤寐思服的境界。拿到 ACE 总监贺喜函的那个晚上我以为我会激动得失眠，却不想一夜好眠，或许是这一程走得辛苦，我已经没有精力再去激动一番了。

这两年我的职业生涯也发生了质的飞跃。我加入了云和恩墨，一手创办了恩墨学院，真正把培训作为了自己的一份事业。

恩墨学院从无到有，从无名到有名，凝结了我的全部心血。经常披星戴月地辅导学员训练，看着日光灯下，大家奋斗的身影，听着键盘噼啪作响，我常常恍惚，仿佛回到了高三那段疲惫而充实的岁月。不过此时，我的身份不再是学生而是老师。我熟悉每个人的技术功底，我会敦促每个人坚持练习，无论多晚我都会在手机边守候着大家的成绩，也正因如此，恩墨战队才有今天这样傲人的成绩。

虽然人生常如浮萍，境遇迥然，无法自控，但我很庆幸我没有走太多弯路，在 Oracle 的坦途上一路前行。从一个 Oracle 爱好者，到一名 Oracle 从业者，并最终成为 Oracle 的传道者，我很满足。我自觉年纪尚轻还没到总结前尘的时候，写下这些只为纪念那十年拼搏奋斗的岁月，缅怀那曾经热血沸腾的青春，以先行者为鉴，与后来者共勉。

第 3 章 正确抉择，丰富人生

——我的职业生涯和学习方法（Joel Perez，孙雪 译）

题记

在我每天接触到的技术的人群中，有很多人会询问我关于学习 Oracle 的方法，希望在 Oracle 领域取得建树，或者成为国际技术顾问。我想说做任何事情没有绝对正确的方法，可能因人而异。但的确存在一些适用于大部分人的方法使我们更有效地学习和提升。接下来我将会通过我的职业生涯及一些经历来分享我个人的学习方法。希望对大家有所帮助。

3.1 初闻 Oracle

我 20 岁的时候，在我的祖国委内瑞拉的一所学校教电子相关的课程。在一次课上，我跟一个学生有以下的对话内容。

我的学生：Joel，你听说过一种叫 Oracle 的技术吗？

我：没有，那是什么？

我的学生：这是一个很棒的数据库软件，我的公司总部位于法国巴黎，现在急需大量的高级 DBA 来管理重要的数据库，如果你有兴趣的话，我可以推荐你过去。

我：我现在还不知道 Oracle 是什么，但我觉得如果像巴黎这样一个发达城市，需要 Oracle 相关的技术人员，那么 Oracle 技术应该会在以后会有很好的发展前景。

那天我回家在网上搜索了一些关于 Oracle 技术相关的内容，连续两个小时的搜索，我就对

Oracle 产生了强烈的兴趣，并决定，Oracle 技术将会成为我今后技术生涯的主要方向。

规划与眼界

在我了解了 Oracle 的那一刻，我就决定了我要学习 Oracle。但是选择一个方向，不能因为一时心血来潮或者短暂的兴趣，一定要有长远的打算，要在这个领域做出些成就。

我认为规划和眼界是很重要的。对于任何一个人，无论你在数学，化学或者象棋等各方面表现出极强的天赋，若没有眼界，对自己的人生没有合理的规划，无论是学习 Oracle 还是其他的知识，都不能获得成功。

3.2 选择适合自己的路

那天晚上我买了一些书，首先从介绍 Oracle 是什么和 Oracle 区别于其他数据库的一些特征开始学习。我发现，Oracle 其实是一个很广泛的领域，包含很多方面的知识。那么在学习的过程中，就不能因贪多而浅尝辄止，你必须要选择适合你的方向，然后深入学习。在做任何决定之前，一定要思考清楚，现实中有很多学习 Oracle 的人，因一时兴起做了选择，但由于对未来没有明确的规划，最终不能在这条路上坚持下来。

3.2.1 选择合适的资料书

在学习 Oracle 的时候，选择合适的资料书是很重要的。找到适合自己的书，然后深入研究，一定会有收获。现在的社会，并不缺少技术类的书籍，我记得在我上学的时候一本技术书还不到 50 美元，并不贵，但却很少有人买，更少有人看。

后来随着我对 Oracle 学习的深入，看了很多的书，除了学习知识以外，我也会比较这些书的不同，分析作者的思路，我慢慢发现，对于一个作者来说，要写一本好书是非常难的。他们不仅需要选择有价值的内容，还要花很多的精力思考如何合理组织这些内容，让读者有最佳的体验。每一个作者的努力付出并不是出于商业的目的，事实上也赚不了钱，而是一种技术的分享。所以在此我建议大家，当你选择了一本好的资料书，一定要充分利用它的价值，因为作者为了让你更好地学习，花了很多的心思和精力。

3.2.2 融入圈子

在我们学习成长的道路上，除了自己的努力，我认为把自己融入一种技术的圈子很重要，让你的身边有一些跟你追求同样东西的人，这样的人可以是你的偶像或者一起学习的伙伴，因为"独学而无友，则孤陋而寡闻"。无论我们所学的是什么，总有一些领域的大师，他们在很早之前就走过了我们将要走的路,所以很多我们会面临的困难和挫折,他们都经历过,因此,我们要从他们的教训中获取经验,从他们的经验里获得成长,可以多听取他们的意见建议,从而成长得更快。同时,当我们以行业内某一个大师级别的人为我们的榜样,我们在日常生活和工作中,也会更多地约束自己。另外,很重要的一点是,我们最好能够找到志同道合的学习伙伴,跟我们一起进步。在彼此的分享交流中,能够共同成长。在我刚开始学习 Oracle 的时候,我有一个朋友曾带我去他所在的公司,以简明的方式带我认识了 Oracle 在生活中的应用,让我对 Oracle 技术有了更深刻的认识。虽然他的分享只有短短的一天,但对我后来的职业生涯影响非常大。

3.2.3 关于大学的课程

自从我决定要在 Oracle 方面深入研究学习之后,我回到大学校园上了跟 Oracle 相关的课程。这次回到校园,距离第一次上大学已相隔 15 年,我不得不承认,大学的教学方式发生了很大的变化。过去那些年虽然已经存在很多的企业版的软件(如 EM 工具)可以辅助教学,但跟数据库相关的课程的操作都是手动完成的,也就是说我们当时还是敲命令实现的,因此对于一个初学 Oracle 的人来说,通过上课能够获得很大的能力提升。但现在时代变了,教学方式的改变给学习的人带来了一些方便,但随之而来的弊端也不少。接下来我会详细谈一下当今大学课程的利与弊。

利,主要表现在以下几方面。

(1)当今大学的 Oracle 相关的课程,会通过与 Oracle 公司的相关配合和推动,使得课程内容与市场的需求一致,培养市场化的人才。比如说对于 Oracle RAC 技术,如果你通过官方文档来学习,你会发现要学的内容有很多,也很复杂,但在你大学的 Oracle 课程当中,会将其中他们认为最重要的框架列出来,然后你按照他们的思路来学习。这些框架的学习能够让你更快地投入到工作当中,掌握最核心的技术点。

(2)能够让你快速地掌握一门技术,比如对于一个完全不懂 Oracle 的人,可能只需要 5 天左右的时间就让你了解 Data Guard 的技术原理和相关操作。

(3)这些课程为你的 Oracle 相关认证奠定基础。

(4)在一些国家,修这些课程能够为你的专业表现获得额外加分。

弊,主要表现在以下几方面。

(1)以前的课程练习和授课都是代码完成的,现在的课程很多采用了企业级的一些集成软件(如 EM 工具),虽然操作起来方便,但不容易清晰地理解其内部原理。如果你想要深入学习原理的话,还需要额外找资料学习。

（2）Oracle 的相关课程和认证都是很昂贵的。

我修过很多 Oracle 的课程，这些课程让我对 Oracle 技术的认识更深刻，也加快了我在 Oracle 这条路上的成长。通常情况下，这些课程能教你快速地掌握一门技术，只有在很少数的情况下，才需要为了学习某项特定技术花几个月的时间。因为我上了很多的 Oracle 相关的课程，这帮助我后来成为拉丁美洲在 Oracle 领域最卓越的领袖之一。

3.3 树立目标

最快的成长方式就是寻找合适的参照物，然后不断接近它。在我的职业生涯中，我曾有两个偶像，一个是美国的 Tom Kyte，另一个是印度的 Arup Nanda。我最初了解到 Kyte 是通过一个技术论坛 https://asktom.oracle.com，这是一个分享技术的论坛，我经常会看上面的内容。说到论坛，这也是我认为学习中比较重要的一个环节，我接下来会详细谈我的一些体会。

我内心一直渴望成为他们那样的人。于是我不断研读他们写的书、文章、博客等，这些对我的指导作用很大。我很看中目标的重要性，如果你想在某个领域有所建树，就多向领域内的专家看齐，了解他们的学习方法、思维模式以及处事风格，这样经过潜移默化的影响，你自己也会变得越来越像他们。

3.3.1 参与论坛分享

论坛是获取和传播知识的一个很基本但很有效的平台。在论坛上，每天有成千上万的人提问题，也有成千上万的人准备回答问题，你可以通过论坛的交流积累一定的专业知识，然后在分享中不断深入和广泛地学习，这能够帮助你快速在论坛成名。我是从 13 年前开始接触论坛的，尤其是 Oracle 论坛，刚开始短暂的接触就让我爱上了这种学习方式。因为每天有好多人问各种各样的问题，这迫使我不断地去钻研和深入学习，这在很大程度上巩固了我自己的知识储备。同时当我跟别人交流的时候，思想碰撞，又能够去发现新的知识点，通过这种方式开阔了我的视野，让我勇于探索和创新。我完全投入论坛的时间大概有一年半。在这期间，我积极参与讨论和互动，渐渐成为全球 Oracle 论坛上回答问题最多的人之一。为此，2003 年，Oracle 给我颁发了"OTN 专家"的荣誉称号，我是拉丁美洲第一个在 Oracle 领域被 OTN 授予专家称号的人，后来也成为全球第一批的 ACE。

3.3.2 写作

我觉得写文章是一个人分享知识的有效途径，我曾在读同事的技术文章的短短几分钟的时间都能够学到很多的技术要点。在我学习 Oracle 生涯中，我会选择行业内一些有名作家的作品去读，然后尝试着学着他们的方式去写。期间我最大的感悟就是，当你为着出版文章的目的去学习新的知识，探索新的领域的时候，你会充满动力，比往常更加深入地研究，因为你心里清楚地知道，你所要表达的内容是要给公众，给千千万万的读者看的。因为不希望出错，所以写作过程也是提高技术、增加学习严谨度的重要途径。

像我之前提到的，我有两个偶像，Tom Kyte 和 Arup Nanda，他们曾被认为是 Oracle 行业内最伟大的作家。幸运的是，在过去的 5 年里，我在世界各地传播 Oracle 技术和各类交流活动的时候，都曾与 Tom 共事。受到他们两人的影响，我在积累了一定的知识之后，开始学着写一些文章，最初主要是发表在"OTN 西班牙频道"和"OTN 葡萄牙频道"，我是这两个频道的主要作者，现在借助 OTN 平台，我已经在全球范围内发表技术文章。

3.3.3 参加技术研讨会

现在各类技术峰会和在线研讨会变得越来越流行，通过各类会议的交流和分享，可以扩展自己的知识面、增加行业见识。有些会议是免费的，当然也有收费的。根据我这些年的经验，大部分的演讲者会在发言的前 45 分钟里将自己对某个领域或者某个技术点的认识和研究成果尽量细致地展示，甚至会包含稍夸大的成分去渲染，以此来达到吸引听众的目的。而后面的 45 分钟，则会描述一些自己的职业生涯的经历，表达一种职业的情怀。但无论哪一部分内容，因为他们丰富的经验和深入的研究，总会让作为听众的你获益匪浅。所以我比较推荐大家多参加一些技术研讨会。

几年前我就开始在这类会议中发表演讲，目前为止我已经在 30 多个国家做过技术演讲，无论是作为演讲者还是作为听众，在这类会议中我总能学到很多的东西借此丰富自己。

3.4 关于认证

前面向大家推荐了一些我个人认为有价值的学习方法和途径，接下来我会谈一谈关于 Oracle 技术认证的事情。对于学一门技术到底要不要考相关认证，这个话题一直都是智者见智，并没有绝对正确的观点。我只是分享一下我个人的一些看法。

我印象深刻的一次经历，有个业务需要持有 Oracle 证书就可以接，并且薪酬是 250000 美元。从能力上讲，我觉得我可能并不比考了认证的人差，但仅仅因为没有证书，我就有些怀疑和犹豫，不知道自己是否配得去接这个业务机会。

我认为考技术认证是非常值得并且有必要的事情。说到费用，各地可能存在差异，但我觉得比起考认证的益处，这远不足以阻拦我们。这不仅是对你所学知识的肯定，也在很大程度上反映出你在某方面的能力。以下是考技术认证的几个优点。

（1）通过认证可以检验你是否掌握了 Oracle 希望你掌握的东西，而这些都是 Oracle 领域最核心最有价值的内容，也是你从一个学习的角色转为从事 Oracle 技术的工作人员所必须具备的能力，因此，从某种意义上说，通过认证能检验你是否真的准备好了跨入 Oracle 领域。

（2）在你认证备考的过程中，会得到很多的资料，这些资料的价值远不止是让你通过认证。对于学技术的你来说，这些资料一生都是有用的，值得你不断研究和深入探索。

（3）认证让你的简历更吸引眼球。事实上，写一大堆你能做什么，你做了什么，可能还不如几个很出名的证书更有意义。

（4）拿 Oracle 的认证来说，OCM 的认证不仅需要掌握很多的技术原理，同时还需要大量的实践，备考过程更是充满压力和挑战。在这样的环境中，不仅仅让你的技术得到提升，也会锻炼你的耐力、毅力，让你为今后学习路上的各种困难做好准备。

我习惯将 Oracle ACED 或者 OCM Maximum Availability， Cloud 12c/11g 作为我主要的头衔，因为这对我来说，不仅仅是一种认证，而是代表着我一路走来的付出与收获。

迄今为止，我从事 Oracle 行业已经 17 年了，期间我参加过各类会议、商务活动等。我个人很喜欢旅行，因为兴趣和工作的缘故，我目前走过了 85 个国家，这并不是一件容易的事情。我想我坚持下来的理由就是，我热爱我所做的，所以我能够完全将自己的生活和工作融为一体，在工作中收获快乐，丰富人生。我自豪地认为，从最初踏入 Oracle 领域到现在，我一直做正确的事和正确的决定。因此今天把我的经历分享给你，不是告诉你我有多厉害，而是希望你通过我的经历去思考自己的人生，选择自己所爱的方向，然后坚定不移地走下去。愿你有一个丰富而充实的人生！

第二篇

知 识 基 础

- 第 4 章　Oracle 数据库的跟踪和分析方法
- 第 5 章　兴趣、思考与实践
- 第 6 章　使用 XTTS 技术进行 U2L 跨平台数据迁移
- 第 7 章　Oracle 的参数和参数文件

第 4 章　Oracle 数据库的跟踪和分析方法

——盖国强

在今天的技术领域，DevOps 已经成为最热门的话题之一，DevOps 是开发和运维一体化的实践趋势，也是运维掌握一定的开发能力，推动和协助开发进行适应高效运维的渐进变革。

在我的技术生涯中，对 Oracle 数据库的接触最多，感受也最深。如果说要将最值得推荐的技能展示给大家，那么我想推荐的就是 Oracle 跟踪方法。事实上，通过跟踪能够实现的也正是不断了解、接近开发的思路和方法，从而有助于运维中的问题诊断、排查和解决。

我在一个活动上分享过一段话，摘录在这里，作为我这一章内容的开始。

"早上我听到一句话印象深刻，叫"隐藏的权利感"，我想把这句话应用到数据库，表达一下我的观点。

Oracle 数据库，虽然是一个商用数据库不开源，但是它又是非常开放的一个产品，Oracle 几乎所有的内部操作，不管是调优的过程还是数据库的各种内部操作，都是可跟踪解析的。比如 Oracle 数据库的启动和关闭过程，全程是可跟踪的。它的启动关闭会解析成多少个递归操作，我们全都可以跟踪出来。

所以我们做 Oracle DBA 的工作时，面对任何事情我们都会非常有信心。Oracle 开放了各种接口，方法和手段给我们，只要我们去分析研究，就能够把一个问题的 Root Cause 找出来，接近 Root Cause 就离解决问题不远了。

一个数据库只有变得更加开放接口，更加开放 DeBug 功能的，才能让我们在研究这个数据库的时候也可以找到更多的乐趣。我觉得这里面找到的乐趣就是我讲的，是隐藏的权利感。

就是我不动声色,但是我知道我在处理接触这个数据库的时候,我有非常强的把控力,我能撼动和解决几乎所有的问题。我觉得这一点对于技术人员是非常重要的。"

Oracle 数据库的这些基本跟踪方法,伴随着我的技术成长和排忧解难的职业历程,以下详细的通过案例进行解析。

4.1 SQL_TRACE 及 10046 事件

最常用的跟踪方式是通过初始化参数 SQL_TRACE 或者设置 10046 事件。首先从文档了解一下 SQL_TRACE 基本介绍,以使大家能对这个工具有所了解,并熟悉其使用方法。

4.2 SQL_TRACE 说明

先来关注一下 Oracle 12c 官方文档对 SQL_TRACE 的说明,如表 4-1 所示。

表 4-1

参 数 类 型	布 尔 型
默认值	false
修改方式	ALTER SESSION ,ALTER SYSTEM
PDB 修改	Yes
取值范围	true | false

通过设置 SQL_TRACE 可以启用或禁用 SQL 跟踪工具,设置 SQL_trace 为 true 可以收集信息用于性能优化或问题诊断;DBMS_SYSTEM 包也可以用于实现类似的功能。

以下警告在不同版本的文档中几乎没有任何变化。

> **警告** 设置初始化参数 SQL_TRACE 为 true 会对整个实例产生严重的性能影响,所以在产品环境中如非必要,确保不要设置这个参数。如果只是对特定的 session 启用跟踪,可以使用 ALTER SESSION 或 DBMS_SYSTEM.SET_SQL_TRACE_IN_SESSION 来设置。如果必须在数据库级启用 SQL_TRACE,你需要保证以下条件以最小化性能影响。
>
> (1)至少保证有 25% 的 CPU idle。
>
> (2)为 USER_Dump_DEST 分配足够的空间。
>
> (3)条带化磁盘以减轻 IO 负担。

在 12c 中文档中提示:不再支持 SQL_TRACE 参数,推荐使用 DBMS_MONITOR 和 DBMS_SESSION 包来替代其功能,该参数作为向后兼容而保留,其原有功能仍然存在。而事实上,DBA 的工作中,SQL_TRACE 很少被使用,更多的是 10046 事件。

自 Oracle 10g 开始，SQL_TRACE 参数才成为动态参数，可以在全局动态启用，在实践中除了研究目的，很少需要如此在全局设置。

```
SQL> alter system set SQL_trace=true;
System altered.
```

大多数时候我们使用 SQL_trace 跟踪当前会话，通过跟踪当前会话可以发现当前操作的后台递归活动（这在研究数据库新特性时尤其有效），研究 SQL 执行，发现后台错误等。

在 session 级启用和停止 SQL_trace 方式如下所示。

```
SQL> alter session set SQL_trace=true;  --启用当前 session 的跟踪：
SQL> select count(*) from dba_users; --此时的 SQL 操作将被跟踪
  COUNT(*)
----------
        34
SQL> alter session set SQL_trace=false;  --结束跟踪
SQL> select value from v$diag_info where name='Default Trace File';
```

在很多时候需要跟踪其他用户的进程，而不是当前用户，这可以通过 Oracle 提供的系统包 DBMS_SYSTEM 的 SET_SQL_TRACE_IN_SESSION 过程来完成，该过程调用提供三个参数。

```
PROCEDURE SET_SQL_TRACE_IN_SESSION
Argument Name       Type            In/Out Default?
------------------- --------------- --------------------
SID                 NUMBER          IN
SERIAL#             NUMBER          IN
SQL_TRACE           BOOLEAN         IN
```

通过查询 v$session 可以获得会话的 sid、serial# 等信息，获得会话信息之后就可以设置跟踪。

```
SQL> select sid,serial#,username from v$session
  2  where username is not null;
       SID    SERIAL# USERNAME
---------- ---------- ------
         8       2041 SYS
         9        437 EYGLE
SQL> exec dbms_system.set_SQL_trace_in_session(9,437,true)
SQL> exec dbms_system.set_SQL_trace_in_session(9,437,false)
```

DBMS_SYSTEM 包功能非常强大，是重要的跟踪手段之一。

SQL_TRACE 还可以通过如下方式针对特定的 SQL 启用跟踪，指定 SQL_ID 的 SQL 会被按照指定条件跟踪其执行过程，这在分析特定 SQL 时非常有效。

```
ALTER SYSTEM SET EVENTS 'SQL_trace [SQL:&&SQL_id] bind=true, wait=true';
```

4.2.1　DBMS_SYSTEM 跟踪案例

以下这个案例是我走上 DBA 道路第一个帮助朋友解决的问题。这个问题的解决过程，就是利用自己获得的已知知识，通过思考、跟踪去发现问题，并给出解决方案，对我的成长意义非凡。

4.2 SQL_TRACE 说明

这个应用是一个后台新闻发布系统，前端展现是一个网站。JAVA 开发，通过中间件连接池连接数据库。当时系统症状是访问新闻页极其缓慢，后台发布管理具有同样的问题。通常需要数十秒才能返回。

处理这个问题时，通过前台或者应用代码去分析会变得十分复杂，我想到的第一个办法就是启用跟踪，然后通过分析跟踪文件找出瓶颈所在。诊断时间在晚上，在无集中用户访问的情况下，让用户在前台进行相关页面的访问，同时进行进程跟踪。

首先通过查询 v$session 视图，获取进程信息。

```
SQL> select sid,serial#,username from v$session where username is not null;
    SID   SERIAL#  USERNAME
---------- ---------- ----------
      7     284   IFLOW
     11     214   IFLOW
     12     164   SYS
     16    1042   IFLOW
```

然后对相应的应用会话启用 SQL_trace 跟踪如下。

```
SQL> exec dbms_system.set_SQL_trace_in_session(7,284,true)
SQL> exec dbms_system.set_SQL_trace_in_session(11,214,true)
SQL> exec dbms_system.set_SQL_trace_in_session(16,1042,true)
```

应用执行一段时间后，关闭 SQL_trace。

```
SQL> exec dbms_system.set_SQL_trace_in_session(7,284,false)
SQL> exec dbms_system.set_SQL_trace_in_session(11,214,false)
SQL> exec dbms_system.set_SQL_trace_in_session(16,1042,false)
```

找到跟踪生成的跟踪文件,然后通过 Oracle 提供的格式化工具——tkprof 对 trace 文件进行格式化处理，筛查其中消耗时间部分。通过检查，发现类似以下语句的是可疑的。

```
********************************************************************************
select auditstatus,categoryid,auditlevel from
 categoryarticleassign a,category b where b.id=a.categoryid and articleId=
 20030700400141 and auditstatus>0

call     count    cpu    elapsed    disk     query    current    rows
-------  ------  -------  --------  -------  -------  ---------  --------
Parse       1    0.00     0.00        0        0         0         0
Execute     1    0.00     0.00        0        0         0         0
Fetch       1    0.81     0.81        0     3892         0         1
-------  ------  -------  --------  -------  -------  ---------  --------
total       3    0.81     0.81        0     3892         0         1
********************************************************************************
```

这里的查询显然是根据 articleId 进行新闻读取的。但是注意到逻辑读有 3892，这是较高的一个数字，这个内容引起了我的注意。

接下来的类似查询跟踪得到的执行计划显示，全表访问被执行。

```
select auditstatus,categoryid from
 categoryarticleassign where articleId=20030700400138 and categoryId in ('63',
  '138','139','140','141','142','143','144','168','213','292','341''-1')
call     count      cpu       elapsed      disk      query     current      rows
-------  -------   --------   --------   --------   --------   --------   --------
Parse        1      0.00       0.00         0          0          0          0
Execute      1      0.00       0.00         0          0          0          0
Fetch        1      4.91       4.91         0         2835        7          1
-------  -------   --------   --------   --------   --------   --------   --------
total        3      4.91       4.91         0         2835        7          1

Rows     Row Source Operation
-------  ---------------------------------------------------
      1  TABLE ACCESS FULL CATEGORYARTICLEASSIGN
```

登录数据库,检查相应表结构,看是否存在有效的索引,以下输出中的 IDX_ARTICLEID 是基于 ARTICLEID 创建的,但是在以上查询中都没有被用到。

```
SQL> select index_name,table_name,column_name from user_ind_columns
  2  where table_name=upper('categoryarticleassign');
INDEX_NAME                     TABLE_NAME                     COLUMN_NAME
-----------------------------  -----------------------------  -----------------
IDX_ARTICLEID                  CATEGORYARTICLEASSIGN          ARTICLEID
IND_ARTICLEID_CATEG            CATEGORYARTICLEASSIGN          ARTICLEID
IND_ARTICLEID_CATEG            CATEGORYARTICLEASSIGN          CATEGORYID
IDX_SORTID                     CATEGORYARTICLEASSIGN          SORTID
PK_CATEGORYARTICLEASSIGN       CATEGORYARTICLEASSIGN          ARTICLEID
PK_CATEGORYARTICLEASSIGN       CATEGORYARTICLEASSIGN          CATEGORYID
PK_CATEGORYARTICLEASSIGN       CATEGORYARTICLEASSIGN          ASSIGNTYPE
```

检查下表结构。

```
SQL> desc categoryarticleassign
 Name                    Null?     Type
 ---------------------   --------  ---------------
 CATEGORYID              NOT NULL  NUMBER
 ARTICLEID               NOT NULL  VARCHAR2(14)
 ASSIGNTYPE              NOT NULL  VARCHAR2(1)
 AUDITSTATUS             NOT NULL  NUMBER
 SORTID                  NOT NULL  NUMBER
 UNPASS                            VARCHAR2(255)
```

分析表结构发现了问题所在,因为 ARTICLEID 是个字符型数据,查询中给入的 articleId=20030700400141 是一个数字值,Oracle 执行查询时需要将 ARTICLEID 转换为数字与给定值进行比较,发生潜在的数据类型转换,这就导致了索引不能被采用,产生了全表扫描的执行计划,在客户的系统中大量类似如下的 SQL 在通过全表扫描产生大量的 IO 操作。

```
SQL> select auditstatus,categoryid
  2  from categoryarticleassign where articleId=20030700400132;
AUDITSTATUS CATEGORYID
----------- ----------
          9         94
          0        383
```

4.2 SQL_TRACE 说明

```
         0        695
Execution Plan
----------------------------------------------------------
   0      SELECT STATEMENT Optimizer=CHOOSE (Cost=110 Card=2 Bytes=38)
   1    0   TABLE ACCESS (FULL) OF 'CATEGORYARTICLEASSIGN' (Cost=110 Card=2 Bytes=38)
```

在这里很有必要解释一下 Oracle 数据库的数据转换，以下一段内容主要来自官方文档的介绍（依据 Oracle 12c 版本，本章末有一节附上了详细的译文）。

通常一个表达式不能包含不同的数据类型，例如一个表达式不能计算 5x10 之后再加上 'JAMES'，数字和字符无法进行联合的计算。然而，Oracle 支持显示和隐式的数据类型转换，可以将一种数据类型转换为另一种，从而使得某些表达式的运算可以正确执行。显示转换是指通过函数明确指定的数据类型转换，而隐式转换则指未明确指定，依赖 Oracle 自动进行的数据类型转换。

基于以下几个原因，Oracle 推荐使用显示类型转换而不是依赖隐式转换。

（1）使用显示转换使得 SQL 含义更容易被理解。

（2）隐式数据类型转换会产生负面的性能影响，尤其是当列值被转换成其他常量的数据类型时。

（3）隐式转换的行为依赖每次转换时的环境如数据库参数设置等，其行为可能多变而不确定，例如将 datetime 值隐式转换为 VARCHAR2,这个转换的格式和 NLS_DATE_FORMAT 参数有关，可能产生不可预期的结果。

（4）隐式转换的算法会因数据库版本而不同，而显示转换则可以预期。

（5）如果隐式转换发生在索引列，则 Oracle 可能用不到索引而影响性能。

在现实的开发环境中，绝大多数隐式转换是开发者无意中引入的，但却导致了大量的性能问题。这个案例就处于这样的场景之中。

要知道，**在字符和数字进行比对时，Oracle 总是将字符转换为数字进行比对**。解决本例问题的方法很简单，只须在传入参数两侧各增加一个单引号使其作为字符传入，即可解决这个问题。重新测试类似的查询，可以发现 Query 模式逻辑读降低为 2，占用 CPU 时间也大大减少。

```
********************************************************************
select unpass from
 categoryarticleassign where articleid='20030320000682' and categoryid='113'

call     count       cpu    elapsed       disk      query    current       rows
------- ------  -------- ---------- ---------- ---------- ----------  ----------
Parse        1      0.00       0.00          0          0          0           0
Execute      1      0.00       0.00          0          0          0           0
Fetch        1      0.00       0.00          0          2          0           0
------- ------  -------- ---------- ---------- ---------- ----------  ----------
total        3      0.00       0.00          0          2          0           0
```

```
Rows     Row Source Operation
-------  ---------------------------------------------------
      0  TABLE ACCESS BY INDEX ROWID CATEGORYARTICLEASSIGN
      1    INDEX RANGE SCAN (object id 3080)
```
****

至此,这个问题得到了完满的解决。但是关于隐式转换的故事还远远没有结束。

在从事数据库工作多年之后,我们发现大多数 DBA 仍然在面对我们 10 多年前面对的同样问题,仍然频繁地在解决系统中的隐式转换、索引失效、全表扫描问题。重复的工作就必须找出更高效的、自动化的手段去消解,这也是今天 DevOps 时代的要义之一。

我们的理念就是,从遇到的问题总结规则,聚规则而成规范,由规范衍生工具,通过工具替代人力。

如果从以上案例总结规范,至少有以下两条需要开发去注意改进。

(1)**使用绑定变量**——绑定变量可以减少硬解析,带来性能上的改进,这是最基本的开发实践。

(2)**避免隐式转换**——隐式转换可能带来索引失效影响性能,也会产生不可预期的程序效果,应当尽量避免。

一个 DBA 如果能够从实践中不断积累、提炼、上升,那么就能够在企业技术架构中承载更重要的使命和职责。

以上整个问题的改善,事实上就是在运维 Ops 的过程中,发现开发 Dev 中存在的问题,如果能够进而通过运维去促进开发,提升运维,这就是 DevOps 的使命和范畴了。

在现实环境中,我们在很多系统中都看到,大量的性能问题都是由于简单的疏忽导致的,而且由于问题的隐蔽性等,这些问题一旦在线上爆发出来,会给诊断优化带来相当的难度,同时会影响业务的正常运行,所以完善的规范和良好的编码对于一个系统来说是至关重要的。DevOps 在 Oracle 数据库开发中的最佳实践至少可以包括以下两点。

(1)**基于运维的开发培训**——通过运维中的实践总结,将规则方法推进到开发端,持续改进开发质量。

(2)**优化前置的 SQL 审核**——将事后救火变更为事前审核,在开发测试阶段发现和解决问题。

治本永远强于治标。

4.2.2 系统递归调用的跟踪

很多时候在进行数据库 DDL 操作时,Oracle 会在后台将这些操作转换为一系列的 DML 和

4.2 SQL_TRACE 说明

查询操作。如果其中的某个步骤出现异常，通常其错误提示不容易判定问题所在。这时候跟踪的作用就体现出来。

以下错误是在一次 DROP USER 中出现的，单从类似这样的提示来看，很多时候是没有丝毫用处的。

```
ORA-00604: error occurred at recursive SQL level 1
ORA-00942: table or view does not exist .
```

关于 recursive SQL 错误，我们有必要做个简单说明。

当我们发出一条简单的 DDL 命令以后，Oracle 数据库要在后台解析这条命令，并转换为 Oracle 数据库的一系列后台操作。这些后台操作统称为递归 SQL。

比如 create table 这样一条简单的 DDL 命令，Oracle 数据库在后台，实际上要把这个命令转换为对于 obj$、tab$、col$ 等底层表的插入操作。对于 drop table 操作，则是在这些系统表中进行反向删除操作。

在面对这样的问题时，通过 SQL_trace 进行后台跟踪，就可以进一步了解 Oracle 数据库的后台操作，找出问题点。

```
SQL> alter session set SQL_trace=true;
SQL> drop user wapcomm;
ORA-00604: error occurred at recursive SQL level 1
ORA-00942: table or view does not exist .
SQL> alter session set SQL_trace=false;
```

格式化（使用 tkprof）跟踪文件后，我们获得以下输出（摘录部分）。

```
********************************************************************************
The following statement encountered a error during parse:
DELETE FROM SDO_GEOM_METADATA_TABLE  WHERE SDO_OWNER = 'WAPCOMM'
Error encountered: ORA-00942
********************************************************************************

drop user wapcomm

********************************************************************************
delete from user_history$
where
 user# = :1             -----后台的递归删除操作…

Rows     Row Source Operation
-------  ---------------------------------------------------
      1  DELETE USER_HISTORY$
      1   TABLE ACCESS FULL USER_HISTORY$

********************************************************************************
declare
  stmt varchar2(200);
BEGIN
```

```
       if dictionary_obj_type = 'USER' THEN
         stmt := 'DELETE FROM SDO_GEOM_METADATA_TABLE ' ||
                 ' WHERE SDO_OWNER = ''' || dictionary_obj_name || ''' ';
         EXECUTE IMMEDIATE stmt;
       end if;
    end;

    call     count       cpu    elapsed       disk      query    current       rows
    ------- ------  -------- ---------- ---------- ---------- ----------  ----------
    Parse        1      0.00       0.00          0          0          0           0
    Execute      1      0.00       0.00          0          0          2           0
    Fetch        0      0.00       0.00          0          0          0           0
    ------- ------  -------- ---------- ---------- ---------- ----------  ----------
    total        2      0.00       0.00          0          0          2           0
    ********************************************************************************
```

使用 TKPROF 格式化以后，Oracle 把错误信息首先呈现出来。ORA-00942 错误是由于 SDO_GEOM_METADATA_TABLE 表/视图不存在所致，问题由此可以定位。对于这一类的错误，定位问题以后解决的方法就要依据具体问题原因而定了。同时可以通过 Oracle 的官方支持站点 MOS（My Oracle Support）去定位是否 Bug 引起的，是否已经有明确的解决方案。对于本例以关键字 SDO_GEOM_METADATA_TABLE 检索可以确认这是一个 Bug，并且给出了解决方案。

```
Problem Description
-------------------
The Oracle Spatial Option has been installed and you are encountering the following errors
while trying to drop a user, who has no spatial tables, connected as SYSTEM:

ERROR at line 1:
ORA-00604: error occurred at recursive SQL level 1
ORA-00942: table or view does not exist
ORA-06512: at line 7

A 942 error trace shows the failing SQL statement as:
DELETE FROM SDO_GEOM_METADATA_TABLE WHERE SDO_OWNER = '<user>'

Solution Description
--------------------
(1) Create a synonym for SDO_GEOM_METADATA_TABLE under SYSTEM which points to
MDSYS.SDO_GEOM_METADATA_TABLE.
(2) Now the user can be dropped connected as SYSTEM.
```

对于本例，为 MDSYS.SDO_GEOM_METADATA_TABLE 创建一个同义词即可解决，是相对简单的情况。这是一个 Oracle 早期版本的问题，只是这样的分析方法应当被 DBA 掌握。

跟踪的目标就是，找出根源，解决问题。只要能够找出问题的根源，或者接近根源，那么距离解决问题也就不远了。我推荐的提问方式也是，当遇到问题时，通过跟踪、日志分析更进一步，而不是只依据浅层的表象提出。

4.3　10046 与等待事件

10046 事件是 Oracle 提供的内部跟踪事件,是对 SQL_TRACE 的增强,通过 10046 可以通知 Oracle 内核执行 SQL_TRACE 类的跟踪操作。如果我们需要获得更多的跟踪信息,就需要用到 10046 事件,而在实际工作中最常用的就是 10046 事件。

自 11g 开始,10046 获得了更多的增强,包括明确的在设置中指定 sql_trace,类似'sql_trace wait=false, bind=true'这样的设定。

在常规的使用方法下,10046 事件的以下几个跟踪级别最为常用(后面的参数是 11g 的新设置中可用的参数设定)。

```
1 - 启用标准的 SQL_TRACE 功能,等价于 sql_trace
4 - Level 1 加上绑定值(bind values)  [ bind=true ]
8 - Level 1 + 等待事件跟踪 [ wait=true ]
12 - Level 1 + Level 4 + Level 8
```

从 11g 开始增加了以下两个跟踪级别。

```
16 - 为每次 SQL 执行生成 STAT 信息输出 [ plan_stat=all_executions ]
32 - 不转储执行统计信息         [ plan_stat=never ]
```

从 11.2.0.2 开始增加以下级别。

```
64 - 自适应的 STAT 转储       [ plan_stat=adaptive ]
```

我们知道,在进行数据库问题诊断及性能优化时,经常需要查询的几个重要视图包括 v$session_wait、v$system_event 等,这些视图中主要记录的就是等待事件。

通过调整以降低等待,是提高性能的一个方法。这些等待事件来自所有数据库操作,对于不同进程的等待可以通过动态性能视图 v$session_wait 等来查询;对于数据库全局等待可以通过 v$system_event 等视图来获得。

同样的,通过 10046 事件对具体 session 进行跟踪,可以获得每个 session 的执行情况及等待事件、统计信息等详细信息,输出到外部跟踪文件以进行分析。

类似 sql_trace,10046 事件可以在全局设置,也可以在 session 级设置。在全局设置,可以在参数文件中增加 Event 参数,此设置对所有用户的所有进程生效,包括后台进程,所以同样除了研究目的应当很少被使用。

```
event="10046 trace name context forever,level 12"
```

通过 alter session 的方式修改,需要 alter session 的系统权限。

```
SQL> alter session set events '10046 trace name context forever, level 12'; -- 启用跟踪
SQL> alter session set events '10046 trace name context off'; --停止跟踪
```

11g 之后的可选设置方式类似。

```
alter session set events 'sql_trace wait=true';
```

对其他用户 session 设置事件跟踪，同样可以通过 DBMS_SYSTEM 的 SET_EV 过程来实现。

```
PROCEDURE SET_EV
Argument Name                  Type                    In/Out Default?
------------------------------ ----------------------- ------ --------
SI                             BINARY_INTEGER          IN
SE                             BINARY_INTEGER          IN
EV                             BINARY_INTEGER          IN
LE                             BINARY_INTEGER          IN
NM                             VARCHAR2                IN
```

其中的参数 SI、SE 来自 v$session 视图。

查询获得需要跟踪的 session 信息。

```
SQL> select sid,serial#,username from v$session where username is not null;
       SID    SERIAL# USERNAME
---------- ---------- ------------------------------
         8       2041 SYS
         9        437 EYGLE
```

执行跟踪。

```
SQL> exec dbms_system.set_ev(9,437,10046,8,'');
PL/SQL procedure successfully completed.
```

结束跟踪。

```
SQL> exec dbms_system.set_ev(9,437,10046,0,'');
PL/SQL procedure successfully completed.
```

基于会话级别跟踪产生的文件，可以通过查询 V$DIAG_INFO 视图，找到跟踪文件的名称和位置信息，查看其中的内容。

以下通过一个简单的测试来看看 10046 事件的作用。

```
SQL> create table eygle as select * from dba_objects;
SQL> select file_id,block_id,blocks from dba_extents where segment_name='EYGLE';
   FILE_ID   BLOCK_ID     BLOCKS
---------- ---------- ----------
         1      21601          8
         1      21609          8
         1      21617          8
         1      21625          8
         1      21633          8
         1      23433          8
         1      23441          8
         1      23449          8
         1      23457          8
         1      23465          8
```

4.3 10046 与等待事件

```
SQL> alter session set events '10046 trace name context forever,level 12';
SQL> select count(*) from eygle; --表上无索引,所以此处引发一次全表扫描
  COUNT(*)
----------
      6207
SQL> alter session set events '10046 trace name context off';
```

检查一下 Oracle 生成的跟踪文件，比如关注一下等待事件 – db file scattered read（注意，自 11g 开始，由于 Serial Table Scan 的特性引入，你可能观察到全表扫描以 Direct Path Read 方式执行）。

```
[oracle@eygle udump]$ cat rac1_ora_20695.trc |grep scatt
WAIT #1: nam='db file scattered read' ela= 11657 p1=1 p2=21602 p3=7
WAIT #1: nam='db file scattered read' ela= 1363 p1=1 p2=21609 p3=8
WAIT #1: nam='db file scattered read' ela= 1297 p1=1 p2=21617 p3=8
WAIT #1: nam='db file scattered read' ela= 1346 p1=1 p2=21625 p3=8
WAIT #1: nam='db file scattered read' ela= 1313 p1=1 p2=21633 p3=8
WAIT #1: nam='db file scattered read' ela= 6226 p1=1 p2=23433 p3=8
WAIT #1: nam='db file scattered read' ela= 1316 p1=1 p2=23441 p3=8
WAIT #1: nam='db file scattered read' ela= 1355 p1=1 p2=23449 p3=8
WAIT #1: nam='db file scattered read' ela= 1320 p1=1 p2=23457 p3=8
WAIT #1: nam='db file scattered read' ela= 884 p1=1 p2=23465 p3=5
```

大家注意这里的等待事件'db file scattered read'，意味着这里使用了全表扫描来访问数据。其中 p1、p2、p3 分别代表了文件号、起始数据块块号和读取数据块的数量。

各参数的含义也可以从 v$event_name 视图中获得。

```
SQL> select name,PARAMETER1 p1,PARAMETER2 p2,PARAMETER3 p3
  2  from v$event_name where name='db file scattered read';
NAME                                   P1         P2         P3
-------------------------------------- ---------- ---------- ----------
db file scattered read                 file#      block#     blocks
```

在数据库内部，这些等待时间最后都会累计到 v$system_event 动态性能视图中，是数据库性能诊断的一个重要参考。

```
SQL> select event,time_waited from v$system_event
  2  where event='db file scattered read';
EVENT                            TIME_WAITED
-------------------------------- --------------------------
db file scattered read           51
```

这里我们有必要提到另外一个相关的初始化参数 db_file_multiblock_read_count，这个参数代表 Oracle 在执行全表扫描时每次 IO 操作可以读取的数据块的数量。

在前面的测试中，db_file_multiblock_read_count 参数设置为 16，由于 extent 大小为 8 个 block，Oracle 的一次 IO 操作不能跨越 extent，所以前面的全表扫描每次只能读取 8 个 block，进行了 10 次 IO 读取。

看一下进一步的测试。

```
SQL> create tablespace eygle datafile '/dev/raw/raw2' size 100M
  2  extent management local uniform size 256K;
SQL> alter table eygle move tablespace eygle;
SQL> select file_id,block_id,blocks from dba_extents where segment_name='EYGLE';
   FILE_ID   BLOCK_ID     BLOCKS
---------- ---------- ----------
         4          9         32
         4         41         32
         4         73         32
SQL> show parameter read_count
NAME                                 TYPE        VALUE
------------------------------------ ----------- ------------------------------
db_file_multiblock_read_count        integer     16
SQL> alter session set events '10046 trace name context forever,level 12';
SQL> select count(*) from eygle;
  COUNT(*)
----------
      6207
SQL> alter session set events '10046 trace name context off';
SQL> select value TRACE_FILE from v$diag_info where name='Default Trace File';
TRACE_FILE
--------------------------------------------------------------------------------
/opt/oracle/admin/rac/udump/rac1_ora_20912.trc

[oracle]$ cat /opt/oracle/admin/rac/udump/rac1_ora_20912.trc |grep scatt
WAIT #1: nam='db file scattered read' ela= 12170 p1=4 p2=10 p3=16
WAIT #1: nam='db file scattered read' ela= 2316 p1=4 p2=26 p3=15
WAIT #1: nam='db file scattered read' ela= 2454 p1=4 p2=41 p3=16
WAIT #1: nam='db file scattered read' ela= 2449 p1=4 p2=57 p3=16
WAIT #1: nam='db file scattered read' ela= 2027 p1=4 p2=73 p3=13
```

可以看到此时 Oracle 只需要 5 次 IO 操作就完成了全表扫描。通常较大的 db_file_multiblock_read_count 设置可以加快全表扫描的执行。但是需要注意的是，增大 db_file_multiblock_read_count 参数的设置，会使全表扫描的成本降低，在 CBO 优化器下可能会使 Oracle 更倾向于使用全表扫描而不是索引访问，一般建议使用默认值。

4.3.1　通过跟踪理解数据库的初始化

在 DBA 的职业生涯中，会面临众多的挑战，其中最重要的一种情况是数据库无法启动，所以深入理解 Oracle 数据库的初始化非常重要。通过 Oracle 的跟踪手段，可以帮助我们获取这些知识，在我的学习过程中，一直在不断地通过跟踪去研究熟悉的或不熟悉的特性和功能，从而加深自己对于数据库的理解。

对于 Oracle 数据库的初始化，我最初的思考是：数据库的核心信息都是存放在数据文件当中的，但是当数据库尚未打开之前，Oracle 是无法获得这部分数据的。那么 Oracle 是怎样完成这个从数据文件到内存的初始化过程的呢？

首先通过以下步骤对数据库的 OPEN 过程进行跟踪，研究获得的跟踪文件。

```
SQL> startup mount;
SQL> alter session set events='10046 trace name context forever,level 12';
SQL> alter database open;
```

以上通过 10046 跟踪获得一个跟踪文件，跟踪文件里将记录从 mount 到 open 的过程中，Oracle 所执行的后台操作。可以通过 tkprof 工具对跟踪文件进行格式化，使得其中的信息更便于阅读。首先我们来参考跟踪文件的前面部分（我的研究首先从 Oracle 9i 开始，逐渐推演到 Oracle 12c，研究不同版本的引导过程方法完全相同），这是第一个对象的创建。

```
create table bootstrap$
( line# number not null,
obj# number not null,
sql_text varchar2(4000) not null)
storage (initial 50K objno 56 extents (file 1 block 520))
```

注意：在这一步骤中，实际上 Oracle 是在内存中创建 bootstrap$ 的结构，然后从数据文件中读取数据到内存中，完成第一次初始化。在 9i 中，读取的位置是文件 1 的 377 块，自从 11g 之后变更为文件 1 的 520 块。注意此处的 file 1 block 520 子句是内部语句，意味这这些对象的存储位置是固定的，该语法对用户创建对象是不可用的。

从数据库的创建脚本 $ORACLE_HOME/rdbms/admin/sql.bsq 文件中，可以获得 bootstrap$ 表的初始创建语句，直至 12c 这些定义未曾变化（在 12c 中 sql.bsq 分解为一系列的 bsq 文件，dcore.bsq 中记录了下面这段代码）。

```
create table bootstrap$
( line#       number not null,               /* statement order id */
  obj#        number not null,               /* object number */
  sql_text    varchar2("M_VCSZ") not null)   /* statement */
  storage (initial 50K)         /* to avoid space management during IOR I */
//                              /* "//" required for bootstrap */
```

接下来从数据库中查询一下，file 1 block 520 上存储的是什么对象。

```
SQL> select segment_name,file_id,block_id
  2  from dba_extents where block_id=520 and file_id=1;
SEGMENT_NAME      FILE_ID    BLOCK_ID
--------------    -------    --------
BOOTSTRAP$            1         520
```

File 1 Block 520 开始存放的正是 Bootstrap$ 对象。继续查看 Trace 文件的内容，Oracle 进一步执行的是如下操作。

```
select line#, sql_text from bootstrap$ where obj# != :1
```

在创建并从数据文件中装载了 bootstrap$ 的内容之后，Oracle 开始递归的从该表中读取信息，加载数据。那么 bootstrap$ 中记录的是什么信息呢？

在数据库中，bootstrap$ 是一张实际存在的系统表。

```
SQL> desc bootstrap$
 Name                    Null?    Type
 ----------------------- -------- --------------------
 LINE#                   NOT NULL NUMBER
 OBJ#                    NOT NULL NUMBER
 SQL_TEXT                NOT NULL VARCHAR2(4000)
```

来看一下这张表的具体内容。

```
SQL> select * from bootstrap$ where rownum <5;
 LINE#  OBJ# SQL_TEXT
 ------ ----- --------------------------------------------------------
    -1    -1  8.0.0.0.0
     0     0  CREATE ROLLBACK SEGMENT SYSTEM STORAGE (  INITIAL 112K NEXT 1024K MINE
               XTENTS 1 MAXEXTENTS 32765 OBJNO 0 EXTENTS (FILE 1 BLOCK 9))
     8     8  CREATE CLUSTER C_FILE#_BLOCK#("TS#" NUMBER,"SEGFILE#" NUMBER,"SEGBLOCK
               #" NUMBER) PCTFREE 10 PCTUSED 40 INITRANS 2 MAXTRANS 255 STORAGE (  IN
               ITIAL 24K NEXT 1024K MINEXTENTS 1 MAXEXTENTS 2147483645 PCTINCREASE 0
               OBJNO 8 EXTENTS (FILE 1 BLOCK 73)) SIZE 225
     9     9  CREATE INDEX I_FILE#_BLOCK# ON CLUSTER C_FILE#_BLOCK# PCTFREE 10 INITR
               ANS 2 MAXTRANS 255 STORAGE (  INITIAL 64K NEXT 1024K MINEXTENTS 1 MAXE
               XTENTS 2147483645 PCTINCREASE 0 OBJNO 9 EXTENTS (FILE 1 BLOCK 81))
```

以上输出只显示了表中的 4 条记录，大家可以自行研究一下其他记录的内容。从这些语句中可以看出，bootstrap$ 中实际上是记录了一些数据库系统基本对象的创建语句。Oracle 通过 bootstrap$进行引导，进一步创建相关的重要对象，从而启动了数据库。

如果向前追溯，可以继续考察一下 bootstrap$的创建过程。查看一下创建数据库的脚本，可以发现数据库在创建过程中最先运行的是一个叫做 CreateDB.sql 的脚本。这个脚本发出 CREATE DATABASE 的命令，具体类似如下的例子。

```
CREATE DATABASE eygle
MAXINSTANCES 1 MAXLOGHISTORY 1 MAXLOGFILES 5 MAXLOGMEMBERS 3 MAXDATAFILES 100
DATAFILE '/opt/oracle/oradata/eygle/system01.dbf'
SIZE 250M REUSE AUTOEXTEND ON NEXT  10240K MAXSIZE UNLIMITED EXTENT MANAGEMENT LOCAL
DEFAULT TEMPORARY TABLESPACE TEMP TEMPFILE '/opt/oracle/oradata/eygle/temp01.dbf'
SIZE 40M REUSE AUTOEXTEND ON NEXT  640K MAXSIZE UNLIMITED
UNDO TABLESPACE "UNDOTBS1" DATAFILE '/opt/oracle/oradata/eygle/undotbs01.dbf'
SIZE 200M REUSE AUTOEXTEND ON NEXT  5120K MAXSIZE UNLIMITED
CHARACTER SET ZHS16GBK NATIONAL CHARACTER SET AL16UTF16
LOGFILE GROUP 1 ('/opt/oracle/oradata/eygle/redo01.log') SIZE 10240K,
GROUP 2 ('/opt/oracle/oradata/eygle/redo02.log') SIZE 10240K,
GROUP 3 ('/opt/oracle/oradata/eygle/redo03.log') SIZE 10240K;
exit;
```

在这个创建过程中，Oracle 会隐含的调用$ORACLE_HOME/rdbms/admin/sql.bsq 脚本，用于创建数据字典。这个文件的位置受到一个隐含的初始化参数（_init_sql_file）的控制。

```
SQL> @GetParDescrb.sql
Enter value for par: init_sql
NAME            VALUE               DESCRIB
```

```
---------------  ---------------------  ------------------------------------------
_init_sql_file   ?/rdbms/admin/sql.bsq   File containing SQL statements to execute upon
                                         database creation
```

如果在创建过程中，Oracle 无法找到 sql.bsq 文件，则数据库创建将会出错。我们可以测试一下移除 sql.bsq 文件，再看这样一个数据库创建过程。

```
SQL> startup nomount;
SQL> @CreateDB.sql
CREATE DATABASE eygle
*
ERROR at line 1:
ORA-01092: ORACLE instance terminated. Disconnection forced
```

此时日志中会记录以下信息。

```
Fri Aug 18 15:45:49 2006
Errors in file /opt/oracle/admin/eygle/udump/eygle_ora_3632.trc:
ORA-01501: CREATE DATABASE failed
ORA-01526: error in opening file '?/rdbms/admin/SQL.bsq'
ORA-07391: sftopn: fopen error, unable to open text file.
Error 1526 happened during db open, shutting down database
USER: terminating instance due to error 1526
```

这就是 sql.bsq 文件在数据库创建过程中的作用。那么在数据库的引导过程中，又该如何去定位 bootstrap$ 的位置呢？

这就不得不提到了 SYSTEM 表空间了。在系统表空间文件头存在一个重要的数据结构 root dba，我们可以通过转储数据文件头获得这个信息，从生成的 trace 文件中，我们可以获得以下信息（Oracle 12c 环境信息摘录）。

```
V10 STYLE FILE HEADER:
    Compatibility Vsn = 202375680=0xc100200
    Db ID=2903506423=0xad0ffdf7, Db Name='PRODCDB'
    Activation ID=0=0x0
    Control Seq=60695=0xed17, File size=103680=0x19500
    File Number=1, Blksiz=8192, File Type=3 DATA
Tablespace #0 - SYSTEM  rel_fn:1
Creation   at   scn: 0x0000.00000007 07/07/2014 05:38:57
Backup taken at scn: 0x0000.00000000 01/01/1988 00:00:00 thread:0
 reset logs count:0x35f7cd7a scn: 0x0000.0018531f
 prev reset logs count:0x32cc9b67 scn: 0x0000.00000001
 recovered at 03/02/2016 12:59:30
 status:0x2004 root dba:0x00400208 chkpt cnt: 941 ctl cnt:940
```

root dba 仅在 SYSTEM 表空间的文件头存在，用于定位数据库引导的 bootstrap$ 信息。Root dba 存储的是用 16 进制表示的二进制数，其中包含 10 位的文件号以及 22 位的数据块号，将 0x00400208 转换为二进制就是 0000 0000 0100 0000 0000 0010 0000 1000，前 10 位为 1，代表文件号为 1，后 22 位转换为 10 进制为 520，代表数据文件 1 上的 520 号数据块。

当然在数据库中无须如此复杂，Oracle 提供工具用于数据块及文件号的转换。

```
SQL> variable file# number
SQL> execute :file#:=dbms_utility.data_block_address_file(to_number('400208','xxxxxxx'));
PL/SQL procedure successfully completed.
SQL> variable block# number
SQL> execute :block#:=dbms_utility.data_block_address_block(to_number('400208','xxxxxxx'))
PL/SQL procedure successfully completed.
SQL>print file#
  FILE#
----------
     1
SQL> print block#
   BLOCK#
----------
     520
```

现在可以全面的来回顾一下数据库的内部引导过程，通过 10046 事件可以跟踪一下数据库的打开过程，使用前面曾经提到过的步骤。

```
oracle@enmocoredb admin]$ SQLplus / as sysdba
SQL*Plus: Release 12.2.0.0.3 Production on Thu Aug 4 15:26:49 2016
Copyright (c) 1982, 2016, Oracle.  All rights reserved.
SQL> shutdown immediate;
SQL> startup mount;
Database mounted.
SQL> alter session set events='10046 trace name context forever,level 12';
SQL> alter database open;
SQL> shutdown immediate;
```

从跟踪文件（以下跟踪文件来自 Oracle 12.2 版本）中我们可以获得以下重要信息。

```
=====================
PARSING IN CURSOR #0x7fdf6c93ae70 len=19 dep=0 uid=0 oct=35 lid=0 tim=2699657623431
hv=1907384048 ad='0x61ce6470' SQLid='a01hp0psv0rrh'
alter database open
END OF STMT
PARSE #0x7fdf6c93ae70:c=2999,e=2961,p=0,cr=0,cu=0,mis=1,r=0,dep=0,og=1,plh=0,tim=2699657623426
WAIT #0x7fdf6c93ae70: nam='db file sequential read' ela= 12 file#=1 block#=520 blocks=1 obj#=-1 tim=2699658571220
=====================
PARSING IN CURSOR #0x7fdf6c938d48 len=188 dep=1 uid=0 oct=1 lid=0 tim=2699658571988
hv=4006182593 ad='0x61c4bd28' sqlid='32r4f1brckzq1'
create table bootstrap$ (
  line#         number not null,
  obj#          number not null,
  sql_text      varchar2(4000) not null)
 storage (initial 50K objno 59 extents (file 1 block 520))
END OF STMT
PARSE #0x7fdf6c938d48:c=0,e=658,p=0,cr=0,cu=0,mis=1,r=0,dep=1,og=4,plh=0,tim=2699658571988
WAIT #0x7fdf6c938d48: nam='PGA memory operation' ela= 13 p1=65536 p2=2 p3=0 obj#=-1 tim=2699658572091
EXEC #0x7fdf6c938d48:c=0,e=215,p=0,cr=0,cu=0,mis=0,r=0,dep=1,og=4,plh=0,tim=2699658572285
CLOSE #0x7fdf6c938d48:c=0,e=3,dep=1,type=0,tim=2699658572352
```

```
====================
PARSING IN CURSOR #0x7fdf6c938d48 len=65 dep=1 uid=0 oct=3 lid=0 tim=2699658572771
hv=1762642493 ad='0x61c4a500' sqlid='aps3qh1nhzkjx'
select line#, sql_text from bootstrap$ where obj# not in (:1, :2)
END OF STMT
PARSE #0x7fdf6c938d48:c=0,e=404,p=0,cr=0,cu=0,mis=1,r=0,dep=1,og=4,plh=0,tim=2699658572770
BINDS #0x7fdf6c938d48:

 Bind#0
  oacdty=02 mxl=22(22) mxlc=00 mal=00 scl=00 pre=00
  oacflg=08 fl2=1000001 frm=00 csi=00 siz=24 off=0
  kxsbbbfp=7fdf6c938900  bln=22  avl=02  flg=05
  value=59
 Bind#1
  oacdty=02 mxl=22(22) mxlc=00 mal=00 scl=00 pre=00
  oacflg=08 fl2=1000001 frm=00 csi=00 siz=24 off=0
  kxsbbbfp=7fdf6c9388d0  bln=24  avl=06  flg=05
  value=4294967295
EXEC #0x7fdf6c938d48:c=1000,e=817,p=0,cr=0,cu=0,mis=1,r=0,dep=1,og=4,plh=867914364,tim=2699658573677
WAIT #0x7fdf6c938d48: nam='db file sequential read' ela= 9 file#=1 block#=520 blocks=
1 obj#=59 tim=2699658573738
WAIT #0x7fdf6c938d48: nam='PGA memory operation' ela= 9 p1=65536 p2=1 p3=0 obj#=
59 tim=2699658573827
WAIT #0x7fdf6c938d48: nam='db file scattered read' ela= 19 file#=1 block#=521 blocks=
3 obj#=59 tim=2699658573931
```

从等待事件上可以明确看到,单块读读取了文件 1 的第 520 个数据块,这也正是引导块的定位过程。

```
WAIT #0x7fdf6c93ae70: nam='db file sequential read' ela= 12 file#=1 block#=520 blocks=1 obj#=-1 tim=2699658571220
```

了解了 SYSTEM 表空间的重要作用,也就可以理解,为什么系统表空间的文件头损坏,或者如果启动对象的数据块损坏后,Oracle 数据库就将无法启动。

我们曾经见过很多案例,很多用户的数据库运行在非归档模式下,又没有备份,最后当 SYSTEM 表空间出现故障后,数据库就无法打开了,这是最为严重的情况,通常是没有办法恢复数据的。

所以我们经常反复建议,**SYSTEM** 表空间极其重要,备份重于一切,希望通过我们的不断呼吁,数据库的安全能够更加引起重视,用户的数据能够更加安全。

数据库的引导过程还可以通过 GDB 工具在 Linux、UNIX 上进行跟踪,分步骤来观察这个启动过程,以下输出可以帮助读者进一步了解这些内部操作。

首先将数据库启动到 Mount 状态,找到进程 SPID。

```
SQL> startup mount;
ORACLE instance started.
Database mounted.
SQL> select spid from v$process where addr in (select paddr from v$session where sid=(select distinct sid from v$mystat));
```

```
SPID
-----
1518
```

然后通过 gdb 跟踪这个进程。

```
localhost:~ oracle$ gdb $ORACLE_HOME/bin/oracle 1518
GNU gdb 6.3.50-20050815 (Apple version gdb-1518) (Sat Feb 12 02:52:12 UTC 2011)
Attaching to program: `/oracle/product/10.2.0/bin/oracle', process 1518.
Reading symbols for shared libraries .+++++++++++ done
0x00007fff80616984 in read ()
(gdb)
```

然后跟踪两个内部指令。

```
(gdb) break kcrf_commit_force
Breakpoint 1 at 0x1025a2d4c
(gdb) break kqlobjlod
Breakpoint 2 at 0x1006c78b4
```

此时执行数据库 OPEN 操作会被挂起。

```
SQL> alter database open;
```

然后重新开启一个 SQL*Plus 进程，查询此时数据库加载的 ROWCACHE 对象。

```
SQL> select parameter,count,gets from v$rowcache where count!=0;

no rows selected
```

然后继续执行，我们看到在第三个步骤之后，数据库加载了一个 ROW Cache 对象。

```
(gdb) c
Continuing.

Breakpoint 1, 0x00000001025a2d4c in kcrf_commit_force ()
(gdb) c
Continuing.

Breakpoint 1, 0x00000001025a2d4c in kcrf_commit_force ()
(gdb) c
Continuing.

Breakpoint 2, 0x00000001006c78b4 in kqlobjlod ()

SQL> select parameter,count,gets from v$rowcache where count!=0;
PARAMETER                      COUNT      GETS
------------------------------ ---------- ----------
dc_objects                     1          1
```

这个对象是什么呢？

```
SQL> select address,cache_name,existent,lock_mode,saddr,substr(key,1,40) keystr from v$rowcache_parent;
```

```
ADDRESS          CACHE_NAME           E  LOCK_MODE  SADDR           KEYSTR
---------------- -------------------- -  ---------- --------------- -----------------------
00000001942E9080 dc_objects           N  3          0000000194782EB0 000000000A00424F4F5453545241502400000000
```

解析其 KEY 值，正是 bootstrap$，这就是数据库初始化时加载的第一个对象。

```
SQL> select dump('BOOTSTRAP$',16) from dual;
Dump('BOOTSTRAP$',16)
------------------------------------------
Typ=96 Len=10: 42,4f,4f,54,53,54,52,41,50,24
```

然后数据库将递归查询该对象中的数据，向内存中加载其他对象。更进一步。

```
(gdb) c
Continuing.
Breakpoint 2, 0x00000001006c78b4 in kqlobjlod ()

ADDRESS          CACHE_NAME           E  LOCK_MODE  SADDR           KEYSTR
---------------- -------------------- -  ---------- --------------- -----------------------
00000001942E30D8 dc_tablespaces       N  0 00        000000000000000000000000000000000000000
00000001942DE9B8 dc_rollback_segments Y  0 00        000000000000000000000000000000000000000
00000001942E9080 dc_objects           Y  0 00        000000000A00424F4F5453545241502400000000
00000001942DE2D0 dc_objects           N  3 0000000194782EB0 000000000600435F4F424A230000000000000000
00000001942E3340 dc_object_ids        Y  0 00        3800000000000000000000000000000000000000
```

这里可以看到数据库加载了回滚段信息，首先加载的是 SYSTEM 的回滚段，转储 Row Cache 信息之后，就可以看到这些详细的内容。

```
SQL> ALTER SESSION SET EVENTS 'immediate trace name row_cache level 10';
```

这里得到的 BUCKET 37 包含了回滚段信息。

```
BUCKET 37:
  row cache parent object: address=0x1942de9b8 cid=3(dc_rollback_segments)
  hash=5fed2a24 typ=9 transaction=0x0 flags=000000a6
  own=0x1942dea88[0x1942dea88,0x1942dea88] wat=0x1942dea98[0x1942dea98,0x1942dea98] mode=N
  status=VALID/INSERT/-/FIXED/-/-/-/-/-
  data=
  00000000 00000000 00000001 00000009 59530006 4d455453 00000000 00000000
  00000000 00000000 00000000 00000000 00000003 00000000 00000000 00000000
  00000000 00000000 00000000 00000000
  BUCKET 37 total object count=1
ROW CACHE HASH TABLE: cid=4 ht=0x192b2e8a8 size=256
```

其中 53595354454d 正是 SYSTEM 回滚段。

```
SQL> select dump('SYSTEM',16) from dual;
Dump('SYSTEM',16)
------------------------------
Typ=96 Len=6: 53,59,53,54,45,4d
```

而另外一个 BUCKET 上正是 BOOTSTRAP$ 对象。

```
BUCKET 43170:
  row cache parent object: address=0x1942e9080 cid=8(dc_objects)
  hash=f3d1a8a1 typ=11 transaction=0x0 flags=000000a6
  own=0x1942e9150[0x1942e9150,0x1942e9150] wat=0x1942e9160[0x1942e9160,0x1942e9160] mode=N
  status=VALID/INSERT/-/FIXED/-/-/-/-/-
  set=0, complete=TRUE
  data=
  00000000 4f42000a 5453544f 24504152 00000000 00000000 00000000 00000000
  00000000 00000001 00000000 00000000 00000000 00000000 00000000 00000000
  00000000 00000000 00000000 00000000 00000000 00000000 00000000 00000000
  00000000 00000000 00000000 00000038 00000001 00000038 066f7802 030d1411
  11066f78 78030d14 1411066f 0001030d 00000000 00000000 00000000 00000000
  00000000 00000000
BUCKET 43170 total object count=1
```

这就是数据库启动过程中,BOOTSTRAP$的加载与引导过程。由上面的讨论我们可以知道 bootstrap$表的重要,如果 bootstrap$表发生损坏,数据库将无法启动。

4.3.2 远程支持之 10046 事件

通过 10046 的跟踪可以快速获得完整的后台信息,在远程支持时尤其有效。以下一个案例来自朋友发来的案例请求,他们的优化工具在客户的环境中执行某个 SQL 查询时,需要 10 分钟时间才能出结果,这是无法接受的,而同样的查询在其他环境上都可以快速的获得输出结果。

首先我请朋友获得了一个 10046 跟踪文件,通过 tkprof 格式化之后,这个 SQL 的输出结果展现出来。一段时间消耗代码如下所示。

```
SELECT B.OWNER,B.CONSTRAINT_NAME,B.CONSTRAINT_TYPE,B.OWNER,B.TABLE_NAME,
  A.OWNER,A.CONSTRAINT_NAME,A.OWNER,A.TABLE_NAME
FROM
  SYS.DBA_CONSTRAINTS A, SYS.DBA_CONSTRAINTS B, SYS.DBA_OBJECTS O WHERE
  A.CONSTRAINT_TYPE='R' AND A.R_CONSTRAINT_NAME=B.CONSTRAINT_NAME AND
  A.R_OWNER=B.OWNER  AND O.OWNER=A.OWNER AND O.OBJECT_NAME=A.TABLE_NAME AND
  O.OBJECT_TYPE='TABLE'  AND B.OWNER='BOM' AND B.TABLE_NAME=
  'BOM_DEPARTMENT_CLASSES' ORDER BY B.OWNER, B.TABLE_NAME

call     count       cpu    elapsed       disk      query    current        rows
------- ------  -------- ---------- ---------- ---------- ----------  ----------
Parse        1      0.12       0.16          0          0          0           0
Execute      1      0.00       0.00          0          0          0           0
Fetch        1    403.63     607.11       9585   15568040          0           0
------- ------  -------- ---------- ---------- ---------- ----------  ----------
total        3    403.75     607.27       9585   15568040          0           0
```

该段 SQL 的 Elapsed 时间超过了 600 秒,逻辑读也非常高,对于一个优化工具来说显然是不可接受的。

接下来的跟踪文件中显示了 SQL 的执行计划(仅保留了核心部分)。

4.3 10046 与等待事件

```
ROWS        Row Source Operation
-------     ---------------------------------------------------
1935557     VIEW
1935557     UNION-ALL PARTITION
1935557     FILTER
1935557     NESTED LOOPS
2863        TABLE ACCESS BY INDEX ROWID USER$
2863        INDEX UNIQUE SCAN I_USER1 (object id 44)
1935557     TABLE ACCESS FULL OBJ$
```

以上执行计划中,最可疑的部分是对于 OBJ$ 的全表扫描,这个环节的行数返回有 1、935、557 行。通过 NL 执行,首先怀疑这里的执行计划选择错误,如果选择索引,执行性能肯定会有极大的不同。

可是 10046 的跟踪事件显示的信息有限不能够准确定位错误的原因,我请朋友通过 10053 事件来生成一个执行计划的跟踪,10053 使用极为简便,通过如下方式就可以捕获 SQL 的解析过程。

```
alter session set events '10053 trace name context forever,level 1';
explain plan for <select_query>;
```

举例如下所示。

```
SQL> alter session set events '10053 trace name context forever,level 1';
Session altered.

SQL> explain plan for select count(*) from obj$;
Explained.
```

然后在 udump 目录下就可以找到 10053 生成的跟踪文件,在该文件中,找到了查询相关表的统计信息,其中 OBJ$ 的信息如下所示,其中 CDN(CarDiNality)指表中包含的记录数量,此处显示 OBJ$ 表中有 24 万左右的记录,使用了 2941 个数据块。

```
***********************
Table stats    Table: OBJ$   Alias: SYS_ALIAS_1
  TOTAL ::  CDN: 245313  NBLKS: 2941  AVG_ROW_LEN: 79
Column:       OWNER# Col#: 3     Table: OBJ$   Alias: SYS_ALIAS_1
    NDV: 221       NULLS: 0     DENS: 4.5249e-03  LO: 0  HI: 259
  NO HISTOGRAM: #BKT: 1 #VAL: 2
-- Index stats
  INDEX NAME: I_OBJ1  COL#: 1
    TOTAL ::  LVLS: 1  #LB: 632   #DK: 245313  LB/K: 1  DB/K: 1  CLUF: 4184
  INDEX NAME: I_OBJ2  COL#: 3 4 5 12 13 6
    TOTAL ::  LVLS: 2  #LB: 1904  #DK: 245313  LB/K: 1  DB/K: 1  CLUF: 180286
  INDEX NAME: I_OBJ3  COL#: 15
    TOTAL ::  LVLS: 1  #LB: 19    #DK: 2007    LB/K: 1  DB/K: 1  CLUF: 340
_OPTIMIZER_PERCENT_PARALLEL = 0
***********************************
```

而在前面的 10046 跟踪信息中,显示 OBJ$ 包含大约 200 万条记录,这是非常巨大的不同,对于 USER$ 表,显示具有 2863 条记录,而统计信息中显示仅有 253 条记录。

```
**************************
Table stats    Table: USER$   Alias: U
TOTAL ::  CDN: 253  NBLKS: 16  AVG_ROW_LEN: 82
Column:       USER# Col#: 1    Table: USER$  Alias: U
NDV: 253      NULLS: 0      DENS: 3.9526e-03 LO: 0  HI: 261
NO HISTOGRAM: #BKT: 1 #VAL: 2
-- Index stats
INDEX NAME: I_USER#  COL#: 1
TOTAL ::  LVLS: 0   #LB: 1 #DK: 258  LB/K: 1  DB/K: 1  CLUF: 13
INDEX NAME: I_USER1  COL#: 2
TOTAL ::  LVLS: 0   #LB: 1 #DK: 253  LB/K: 1  DB/K: 1  CLUF: 87
**************************
```

这说明数据字典中记录的统计信息与真实情况不符合，导致了 SQL 选择了错误的执行计划，在使用 CBO 的 Oracle9i 数据库中，这种情况极为普遍，通过删除表的统计信息，或者重新收集正确的统计信息，可以使 SQL 执行恢复到正常合理的范畴内。在 Oracle10g 开始的自动统计信息收集，就是为了防止出现统计信息陈旧的现象。

以下是通过 dbms_stats 包清除和重新收集表的统计信息的简单参考。

```
SQL> exec dbms_stats.delete_table_stats(user,'OBJ$');
PL/SQL procedure successfully completed.
SQL> exec dbms_stats.gather_table_stats(user,'OBJ$');
PL/SQL procedure successfully completed.
```

基于这样的判断我们建议客户做出修正，最后的客户反馈结果是：按照你的建议，在更新 OBJ$ 的统计信息后，该语句的执行时间由 10 分钟减少到了 2 分钟。接下来，按照类似的思路，又更新了 USER$、CON$、CDEF$ 表的统计信息，这时，语句的执行时间减少到了零点几秒。至此，问题解决。

4.3.3 通过 10046 事件跟踪解决未知问题

掌握了 Oracle 数据库最为重要的跟踪方法，就可以在遇到问题时，快速定位根源。而找到问题根源，距离解决问题也就不远了——不论这些问题是已知的还是未知的。

以下一个案例来自于 Oracle Database 12.2 的版本，在数据库启动时遇到错误，数据库无法启动，抛出的异常是 ORA-00600 908 错误，关于这个错误在 MOS 上也没有说明，这属于最早发现的 12.2 的问题。

```
[oracle@enmocoredb ~]$ SQLplus / as sysdba
SQL*Plus: Release 12.2.0.0.3 Production on Mon Aug 1 09:59:54 2016
Copyright (c) 1982, 2016, Oracle.  All rights reserved.
Connected to an idle instance.

SQL> startup
ORACLE instance started.
Database mounted.
```

```
ORA-00603: ORACLE server session terminated by fatal error
ORA-01092: ORACLE instance terminated. Disconnection forced
ORA-00604: error occurred at recursive SQL level 1
ORA-00600: internal error code, arguments: [908], [], [], [], [], [], [], [], [], [], [], []
ORA-07445: exception encountered: core dump [ksupdbsesinc()+866] [SIGSEGV]
[ADDR:0x13650] [PC:0x6170C62] [Address not mapped to object] []
Process ID: 24446
Session ID: 30 Serial number: 42586
```

进一步的检查告警日志文件，获取后台记录的更详细信息，在这个案例中后台的异常日志和前台抛出的一致，包括跟踪日志，很难获得更明确的判断线索。

```
2016-08-01T10:02:56.913024+08:00
Errors in file /u01/app/oracle/diag/rdbms/eygle/eygle/trace/eygle_ora_24446.trc:
ORA-00604: error occurred at recursive SQL level 1
ORA-00600: internal error code, arguments: [908], [], [], [], [], [], [], [], [], [], [], []
ORA-07445: exception encountered: core dump [ksupdbsesinc()+866] [SIGSEGV] [ADDR:0x13650]
[PC:0x6170C62] [Address not mapped to object] []
Error 604 happened during db open, shutting down database
```

为了进一步分析这个问题，我们在 Open 数据库的阶段启用跟踪，以寻找最后出现问题的步骤，这一方法在实践中非常有效。

```
SQL> startup mount;
ORACLE instance started.
Database mounted.
SQL> alter session set events '10046 trace name context forever,level 12';
Session altered.

SQL> alter database open;
ORA-00600: internal error code, arguments: [908], [], [], [], [], [], [], [],[], [], [], []
```

现在可以在后台找到这个跟踪文件，通过定位最后出现问题的部分，获得线索。以下一段输出是数据库启动报错前最后执行的一段递归 SQL。

```
=====================
PARSING IN CURSOR #0x7f19251432d8 len=123 dep=1 uid=0 oct=3 lid=0 tim=2423484984490
hv=1601912009 ad='0x6181e130' SQLid='65m6cgpgrqg69'
 select /*+ NO_PARALLEL(c) */ c.con_id# from cdb_service$ c where lower(c.name) = lower(:1)
or lower(c.name) = lower(:2)
END OF STMT
PARSE #0x7f19251432d8:c=0,e=293,p=0,cr=0,cu=0,mis=1,r=0,dep=1,og=4,plh=0,tim=2423484984490
BINDS #0x7f19251432d8:

Bind#0
 oacdty=01 mxl=32(04) mxlc=00 mal=00 scl=00 pre=00
 oacflg=18 fl2=0001 frm=01 csi=852 siz=32 off=0
 kxsbbbfp=7f19266bdf70  bln=32  avl=04  flg=05
 value="enmo"
Bind#1
 oacdty=01 mxl=32(04) mxlc=00 mal=00 scl=00 pre=00
 oacflg=18 fl2=0001 frm=01 csi=852 siz=32 off=0
```

```
kxsbbbfp=7f19266bdf38  bln=32  avl=04  flg=05
value="enmo"
EXEC #0x7f19251432d8:c=1000,e=594,p=0,cr=0,cu=0,mis=1,r=0,dep=1,og=4,plh=2885951592,tim=2423484985147
FETCH #0x7f19251432d8:c=0,e=24,p=0,cr=3,cu=0,mis=0,r=1,dep=1,og=4,plh=2885951592,tim=2423484985192
STAT  #0x7f19251432d8 id=1 cnt=1 pid=0 pos=1 obj=434 op='TABLE ACCESS FULL CDB_SERVICE$
(cr=3 pr=0 pw=0 str=1 time=23 us cost=2 size=16 card=1)'
CLOSE #0x7f19251432d8:c=0,e=41,dep=1,type=0,tim=2423484985251
kswscrs: error service is already defined in pdb 3.
       : current pdb id=1
```

注意最后出现的提示：kswscrs: error service is already defined in pdb 3，current pdb id=1。这个提示给了我们一个重要线索，错误出现的原因是服务名已经在 PDB 3 中被定义和使用，因此出现了冲突。

再来看看最后执行的这个递归 SQL，该 SQL 从 cdb_service$ 表来取得服务名，进行验证。

```
select /*+ NO_PARALLEL(c) */ c.con_id# from cdb_service$ c where lower(c.name) = lower(:1)
or lower(c.name) = lower(:2)
```

两个绑定变量的输入参数是：value="enmo" 。

找到了这样一个方向，进一步的检查数据库参数，发现在初始化参数中的确设置了一个服务名 enmo，根据错误应该是这个服务名和 PDB 自动注册产生了冲突。

```
SQL> startup mount;
ORACLE instance started.
Database mounted.
SQL> show parameter service
NAME                 TYPE       VALUE
-------------------- ---------- -----------------
service_names        string     yhem,eygle,enmo
```

接下来尝试去掉这个服务名，重新启动数据库，数据库成功启动。

```
SQL> alter system set service_names='yhem,eygle';
System altered.
SQL> shutdown immediate;
ORA-01109: database not open
Database dismounted.
ORACLE instance shut down.
SQL> startup
ORACLE instance started.
Database mounted.
Database opened.
SQL> show parameter service
NAME                 TYPE       VALUE
-------------------- ---------- -------------
service_names        string     yhem,eygle
```

在熟悉了 Oracle 的跟踪方法之后，就可以据此不断深入理解 Oracle 数据库的工作原理，在实践中应对各种异常，并直指根源，找到解决方案。

4.3 10046 与等待事件

在 12c 的多租户环境，修改 service_names 参数应该非常谨慎，对于该参数的修改会直接反映到监听器的动态注册，甚至会覆盖 PDB 的动态注册，影响服务。以下过程展示这一知识点，在运维工作中尤其应当引起大家的注意和关注。

测试环境的 CDB 中存在两个 PDB，分别是 ENMO 和 YHEM2，同时 service_names 参数中存在两个服务名设定，分别是 yhem 和 eygle，这样数据库存在了四个服务名。

```
SQL> col name for a30
SQL> select con_id,dbid,name from v$pdbs;
    CON_ID     DBID       NAME
---------- ---------- --------------------
     2      612507346  PDB$SEED
     3      965292808  ENMO
     4     2839503056  YHEM2
SQL> show parameter service_names
NAME                TYPE       VALUE
--------------      ---------  ------------------
service_names       string     yhem,eygle
```

在监听器中的注册的服务名如下（去除了不必要内容）。

```
[oracle@enmocoredb ~]$ lsnrctl status
LSNRCTL for Linux: Version 12.2.0.0.3 - Production on 03-AUG-2016 11:08:46
Copyright (c) 1991, 2016, Oracle.  All rights reserved.

Connecting to (ADDRESS=(PROTOCOL=tcp)(HOST=)(PORT=1521))
STATUS of the LISTENER
------------------------
Services Summary...
Service "enmo" has 1 instance(s).
  Instance "eygle", status READY, has 1 handler(s) for this service...
Service "eygle" has 1 instance(s).
  Instance "eygle", status READY, has 1 handler(s) for this service...
Service "yhem" has 1 instance(s).
  Instance "eygle", status READY, has 1 handler(s) for this service...
Service "yhem2" has 1 instance(s).
  Instance "eygle", status READY, has 1 handler(s) for this service...
The command completed successfully
```

如果通过如下方式修改了参数，去除部分服务名。

```
SQL> alter system set service_names='yhem';
System altered.
```

此时的监听状态会随之改变，包括 PDB 自动注册在内的服务名被一起清除。

```
[oracle@enmocoredb ~]$ lsnrctl status
LSNRCTL for Linux: Version 12.2.0.0.3 - Production on 03-AUG-2016 11:10:33
Copyright (c) 1991, 2016, Oracle.  All rights reserved.
Services Summary...
Service "eygle" has 1 instance(s).
  Instance "eygle", status READY, has 1 handler(s) for this service...
```

```
Service "yhem" has 1 instance(s).
  Instance "eygle", status READY, has 1 handler(s) for this service...
The command completed successfully
```

如果遇到这种状况，这意味着所有通过服务名访问 PDB 的请求都会无法获取连接。解决这个问题的方法是，将 PDB 的服务名重新通过 service_names 参数进行显示的设置，就可以临时恢复这个问题。

```
SQL> alter system set service_names='yhem,yhem2,enmo';
SQL> !lsnrctl status
LSNRCTL for Linux: Version 12.2.0.0.3 - Production on 03-AUG-2016 15:19:49
Copyright (c) 1991, 2016, Oracle.  All rights reserved.

Connecting to (ADDRESS=(PROTOCOL=tcp)(HOST=)(PORT=1521))
Services Summary...
Service "enmo" has 1 instance(s).
  Instance "eygle", status READY, has 1 handler(s) for this service...
Service "eygle" has 1 instance(s).
  Instance "eygle", status READY, has 1 handler(s) for this service...
Service "yhem" has 1 instance(s).
  Instance "eygle", status READY, has 1 handler(s) for this service...
Service "yhem2" has 1 instance(s).
  Instance "eygle", status READY, has 1 handler(s) for this service...
The command completed successfully
```

但是注意，在 12.2 的已知问题未作为 Bug 修复之前，如果带着这个 service_names 参数设置重启数据库就会遇到前文所说的问题。

4.3.4　通过 10046 解决数据库 RAC 集群案例

在某次客户案例中，遇到了数据库无法启动的一则案例，根据现场工程师的反馈，获得了启动时的 10046 跟踪。在跟踪文件中，末端呈现的是大量的 TT 锁等待（这是一个 10.2.0.4 版本的数据库）。

```
    WAIT #2: nam='enq: TT - contention' ela= 488293 name|mode=1414791174 tablespace ID=0 operation=0 obj#=-1 tim=6632936391
    WAIT #2: nam='enq: TT - contention' ela= 488294 name|mode=1414791174 tablespace ID=0 operation=0 obj#=-1 tim=6633424737
    WAIT #2: nam='enq: TT - contention' ela= 488292 name|mode=1414791174 tablespace ID=0 operation=0 obj#=-1 tim=6633913072
    WAIT #2: nam='enq: TT - contention' ela= 488292 name|mode=1414791174 tablespace ID=0 operation=0 obj#=-1 tim=6634401394
    WAIT #2: nam='enq: TT - contention' ela= 488292 name|mode=1414791174 tablespace ID=0 operation=0 obj#=-1 tim=6634889716
    WAIT #2: nam='enq: TT - contention' ela= 488293 name|mode=1414791174 tablespace ID=0 operation=0 obj#=-1 tim=6635378039
    WAIT #2: nam='enq: TT - contention' ela= 488292 name|mode=1414791174 tablespace ID=0 operation=0 obj#=-1 tim=6635866366
```

找到最后执行的递归 SQL，从这一点开始分析，主要信息显示如下。

4.3 10046 与等待事件

```
=====================
  PARSING IN CURSOR #3 len=348 dep=1 uid=0 oct=3 lid=0 tim=5645840078 hv=2512561537
ad='b5f8e850'
select name,intcol#,segcol#,type#,length,nvl(precision#,0),decode(type#,2,nvl(scale,-127/*MAXSB1MINAL*/),
178,scale,179,scale,180,scale,181,scale,182,scale,183,scale,231,scale,0),null$,fixedstorage,nvl(deflengt
h,0),default$,rowid,col#,property, nvl(charsetid,0),nvl(charsetform,0),spare1,spare2,nvl(spare3,0)  from
col$ where obj#=:1 order by intcol#
END OF STMT
EXEC #3:c=0,e=118,p=0,cr=0,cu=0,mis=0,r=0,dep=1,og=4,tim=5645840075
FETCH #3:c=0,e=52,p=0,cr=3,cu=0,mis=0,r=1,dep=1,og=4,tim=5645840215
STAT #3 id=1 cnt=130 pid=0 pos=1 obj=0 op='SORT ORDER BY (cr=35 pr=0 pw=0 time=732 us)'
STAT #3 id=2 cnt=130 pid=1 pos=1 obj=21 op='TABLE ACCESS CLUSTER COL$ (cr=35 pr=0 pw=0
time=330 us)'
STAT #3 id=3 cnt=11 pid=2 pos=1 obj=3 op='INDEX UNIQUE SCAN I_OBJ# (cr=22 pr=0 pw=0 time=102 us)'
WAIT #2: nam='DFS lock handle' ela= 450 type|mode=1413545989 id1=4 id2=0 obj#=-1
tim=5645841105
WAIT #2: nam='DFS lock handle' ela= 445 type|mode=1413545989 id1=2 id2=0 obj#=-1
tim=5645841592
WAIT #2: nam='enq: US - contention' ela= 443 name|mode=1431502854 undo segment #=0 0=0
obj#=-1 tim=5645842085
WAIT #2: nam='gc current block 2-way' ela= 1420 p1=1 p2=9 p3=16777231 obj#=-1 tim=5645843634
WAIT #2: nam='gc current grant busy' ela= 516 p1=1 p2=9 p3=33619983 obj#=-1 tim=5645844244
WAIT #2: nam='rdbms ipc reply' ela= 637 from_process=17 timeout=900 p3=0 obj#=-1
tim=5645844928
WAIT #2: nam='enq: TT - contention' ela= 488657 name|mode=1414791174 tablespace ID=0
operation=0 obj#=-1 tim=5646333647
WAIT #2: nam='enq: TT - contention' ela= 488293 name|mode=1414791174 tablespace ID=0
operation=0 obj#=-1 tim=5646821976
```

等待事件是分析的起点，而真正能够理解等待事件后面隐藏的深意是 DBA 不断成长的重要标志。

第一个需要分析清楚的是 DFS lock handle，该事件出现了两次。

```
WAIT #2: nam='DFS lock handle' ela= 450 type|mode=1413545989 id1=4 id2=0 obj#=-1
tim=5645841105
WAIT #2: nam='DFS lock handle' ela= 445 type|mode=1413545989 id1=2 id2=0 obj#=-1
tim=5645841592
```

DFS 是 Distributed File System 的缩写，最早在 Oracle 并行服务器（OPS）时代引入，指分布式集群文件系统，但是今天这个 DFS 涵盖的意义更加广泛，这里代表着需要获取一个全局锁。之所以加上 Handle，是因为很多锁的获取是分为两个步骤，首先获得 lock handle 上的锁定。这里的 DFS 给我们的提示是，这是一个 RAC 环境，而事实上，另外一个节点还在正常运行，当前故障节点无法启动。

由于 Oracle 的锁类型很多，这个事件的一个参数是 type|mode，通过这个参数合并编码了锁类型和请求模式。可以通过 SQL 将这两个元素解码出来。

```
select chr(bitand(&&p1,-16777216)/16777215) ||
  chr(bitand(&&p1,16711680)/65535) type, mod(&&p1, 16) md
from dual;
```

```
TY      MD
--      ----------
TA      5
```

查询得到的结论是 TA 锁,请求模式 5。如果不熟悉 TA 这个锁类型,可以通过数据字典查询一下。

```
SQL> exec print_table('select * from v$lock_type where type=''TA''');
TYPE              : TA
NAME              : Instance Undo
ID1_TAG           : operation
ID2_TAG           : undo segment # / other
IS_USER           : NO
IS_RECYCLE        : NO
DESCRIPTION       : Serializes operations on undo segments and undo tablespaces
CON_ID            : 0
-----------------
```

以上 print_table 引用了一个 Tom 的脚本,其内容如下所示。

```
create or replace procedure print_table( p_query in varchar2 )
AUTHID CURRENT_USER
is
    l_theCursor     integer default dbms_sql.open_cursor;
    l_columnValue   varchar2(4000);
    l_status        integer;
    l_descTbl       dbms_sql.desc_tab;
    l_colCnt        number;
begin
    execute immediate
    'alter session set
    nls_date_format=''dd-mon-yyyy hh24:mi:ss'' ';

    dbms_sql.parse( l_theCursor, p_query, dbms_sql.native );
    dbms_sql.describe_columns
    ( l_theCursor, l_colCnt, l_descTbl );

    for i in 1 .. l_colCnt loop
        dbms_sql.define_column
        (l_theCursor, i, l_columnValue, 4000);
    end loop;

    l_status := dbms_sql.execute(l_theCursor);

    while ( dbms_sql.fetch_rows(l_theCursor) > 0 ) loop
        for i in 1 .. l_colCnt loop
            dbms_sql.column_value
            ( l_theCursor, i, l_columnValue );
            dbms_output.put_line
            ( rpad( l_descTbl(i).col_name, 30 )
              || ': ' ||
              l_columnValue );
        end loop;
```

```
          dbms_output.put_line( '-----------------' );
      end loop;
      execute immediate
          'alter session set nls_date_format=''dd-MON-rr'' ';
exception
    when others then
      execute immediate
          'alter session set nls_date_format=''dd-MON-rr'' ';
      raise;
end;
/
```

我们看到 TA 锁是 Instance Undo——实例重做锁，其描述表明是和 Undo 段、Undo 表空间相关的串行操作。

进一步锁定信息在下一个事件体现出来，成为显性的 US 锁等待。

```
WAIT #2: nam='enq: US - contention' ela= 443 name|mode=1431502854 undo segment #=0 0=0
obj#=-1 tim=5645842085
```

这里的 name|mode 转换出来，请求的是模式 6——X 的排他锁。

```
select chr(bitand(&&p1,-16777216)/16777215) ||
chr(bitand(&&p1,16711680)/65535) type,
mod(&&p1, 16) md from dual dual;
Enter value for p1: 1431502854
TY    MD
--  ----------
US     6
```

再进一步，Oracle 通过 GC Current 读向远程请求了文件 1 的第 9 个 Block，但是在远程实例中这个请求没有被授予，呈现 Busy 的状态，也就是排它锁不能被获取。

```
WAIT #2: nam='gc current block 2-way' ela= 1420 p1=1 p2=9 p3=16777231 obj#=-1 tim=5645843634
WAIT #2: nam='gc current grant busy' ela= 516 p1=1 p2=9 p3=33619983 obj#=-1 tim=5645844244
```

再向下，数据库以 TT 锁等待处于无尽循环等待，数据库无法启动。

```
  WAIT #2: nam='enq: TT - contention' ela= 488657 name|mode=1414791174 tablespace ID=0
operation=0 obj#=-1
  WAIT #2: nam='enq: TT - contention' ela= 488293 name|mode=1414791174 tablespace ID=0
operation=0 obj#=-1
```

那么 TT 锁是什么？通过以下查询，可以看到这同样是串行表空间操作的 DDL 锁。

```
SQL> set serveroutput on
SQL> exec print_table('select * from v$lock_type where type=''TT''');
TYPE                : TT
NAME                : Tablespace
ID1_TAG             : tablespace ID
ID2_TAG             : operation
IS_USER             : NO
IS_RECYCLE          : NO
```

```
DESCRIPTION                    : Serializes DDL operations on tablespaces
CON_ID                         : 0
-----------------
```

整个问题的脉络梳理清楚了，还剩下一个问题，为什么在数据库的启动过程中要在回滚段上获取串行 DDL 排他锁？文件 1 的 Block 9 给我们了一个线索，存活节点不能给予的锁定块到底是什么？

```
SQL> exec print_table('select segment_name,segment_type from dba_extents where file_id=1
and block_id=9');
SEGMENT_NAME      SEGMENT_TYPE
--------------    ----------------
SYSTEM            ROLLBACK
```

我们看到了输出结果，系统回滚段，那么数据库启动过程为何要在 SYSTEM 回滚段上获取排他锁呢？

回顾 BOOTSTRAP$ 中的对象，数据库初始化中顺序执行的第一个语句就是 CREATE ROLLBACK SEGMENT SYSTEM STORAGE，也就是分配系统回滚段。

```
CREATE ROLLBACK SEGMENT SYSTEM STORAGE
( INITIAL 112K NEXT 1024K MINEXTENTS 1 MAXEXTENTS 32765 OBJNO 0 EXTENTS
(FILE 1 BLOCK 9))
```

正是因为这样一个 DDL 操作需要获得排他锁，数据库在启动过程中才需要向存活实例发出锁请求，无法获得就只有等待。

这里需要注意，系统回滚段从 11g 开始，向后偏移至第 128 个 Block。

```
CREATE ROLLBACK SEGMENT SYSTEM STORAGE ( INITIAL 112K NEXT 56K MINEXTENTS 1 MAXEXTENTS
32765 OBJNO 0 EXTENTS (FILE 1 BLOCK 128))
```

那么对方实例为什么无法给予授权？那一定是因为对方节点同样存在锁定。我们从存活实例上查询 V$LOCK 分析其当前锁定，发现同样存在大量的 TT 锁等待，而连接时间最长的会话 CTIME 已经长达 8066215 秒。

```
SQL> select * from gv$lock where type='TT';
   INST_ID ADDR             KADDR                   SID TY        ID1 ID2 LMODE REQUEST     CTIME BLOCK
---------- ---------------- ---------------- ---------- --        --- --- ----- -------  -------- -----
         1 07000002D8867018 07000002D8867038       2112 TT          0   0     0       4     12725     0
         1 07000002D88643B0 07000002D88643D0       2115 TT          0   0     0       4     13055     0
         1 07000002D88707D0 07000002D88707F0       2116 TT          0   0     4       0   8066215     2
         1 07000002D8864020 07000002D8864040       2116 TT          0   0     0       4     13385     0
         1 07000002D88644E0 07000002D8864500       2124 TT          0   0     0       4     10415     0
         1 07000002D88670B0 07000002D88670D0       2134 TT          0   0     0       4     11735     0
         1 07000002D88674D8 07000002D88674F8       2145 TT          0   0     0       4     10745     0
         1 07000002D8868BB0 07000002D8868BD0       2155 TT          0   0     4       0   8314633     2
         1 07000002D88673A8 07000002D88673C8       2155 TT          0   0     0       4     11405     0
```

注意这两个长时间的锁定，其 BLOCK 类型为 2，这在官方文档上有明确的解释（文档 v$lock

视图部分），这样的进程阻塞了 RAC 环境远程的操作。

```
2 - The lock is not blocking any blocked processes on the local node, but it may or may not be blocking processes on remote nodes. This value is used only in Oracle Real Application Clusters (Oracle RAC) configurations (not in single instance configurations).
```

那么这些进程到底在执行什么操作？通过 V$SESSION.SQL_ID 发现这些 SQL 正在执行表空间的剩余空间计算工作。

```
SELECT a.ts_name, bytes/data_sum, data_sum, free_sum
FROM
  (SELECT tablespace_name ts_name,
    SUM(NVL(blocks,0)) data_sum,
    SUM(NVL(bytes,0)) bytes
  FROM dba_data_files
  GROUP BY tablespace_name
  )a,
  (SELECT tablespace_name ts_name,
    SUM(NVL(blocks,0)) free_sum
  FROM user_free_space
  GROUP BY tablespace_name
  )b
WHERE a.ts_name=b.ts_name
ORDER BY a.ts_name
```

通过手工 Kill 终止两个阻塞进程之后，数据库节点可以成功启动。通过 MOS 确认这是一个 Bug，如果对于 DBA_FREE_SPACE 的查询未正常结束可能导致锁定无法释放，产生 TT 队列竞争，其影响范围甚至包括 11.2.0.4 版本，如图 4-1 所示。

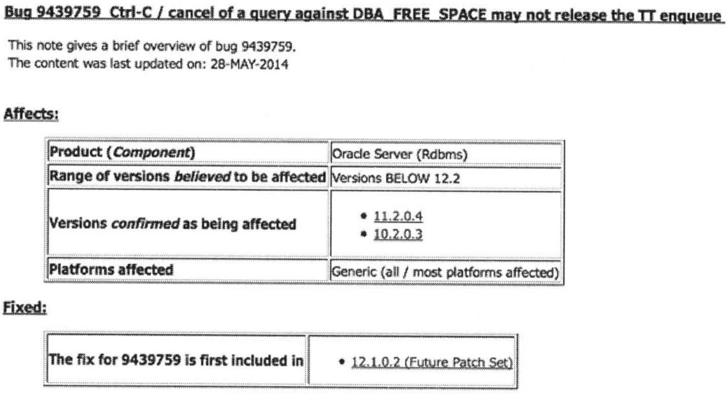
Bug 9439759 Ctrl-C / cancel of a query against DBA_FREE_SPACE may not release the TT enqueue

This note gives a brief overview of bug 9439759.
The content was last updated on: 28-MAY-2014

Affects:

Product (*Component*)	Oracle Server (Rdbms)
Range of versions *believed* to be affected	Versions BELOW 12.2
Versions *confirmed* as being affected	• 11.2.0.4 • 10.2.0.3
Platforms affected	Generic (all / most platforms affected)

Fixed:

The fix for 9439759 is first included in	• 12.1.0.2 (Future Patch Set)

图 4-1

其实 DBA_FREE_SPACE 视图的结构在不同版本之间变化较大，尤其是回收站引入之后。

查询 DBA_FREE_SPACE 可能极慢并且可能触发问题，在监控表空间变化时，应当注意这些问题。

4.4 Oracle 跟踪总结

在我的职业生涯中，每当我接触到一项技术之后，总是希望能够了解隐藏在软件之后的细节。所以我总会迫切地去寻找更多的软件跟踪方法，而一旦掌握了这些方法之后，就能够持续地帮助我去发现原理、解决问题。这样的经验一直伴随着我成长和前行。

这篇文章中，总结的案例跨越了我至今的整个职业历程，涉及的数据库版本从 Oracle9i 到今天的 Oracle 12.2。也正因为这些方法的持续赋能，我更愿意将这作为我最宝贵的经验分享给读者，希望大家能够掌握方法，从中获取长期受益的经验。

4.5 参考资料：数据类型比较规则

本章内容多个案例涉及了 Oracle 的数据比较和隐式转换，因此我们在此翻译了官方文档中的一段文字，作为 Oracle 数据比较的基础知识，供读者参考。

4.5.1 数值类型

数值类型的值的大小跟其数字本身大小的比较是一致的，数字较大的值也较大。负数小于 0 和所有正数，比如，–1 小于 100，–100 小于 –1。

浮点型的数值（不是数字或不能以数字表示）大于任何其他的数值类型的值，并等于它本身。

4.5.2 日期类型

一般认为日期类型的值，出现越晚，值越大。例如，2005 年 3 月 29 日下午 1:35 对应的日期类型的值小于 2006 年 1 月 5 日对应的日期类型的值，但大于 2005 年 1 月 5 日早上 10:09 对应的日期类型的值。

4.5.3 字符类型

字符类型的值的比较有以下两种规则。

（1）基于二进制或语言学排序。

（2）基于填充空格或不填充空格的语义比较。

接下来将会详细介绍这两种规则。

1．二进制或语言学排序

二进制比较是数据库的默认比较规则。数据库字符集中的每一个字符都有一个特定的值，对于任何一个字符类型的值，会通过比较该值中每一个字符在字符集中对应的编码值的和来决定它的大小。和越大，则该字符类型的值越大。Oracle 认为空格小于任何字符，这个结论适用于大部分的字符集。

下面是比较常见的一些字符集。

- **7-bit ASCII**（American Standard Code for Information Interchange）。
- **EBCDIC Code**（Extended Binary Coded Decimal Interchange Code）。
- **ISO 8859/1**（International Organization for Standardization）。
- **JEUC** Japan Extended UNIX。

当你所比较的字符类型对应的数值编码的二进制序列跟字符产生的语法序列不一致的时候，就需要考虑使用语法比较。但语法比较需要满足两个条件，一是 NLS_SORT 参数没有被设置为 BINARY，二是 NLS_COMP 参数的值被设为 LINGUISTIC。使用语法比较的时候，所有的 SQL 排序和比较都会根据 NLS_SORT 参数指定的语法规则来进行。

2．基于填充空格或不填充空格的语义比较

（1）**基于填充空格的语义比较**。如果两个字符类型的值长度不相等，Oracle 会首先在较短的值后面添加空格，使二者长度相等。然后通过比较两个字符类型的值从左到右第一个不同的字符的编码值，编码值大的我们就认为该字符类型的值比较大。如果两个值不存在不同字符，则认为二者相等。也就是说，如果两个字符类型的值的区别仅在于末尾空格的数目，我们认为这两个值相等。

这种基于空格填充的比较规则仅在以下条件下使用：要比较的值既不是 CHAR 或者 NCHAR 类型的表达式，也不是文本或者用户函数返回的结果。

（2）**基于不填充空格的语义比较**。Oracle 通过比较两个字符类型的值的左边起首个不同字母对应的编码值的大小决定大小。编码值越大的，我们认为该字符类型的值也比较大。当两个字符类型的值的长度不等时，我们认为长度较长的值也较大。只有两个字符类型的值长度相等，并且对应位置字符也相等时，我们才认为这两个值是相等的。当要比较的两个值中的一个或者

两个都是 VARCHAR2 或者 NVARCHAR2 类型时，就会用到该比较规则。

使用不同的比较规则比较两个字符类型的值，结果可能不相同。表 4-2 列举了五组字符类型的值分别基于这两种规则时的比较结果。

表 4-2

Blank-Padded	Nonpadded
'ac' > 'ab'	'ac' > 'ab'
'ab' > 'a '	'ab' > 'a '
'ab' > 'a'	'ab' > 'a'
'ab' = 'ab'	'ab' = 'ab'
'a ' = 'a'	'a ' > 'a'

表 4-3 中列举了 ASCII 包含的字符集，表 4-4 列举了 EBCDIC 包含的字符集。区分大小写。对于某些语言，部分字符集中的字符的数值可能并不与其语法的顺序相吻合。

表 4-3

Symbol	Decimal value	Symbol	Decimal value
blank	32	;	59
!	33	<	60
"	34	=	61
#	35	>	62
$	36	?	63
%	37	@	64
&	38	A-Z	65-90
'	39	[	91
(	40	\	92
)	41	]	93
*	42	^	94
+	43	_	95
,	44	`	96
-	45	a-z	97-122
.	46	{	123
/	47	|	124
0-9	48-57	}	125
:	58	~	126

表 4-4

Symbol	Decimal value	Symbol	Decimal value
blank	64	%	108
¢	74	_	109
.	75	>	110
<	76	?	111
(	77	:	122
+	78	#	123
|	79	@	124
&	80	'	125
!	90	=	126
$	91	"	127
*	92	a-i	129-137
)	93	j-r	145-153
;	94	s-z	162-169
ÿ	95	A-I	193-201
-	96	J-R	209-217
/	97	S-Z	226-233

4.5.4　对象类型

对于对象类型的值的比较，通常会采用以下两种函数：MAP 和 ORDER 函数。这两个函数都可以基于对象类型作比较，但二者有很大区别。使用这两种函数必须要在对象的类型中指定要用来比较的方法。

4.5.5　数组和嵌套表类型

1．数据类型的优先级

Oracle 利用数据类型的优先级来决定数据类型的隐式转换。接下来我们会详细描述。Oracle 的数据类型具有以下的优先级顺序。

（1）日期类型和间隔日期类型。

（2）双精度二进制类型。

（3）浮点二进制类型。

(4)数值类型。

(5)字符类型。

(6)其他内置数据类型。

2. 数据转换

通常情况下,在一个表达式中不可能同时包含两种数据类型。比如在一个表达式中,不能同时计算 5*10+'JAMES',但 Oracle 支持使用隐式转换或者显式转换的方式将一个类型转换成另一个类型,然后进行相关运算。

3. 关于数据类型的隐式转换和显式转换

Oracle 建议多使用显式转换而不要太多依赖于隐式转换,主要有以下几点原因。

(1)使用显式转换可以让 SQL 语句更容易理解。

(2)隐式转换会对性能造成一些影响,尤其是当我们需要把一列值通过隐式转换转为常量的时候。

(3)隐式转换可能会依赖环境,并不一定会起作用。例如,使用隐式转换的方式将一个日期类型的值转换为 VARCHAR2 类型,受到 NLS_DATE_FORMAT 参数的影响,并不一定能返回正确的结果。

(4)实现隐式转换的算法会因在软件版本和数据库产品的不同而导致结果有差异,容易出错。而显式转换的结果会比较稳定。

(5)当索引表达式上出现隐式转换时,这时 Oracle 数据库会将索引类型定义为"转换前的类型",因而导致索引不再可用。这会对性能产生很不好的影响。

4. 隐式数据转换

在 Oracle 数据库中,面对类型不一致的情况,只要某种转换方式行得通,Oracle 会自动将一个数据类型转换成另外一种。

表 4-5 列出了 Oracle 数据库中的隐式转化,包含了所有可转换的情况。

4.5 参考资料：数据类型比较规则

表 4-5

	CHAR	VARCHAR2	NCHAR	NVARCHAR2	DATE	DATETIME/INTERVAL	NUMBER	BINARY_FLOAT	BINARY_DOUBLE	LONG	RAW	ROWID	CLOB	BLOB	NCLOB
CHAR	--	×	×	×	×	×	×	×	×	×	×	--	×	×	×
VARCHAR2	×	--	×	×	×	×	×	×	×	×	×	×	×	--	×
NCHAR	×	×	--	×	×	×	×	×	×	×	×	×	×	×	×
NVARCHAR2	×	×	×	--	×	×	×	×	×	×	×	×	×	×	×
DATE	×	×	×	×	--	--	--	--	--	--	--	--	--	--	--
DATETIME/INTERVAL	×	×	×	×	--	--	--	--	--	×	--	--	--	--	--
NUMBER	×	×	×	×	--	--	×	×	×	--	--	--	--	--	--
BINARY_FLOAT	×	×	×	×	--	--	×	--	×	--	--	--	--	--	--
BINARY_DOUBLE	×	×	×	×	--	--	×	×	--	--	--	--	--	--	--
LONG	×	×	×	×	--	×[Foot 1]	--	--	--	--	×	--	×	--	×
RAW	×	×	×	×	--	--	--	--	--	×	--	--	×	×	--
ROWID	--	×	×	×	--	--	--	--	--	--	--	--	--	--	--
CLOB	×	×	×	×	--	--	--	--	--	--	--	--	--	--	×
BLOB	--	--	--	--	--	--	--	--	--	--	×	--	--	--	--
NCLOB	×	×	×	×	--	--	--	--	--	×	--	--	×	--	--

注意，不能直接将 long 类型的值转换为间隔日期类型的，但可以使用 TO_CHAR(interval) 将 long 类型转化为 VARCHAR2 类型。

以下是使用隐式转换的一些规则。

（1）在使用 insert 和 update 操作时，Oracle 会对受影响的列进行类型转换。

（2）在 select from 操作中，Oracle 会将 from 子句中的类型转换为目标类型。

（3）在数值类型的值的先关操作中，Oracle 会首先基于最多符合原则调整数值的精度。在这种情况下，对于数值类型的值的操作结果可能会跟以下我们将会提到的操作结果不一致。

（4）若要比较字符类型和数值型的值，Oracle 会将字符型转换成数值类型。

（5）将字符类型或者 Number 类型的值与浮点类型进行转换可能会产生误差，因为字符类型和 Number 类型使用十进制的精度表达数值大小，而浮点型使用的是二进制的精度。

（6）如果将一个 CLOB 的值转换为字符类型，比如 VARCHAR2 类型，或者将 BLOB 的值转换为 RAW 类型，如果源类型下的值大于转换后的值，则会返回错误。

（7）如果将时间戳类型转换为日期类型，分时和秒时的部分会被直接摒弃掉，而在 Oracle 数据库的早期版本，则会将这些部分近似处理。

（8）将二进制的浮点类型转换为二进制的双精度类型不会产生误差。

（9）将双精度二进制类型转换为浮点型二进制类型的结果可能不准确，会产生误差。因为双精度比浮点型多使用一位，要将多出来的这一位删掉，肯定会产生误差。

（10）当比较字符型和日期类型的值，Oracle 会将字符型转换为日期类型。

（11）当使用带有参数的 SQL 函数或者操作的时候，参数对应的数据类型不能被识别，Oracle 会将参数的类型转换为能够被识别的类型。

（12）在进行运算的时候，Oracle 会把等号右边的类型转换为跟等号左边一样的类型。

（13）在转换中，Oracle 会将非字符类型的值转换为 CHAR 或者 NCHAR。

（14）在算数运算中，为比较字符类型和非字符类型的大小，Oracle 会根据实际情况将任何字符类型的转换为数值，日期或者 rowid 等。如果算数运算中包含 char/nchar 类型与 nchar/nvarchar2 类型，Oracle 会将这些类型都转换为 Number 类型再进行运算。

（15）大部分的 SQL 函数能够识别和接受 CLOB 类型的参数，并且会在 CLOB 类型和字符类型之间做隐式转换。因此，在一些不识别 CLOB 类型或者不能对 CLOB 类型的值做隐式转换的函数中，调用前 Oracle 会先尝试进行转换。如果 CLOB 类型的字段大于 4000 字节，则 Oracle 默认智能转换前面的 4000 字节，也就是说，当 CLOB 字段过大的时候，可能会导致错误产生。

（16）当进行字符类型和 RAW 类型或者 LONG RAW 类型的相互转换的时候，二进制类型的值会被转换为十六进制来表示，用一个十六进制的字符代表 RAW 类型的四位。

（17）CHAR 和 VARCHAR2 类型的转换或者 NCHAR 和 NVARCHAR2 类型的比较可能需要不同的字符集。默认的转换方向是将当前的数据库字符集转换为国际字符集。表 4-6 描述了不同字符类型的隐式转换方向。

表 4-6

	to CHAR	to VARCHAR2	to NCHAR	to NVARCHAR2
from CHAR	--	VARCHAR2	NCHAR	NVARCHAR2
from VARCHAR2	VARCHAR2	--	NVARCHAR2	NVARCHAR2
from NCHAR	NCHAR	NCHAR	--	NVARCHAR2
from NVARCHAR2	NVARCHAR2	NVARCHAR2	NVARCHAR2	--

对于一些未定义类型的值，是不能进行隐式转换的。必须使用 CAST ... MULTISET 语法进行显式转换。

5．隐式数据类型转换实例

（1）文本类型隐式转换

文本类型值"10"属于 CHAR 类型，当它出现在数学表达式中的时候，Oracle 会将它隐式转换为 Number 类型。

```
SELECT salary + '10' FROM employees;
```

（2）字符和数字类型隐式转换

当比较数字类型和字符类型值的大小时，oracle 会隐式将字符类型转换为数字类型，在下面的例子中，oracle 将'200'转换为 200。

```
SELECT last_name FROM employees WHERE employee_id = '200';
```

（3）日期类型隐式转换

在下面的语句中，Oracle 将'24-JUN-06'隐式转换为日期类型的默认格式'DD-MON-YY'。

```
SELECT last_name FROM employees WHERE hire_date = '24-JUN-06';
```

6．显式数据转换

你可以使用 SQL 的转换函数显式地转换数据类型。表 4-7 列举了将一个类型转换为其他类型可以使用的一些函数。

在 Oracle 不能进行隐式转换的情况下，不能指定 RAW 类型或者 LONG RAW 类型。例如 LONG 类型和 LONG RAW 类型不能出现在包含函数或者操作符的表达式中。

表 4-7 Explicit Type Conversions

	to CHAR,VARCHAR2,NCHAR,NVARCHAR2	to NUMBER	to Datetime/Interval	to RAW	to ROWID	to LONG,LONG RAW	to CLOB,NCLOB,BLOB	to BINARY_FLOAT	to BINARY_DOUBLE
from CHAR,VARCHAR2,NCHAR,NVARCHAR2	TO_CHAR(char.) TO_NCHAR(char.)	TO_NUMBER	TO_DATE TO_TIMESTAMP TO_TIMESTAMP_TZ TO_YMINTERVAL TO_DSINTERVAL	HEXTORAW	CHARTOROWID	--	TO_CLOB TO_NCLOB	TO_BINARY_FLOAT	TO_BINARY_DOUBLE
from NUMBER	TO_CHAR(number) TO_NCHAR(number)	--	TO_DATE NUMTOYM-INTERVAL NUMTODS-INTERVAL	--	--	--	--	TO_BINARY_FLOAT	TO_BINARY_DOUBLE
from Datetime/Interval	TO_CHAR(date) TO_NCHAR(datetime)	--	--	--	--	--	--	--	--
from RAW	RAWTOHEX RAWTONHEX	--	--	--	--	--	TO_BLOB	--	--
from ROWID	ROWIDTOCHAR	--	--	--	--	--	--	--	--
from LONG/LONGRAW	--	--	--	--	--	--	TO_LOB	--	--
from CLOB,NCLOB,BLOB	TO_CHAR TO_NCHAR	--	--	--	--	--	TO_CLOB TO_NCLOB	--	--
from CLOB,NCLOB,BLOB	TO_CHAR TO_NCHAR	--	--	--	--	--	TO_CLOB TO_NCLOB	--	--
from BINARY_FLOAT	TO_CHAR(char.) TO_NCHAR(char.)	TO_NUMBER	--	--	--	--	--	--	TO_BINARY_DOUBLE
from BINARY_DOUBLE	TO_CHAR(char.) TO_NCHAR(char.)	TO_NUMBER	--	--	--	--	--	TO_BINARY_FLOAT	--

7. 数据类型转换中的安全考虑

当一个日期类型的值被转换为文本的时候，不管是用隐式转换的方式还是显式转换的方式，如果没有指定格式规范，这个格式由部分全局会话中的参数决定，这些参数包括 NLS_DATE_FORMAT、NLS_TIMESTAMP_FORMAT 或者 NLS_TIMESTAMP_TZ_FORMAT 等，而这些参数都可以通过客户端或者命令行做修改。

在显式转换中，要是没有明确指定转换格式（比如日期类型的值通过动态 SQL 语句执行后被应用），会对数据库的安全造成该很大影响。所谓动态 SQL 语句指的是其语句在提交给数据库执行前，是一些碎片。动态 SQL 语句频繁地与内置 PL/SQL 包或者 DBMS_SQL 中的 PL/SQL 语句关联并快速执行，但这仅仅是动态组织的 SQL 语句被当做参数传递的其中一个例子（在该例子中 start_date 属于日期类型）。

```
EXECUTE IMMEDIATE
'SELECT last_name FROM employees WHERE hire_date > ''' || start_date || '''';
```

上述的例子中，start_date 的值根据会话级别的 NLS_DATE_FORMAT 参数的格式被隐式转换成文本类型。日期类型被转换成文本格式但使用双引号引起来。这样，任何用户都可以通过设置全局的参数格式来决定日期类型的值转换后的格式。如果 SQL 语句是以 PL/SQL 的存储过程执行，这些存储过程的执行可能会受到这些会话参数的影响。如果这些过程以定义者的角色执行，拥有比会话高一点的权限，就可以获得对未授权的敏感数据的访问，造成不安全。

这里需要注意，这些安全风险不仅存在于 SQL 执行的时候。事实上，在中间层的应用上，在利用 OCI 日期函数将日期类型转换为文本类型后重构 SQL 语句的过程中也存在安全风险。如果一个会话的全局参数被用户获得，会对应用产生很大的影响。

数值类型的隐式转换和显式转换还可能受到同类问题的影响。由于依赖于如 NLS_NUMERIC_CHARACTERS 等定义格式和分隔符的参数，如果将分隔符定义为单引号或者双引号，一些基本的 SQL 的输入都会被合并。

第 5 章　兴趣、思考与实践

——我的 DBA 手记（张乐奕）

> **题记**
>
> 在学习 Oracle 技术的过程中，如何掌握正确的方法，推进自己的学习和思考，获得持续的成长是非常重要的。在 Kamus 的 DBA 手记中，我们整理收录了他关于自我学习的总结、动手实践的验证、设计分析的解惑，在这样的系列文章中，我们得以学习和借鉴他的思路精华，并最终对大家有所裨益。

5.1　Kamus 谈技术、学习与分享

随着 Oracle 数据库的发展和广泛使用，整个数据和运维行业的变革也是显而易见的。Oracle 数据库为什么如此受欢迎，这些变革将会产生什么样的影响呢？我们来看 Kamus 在访谈中表达的看法总结。

5.1.1　Oracle 的适用场景

问及当前使用 Oracle 数据库的应用，Kamus 说，可能已经不能问主要有哪些应用采用 Oracle 数据库了，倒是应该反过来问，有哪些应用不选 Oracle 数据库呢？

除了四大银行核心系统（这是 Oracle 数据库没有进入的领域）采用 IBM 大机之外，几乎没有 Oracle 数据库不涉猎的（实际上很多稍小的银行核心系统也是采用 Oracle 数据库），当然一些大客户的系统非常多，也绝不仅仅是只采用 Oracle 一家的产品，比如电信行业的 BI 目前

应该就是被 IBM 垄断的。另外，现在如火如荼的电子商务（Web 网站）领域，也大量采用 MySQL 等开源数据库，甚或是如 MongoDB 这样 NoSQL 数据库，但是通常在重要的订单生成及支付环节，还是在使用 Oracle 数据库。

5.1.2 为什么 Oracle 广受欢迎

这要从 Oracle 数据库的优点说起。Oracle 数据库的优点是见仁见智的问题，每个数据库产品在技术上都有自己的优点，除了技术上的优劣，非技术的原因其实也有很大关系，而 Oracle 数据库之所以受欢迎，是因为其采用了免费下载的方式，迅速被大量技术人员所熟悉并获得了他们的支持，这一点跟 Windows 在国内的发展其实可以相提并论。我一直认为其他厂商也完全应该仿效这种模式，免费下载，免费测试，商业使用再付费。另外，每 2 年一次的大版本升级，也让 Oracle 数据库始终走在技术前沿，不管如何，混个脸熟，这是必须的。

5.1.3 如何实现灾备

关于 Oracle 数据库的灾备技术，Oracle 的 DataGuard，RAC，GoldenGate 已经广泛应用，如果后两者可以算作灾备的方式的话。如今各大型数据库使用灾备是很普遍的，可见数据的安全已经得到了人们的高度重视。

可以分享一则我们救援的数据丢失及恢复的真实案例，案例场景如下：客户准备为没有任何备份的数据库添加 RMAN 备份策略，但是由于空间问题，需要先添加一些磁盘，在 Linux 操作系统上，添加了一块磁盘，然后重启了主机，想当然的将最后一块磁盘格式化了（意思是之前有/dev/sda 和/dev/sdb，添加了新硬盘以后，认为新硬盘必然是/dev/sdc，因此就直接格式化了/dev/sdc，并且创建了文件系统），但是实际上新加的盘在机器重启以后被认成/dev/sda，而原先的 sda 和 sdb 则变成了 sdb 和 sdc，所以格式化/dev/sdc 的命令就直接将一个包含大量数据的 ASM 磁盘格式化了。

没有任何备份，磁盘被格式化，非常悲惨的一次事件。虽然最后我们通过一系列手段将大部分数据救回，但是客户也为此损失了大量财力、人力和时间。

这个案例之所以印象深刻，是因为在实施灾备的过程中破坏了数据，就好比在买保险的路上被车撞了，它告诉我们灾难时刻存在，一分一秒都可能碰上，细心＋技术可以避免一些问题，但是只有完善的灾备才可以救命于水火。试想如果你的机房部署在 2001 年 9 月 11 日的纽约双子大厦，仅此一份，没有异地备份，前一秒可能还在洋洋自得，系统性能调整得多好，业务又得到长足发展，后一秒一切都化为乌有，再细心再有技术也于事无补。

5.1.4 数据库发展对 DBA 工作的影响

数年前，DBA 的工作范畴可能还局限在 Oracle 数据库中，只要登录进 SQL*Plus，所有的工作都能完成了。但是如今，DBA 已经不仅仅是数据库实例管理员了，而已经可以称为数据管理员。没有任何一个软件，可以替代行业知识，可以替代人对数据的熟悉程度，也没有任何一个软件可以用来决策系统架构，所谓"逆水行舟，不进则退"，如果在 DBA 这个岗位工作了 5 年，却还只知道数据库的备份恢复，表空间的整理，索引的调整，那么可能确实要为自己的未来担忧一下了。

同时 Exadata 的出现，意味着 DBA 的角色前所未有的复杂起来，不仅需要懂数据库，还需要懂主机、存储、网络，更需要懂的是应用的特点、数据的分布特点，所以只要能跟随着技术，不断学习，同时不断积累行业知识，DBA 可以做到老，哪怕在中国也是可以做到老的。当然，如果你想 DBA 做到年薪 100 万，恐怕还是要先考虑如何进入管理层。

5.1.5　Oracle 的学习方法

学习任何东西唯一一个需要具备的素质就是兴趣，唯一一个不能缺少的素质也是兴趣。如果当你看到一篇跟你技术相关的并且又很有帮助的文档，但是却完全没有兴奋感的话，那么可以考虑一下，是不是应该转型，不要再做技术了。

DBA 跟其他的技术工种一样，没有什么需要具备的特殊素质，能做的好还是不好，有兴趣没兴趣最重要。Oracle 是值得成为兴趣的行当，绝不应该仅仅当成谋生的手段。

那么如何系统学习 Oracle？

推荐我学习的三部曲，即阅读官方文档 + 自己动手做实验 + 论坛中参与讨论，这是快速提高技术的方法。

- 如果想学习 Oracle，但是从来不知道 docs.oracle.com 这个网址，那么你并没有认真在学习。
- 如果想学习 Oracle，但是自己的机器上都没有 Oracle 数据库，也没有几个虚拟机环境，那么你并没有认真在学习。
- 如果想学习 Oracle，却从不知道有一个 itpub.net 可以讨论 Oracle，也不知道 ACOUG 用户组，倒不能说你没有在认真学习，只是自己一个人学习不觉得孤独吗？

如果你一直在没有兴趣地，不认真地，孤独地学习着 Oracle，那么要不改变，要不放弃。因为不合适的学习方法并不利于个人的成长。

5.1.6　如何成为 ACE

Oracle ACE 实际上也并不仅仅是一个技术上的殊荣，而更多的是考虑社区贡献的多少，一个人即使有很强的技术，但是从来不愿意分享，那么仍然不符合 ACE 的选拔条件。当然实际上这样的人并不会太多，特别是在当今的网络时代，拥有高深技术却不愿意分享的人真的不会太多，所以从这个层面看上去，似乎 Oracle ACE 都是技术高手，但是实际上，Oracle ACE 应该是积极的 Oracle 社区分享者，分享则包括了写自己的 Blog、自己的著作出版、翻译出版、组织社区活动以及积极在社区活动中做主题演讲等等。

如果你明确了这些，关于如何成为 ACE 就是很显而易见的事情。

从认真学习 Oracle 开始，坚持每天学习，坚持总结心得，坚持分享体会（通过网络中的各种方式，比如 blog，比如论坛，比如用户组），大约需要 2 年的时间，可以有所斩获，然后再坚持 2 年，大约就有机会成为 Oracle ACE 了，说上去似乎并不难，但是持之以恒饱含热情地学习并且分享持续 4 年的时间，大约是现在这个浮躁的社会很少人能够做到的，但是如果你做到了，你就成功了。4 年时间在一个人的职业生涯中其实并不算多，只是看你愿不愿意静下心来付出这 4 年。

任何事情都没有捷径，如果最初就抱着走捷径的目的去做一件事情，那么一开始就输了。

5.2　以 12c Identity 类型示范自我探索式学习方法

题记：这篇文章首先我会从几个方面介绍下我的学习思路和方法，供大家参考。然后借助案例分析的过程来分享在技术研究中的一些技巧。作为一个做技术的人，方法很重要。凡事只有掌握了有效的方法，才能在学习的过程中事半功倍。但我并不主张笼统地谈正确的学习方法是什么，因为不同的技术领域，甚至是不同的知识背景，都有其特殊性，那就要求我们在学习的过程中发现知识的特殊性，并找到最适合的学习方法。

首先描述一下**我自己在学习新知识的时候大概是什么状态，什么思路**，因为自认为自己的学习能力还不错，因此也期望这样的学习方法对其他人会有帮助。看这篇文章的时候，你可以同步地想一想如果是你遇到这样的错误，你会怎么处理，怎么发散，怎么研究？

Oracle Database 12c 作为最新一代的 Oracle 数据库产品，已经广为使用，那么，如果学习一个新版本的数据库，应该如何开始呢？

我通常是从 New Features Guide 文档看起，先通览文档的目录，遇到感兴趣的新功能点，就开始做实验来验证这个新功能。当然，这之前需要先把新版本的数据库安装好、新版本的全部文档下载到本地，这样即使你坐在飞机上也有文档可查。

这次我的计划是**实验一下 Identity 类型的字段**，这个字段可以用来作主键，会自动递增，这种类型的字段在 SQL Server 中早就存在，但是 Oracle 直到 12c 才推出这个功能。

通常我不会用 sys 用户进行任何实验（除非是验证 sysdba 的新功能），因此总是会先创建一个我自己的 dba 用户。

在 12c 中创建这个用户首先就遇到了错误（测试环境启用了多租户架构）。

```
SQL> CREATE USER kamus IDENTIFIED BY oracle DEFAULT tablespace users;
ERROR at line 1:
ORA-65096: invalid common USER OR ROLE name
```

对于一个不熟悉的错误，第一件事情不是去 Google，而是**用 oerr 实用程序来看看 Oracle 自己对这个错误是怎么解释的**。为什么我喜欢非 Windows 环境中的 Oracle？oerr 的存在也是很大一个原因。

```
[oracle@dbserver-oel ~]$ oerr ora 65096
65096, 00000, "invalid common user or role name"
// *Cause:  An attempt was made to create a common user or role with a name
//          that wass not valid for common users or roles. In addition to
//          the usual rules for user and role names, common user and role
//          names must start with C## or c## and consist only of ASCII
//          characters.
// *Action: Specify a valid common user or role name.
```

错误信息的解析非常明确地告知"试图创建一个通用用户，必须要用 C## 或者 c##开头"，这时候心里会有疑问，什么是 common user？但是我通常不会先急着去翻文档，而是先把手头的事情做完，也就是先把用户创建上。

```
SQL> CREATE USER c##kamus IDENTIFIED BY oracle DEFAULT tablespace users;
USER created.
SQL> GRANT dba TO c##kamus;
GRANT succeeded.
```

创建 C##KAMUS 用户成功之后，再返回去解决心中的疑问，什么是 common user？在联机文档的左上角搜索关键字 common user，会得到如图 5-1 所示的结果。

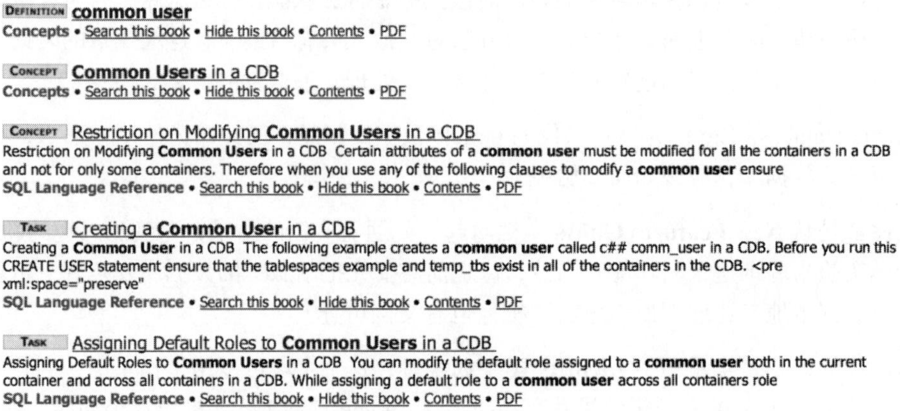

DEFINITION **common user**
Concepts • Search this book • Hide this book • Contents • PDF

CONCEPT **Common Users** in a CDB
Concepts • Search this book • Hide this book • Contents • PDF

CONCEPT Restriction on Modifying **Common Users** in a CDB
Restriction on Modifying **Common Users** in a CDB Certain attributes of a **common user** must be modified for all the containers in a CDB and not for only some containers. Therefore when you use any of the following clauses to modify a **common user** ensure
SQL Language Reference • Search this book • Hide this book • Contents • PDF

TASK Creating a **Common User** in a CDB
Creating a **Common User** in a CDB The following example creates a **common user** called c## comm_user in a CDB. Before you run this CREATE USER statement ensure that the tablespaces example and temp_tbs exist in all of the containers in the CDB. <pre xml:space="preserve">
SQL Language Reference • Search this book • Hide this book • Contents • PDF

TASK Assigning Default Roles to **Common Users** in a CDB
Assigning Default Roles to **Common Users** in a CDB You can modify the default role assigned to a **common user** both in the current container and across all containers in a CDB. While assigning a default role to a **common user** across all containers role
SQL Language Reference • Search this book • Hide this book • Contents • PDF

图 5-1

通常我会先浏览目录，如果看完觉得心中疑问已经解决，就会返回继续做之前的实验，不

5.2　以 12c Identity 类型示范自我探索式学习方法

会再浏览其他的链接；如果想要查询怎么做，比如说如何创建 common user，才会继续去看正文部分。这样的好处是可以**保持专注不至于被过多文档分心**。

但是由于 common user 这个概念几乎是崭新的，所以我很有兴趣继续探索一下：跟 common user 相对的 local user 该如何创建。继续去看正文当然是个方法，但是这里我选择的是直接去看 SQL Language Reference，因为我们知道一定是在 Create User 语法里面会有不同的定义，进入 Create User 语法页面，直接搜索 common user，就可以看到如下这段话。

CONTAINER Clause：

To create a local user in a pluggable database （PDB）, ensure that the current container is that PDB and specify CONTAINER = CURRENT. To create a common user, ensure that the current container is the root and specify CONTAINER = ALL. The name of the common user must begin with C## or c##. If you omit this clause and the current container is a PDB, then CONTAINER = CURRENT is the default. If you omit this clause and the current container is the root, then CONTAINER = ALL is the default.

也就是说我们一定要先登录进一个 PDB，才可以创建本地用户，那么如何知道现在的 SQL*Plus 是登录进了哪个 DB 呢？这个疑问其实是一个很简单的联想，既然需要去一个地方，那么一定有方法知道我现在在什么地方，通过简单地查询文档，可以得知以下的方法。现在确实在 CDB 中。

```
SQL> SHOW con_name
CON_NAME
------------------------------
CDB$ROOT
SQL> SELECT SYS_CONTEXT ('USERENV', 'CON_NAME') FROM DUAL;
SYS_CONTEXT('USERENV','CON_NAME')
----------------------------------------------
CDB$ROOT
```

dbca 建库的时候，有一个新选项是"同时创建 PDB"，我勾选过（对于 dbca 中出现的新选项，如果不是条件不允许，我都会选中进行测试），创建了名字为 pdbtest 的 PDB，那么现在我想尝试登录这个 PDB，去创建一个 local user。

如何登录 PDB？Administrator's Guide 中有专门的一个章节 "Part VI Managing a Multitenant Environment" 来描述如何管理多租户环境，浏览目录就可以直接找到 "Connecting to a PDB with SQL*Plus" 这部分，如下所示。

You can use the following techniques to connect to a PDB with the SQL*Plus CONNECT command:

Database connection using easy connect

Database connection using a net service name

那尝试直接使用 easy connect 来登录 PDB。

```
$ sqlplus sys/oracle@127.0.0.1:15210/pdbtest AS sysdba
SQL*Plus: Release 10.2.0.4.0 - Production ON Sat Jul 6 21:44:42 2013

Copyright (c) 1982, 2007, Oracle. ALL Rights Reserved.
Connected TO:
Oracle DATABASE 12c Enterprise Edition Release 12.1.0.1.0 - 64bit Production
WITH the Partitioning, OLAP, Advanced Analytics AND REAL Application Testing options
```

进行如下操作。

```
SQL> SELECT SYS_CONTEXT ('USERENV', 'CON_NAME') FROM DUAL;
SYS_CONTEXT('USERENV','CON_NAME')
----------------------------------------
PDBTEST
SQL> SELECT NAME,PDB FROM dba_services;
ERROR at line 1:
ORA-01219: DATABASE OR pluggable DATABASE NOT OPEN: queries allowed ON fixed TABLES OR views ONLY
```

PDB 没有 Open？尝试打开。无法使用 startup 命令。原因是我使用了旧版本的 SQL*Plus（如上所示是 10.2.0.4.0）连接到 12c 数据库的 PDB 中，某些新特性不被支持。

```
SQL> startup
ORA-24543: instance startup OR shutdown NOT allowed IN pluggable DATABASE
```

使用 12c 自带的 SQL*Plus 登录，就可以使用 startup 命令将 PDB 打开，使用 SQL*Plus 管理 PDB 的详细命令可以参看文档描述。

或者可以使用如下语句打开 PDB。

```
SQL> SHOW USER
USER IS "SYS"
SQL> startup
Pluggable DATABASE opened.
SQL> SHOW con_name

CON_NAME
---------
PDBTEST

SQL> ALTER PLUGGABLE DATABASE OPEN;
SQL> SELECT NAME, OPEN_MODE, RESTRICTED, OPEN_TIME FROM V$PDBS;
NAME      OPEN_MODE   RES OPEN_TIME
--------- ----------  --- -------------------------
PDBTEST   READ WRITE  NO  06-JUL-13 09.48.57.260 PM
```

到此，可以创建 local user 了。

```
SQL> CREATE USER kamus IDENTIFIED BY oracle;
USER created.
```

5.2 以 12c Identity 类型示范自我探索式学习方法

```
SQL> GRANT dba TO kamus;
GRANT succeeded.
```

那么在一个 PDB 中可以看到多少用户呢？可以看到 CDB 中的用户吗？

这又是一个简单的联想，**学习的过程其实是一个发散再收缩的循环**。看来不可以，只能看到自己的用户，当然这里有很多 common user。可以看到即使是在 PDB 中，cdb_视图也是可以使用的。

```
SQL> SELECT CON_ID,COUNT(*) FROM cdb_users GROUP BY con_id;
CON_ID   COUNT(*)
---------- ----------
     3         38
```

再回到 CDB 中看一下，会是什么情况？可以看到所有容器数据库中的用户都可以查询到。

```
SQL> SELECT CON_ID,COUNT(*) FROM cdb_users GROUP BY con_id;
   CON_ID   COUNT(*)
---------- ----------
        1         36
        2         35
        3         38
```

终于，我可以回到最开始的实验目标上去了，在 PDB 中创建了 T1 表，id 列为 Identity 类型。

```
SQL> CREATE TABLE t1 (id NUMBER GENERATED AS IDENTITY);
TABLE created.
```

根据文档描述，Identity 类型仍然是通过 Sequence 来实现的，那么应该是自动创建了一个 Sequence，果然如此。在你学习的过程中会多此一步来查询一下 Sequence 视图吗？

```
SQL> SELECT SEQUENCE_NAME FROM user_sequences;
SEQUENCE_NAME
--------------
ISEQ$$_91620
```

默认创建的 Sequence，CACHE_SIZE 是 20，开始值是 1，这都跟单独创建的 Sequence 默认值一样。

```
SQL> SELECT * FROM user_sequences;
SEQUENCE_NAME MIN_VALUE  MAX_VALUE  INCREMENT_BY C O CACHE_SIZE LAST_NUMBER P_C S K
------------- ---------- ---------- ------------ - - ---------- ----------- --- - -
ISEQ$$_91620          1 1.0000E+28            1 N N         20           1 N N
```

插入一条数据试一下，报错报错还是报错。所以是 generated always 的 identity 列，如果只有这一列，就没法插入数据。

```
SQL> INSERT INTO t1 VALUES('');
ERROR at line 1:
ORA-32795: cannot INSERT INTO a generated always IDENTITY COLUMN
```

```
SQL> INSERT INTO t1 VALUES(ISEQ$$_91620.NEXTVAL);
ERROR at line 1:
ORA-32795: cannot INSERT INTO a generated always IDENTITY COLUMN
SQL> INSERT INTO t1 VALUES(NULL);
ERROR at line 1:
ORA-32795: cannot INSERT INTO a generated always IDENTITY COLUMN
```

换 GENERATED BY DEFAULT ON NULL 类型试一下，**Wait**，如果删除了表，对应的序列会自动删除吗？理论上应该会，当然还是要测试一下。

```
SQL> DROP TABLE t1;
TABLE dropped.
```

序列还在？

```
SQL> SELECT * FROM user_sequences;
SEQUENCE_NAME  MIN_VALUE  MAX_VALUE  INCREMENT_BY C O CACHE_SIZE LAST_NUMBER P_C S K
-------------  ---------  ---------  ------------ - - ---------- ----------- --- - -
ISEQ$$_91620           1 1.0000E+28             1 N N         20           1     N N
Elapsed: 00:00:00.00
```

再建一张测试表。

```
SQL> CREATE TABLE t2 (id NUMBER GENERATED BY DEFAULT AS IDENTITY);
TABLE created.
```

现在是 2 个序列了。

```
SQL> SELECT * FROM user_sequences;
SEQUENCE_NAME  MIN_VALUE  MAX_VALUE  INCREMENT_BY C O CACHE_SIZE LAST_NUMBER P_C S K
-------------  ---------  ---------  ------------ - - ---------- ----------- --- - -
ISEQ$$_91620           1 1.0000E+28             1 N N         20           1     N N
ISEQ$$_91622           1 1.0000E+28             1 N N         20           1     N N
```

写完整的 Drop 语句试一下。

```
SQL> DROP TABLE t2 cascade CONSTRAINT purge;
TABLE dropped.
```

后面创建的序列已经被自动删除了，之前创建的还在。

```
SQL> SELECT * FROM user_sequences;
SEQUENCE_NAME  MIN_VALUE  MAX_VALUE  INCREMENT_BY C O CACHE_SIZE LAST_NUMBER P_C S K
-------------  ---------  ---------  ------------ - - ---------- ----------- --- - -
ISEQ$$_91620           1 1.0000E+28             1 N N         20           1     N N
```

两者的不同应该是 purge，如果被删除的表还在回收站中，序列是会保留的，因为表还可能从回收站里面再 restore 回来，需要保证序列仍然有效。那么清空回收站实验一下。

```
SQL> purge recyclebin;
Recyclebin purged.
```

5.2 以 12c Identity 类型示范自我探索式学习方法

果然，相应的序列也被删除了。

```
SQL> SELECT * FROM user_sequences;
no ROWS selected
```

再回到正题，创建 T3 表，插入一条数据。

```
SQL> CREATE TABLE t3 (id NUMBER GENERATED BY DEFAULT ON NULL AS IDENTITY);
TABLE created
SQL> INSERT INTO t3 VALUES(NULL);
1 ROW created.
```

序列的 LAST_NUMBER 已经增加为 21。

```
SQL> SELECT * FROM user_sequences;
SEQUENCE_NAME  MIN_VALUE   MAX_VALUE   INCREMENT_BY C O CACHE_SIZE  LAST_NUMBER  P_C S K
-------------  ---------   ---------   ------------ - - ----------  -----------  --- - -
ISEQ$$_91624           1   1.0000E+28             1 N N         20           21  N N
```

后台如何操作的？使用 10046 trace，再插入几条数据。

```
SQL> INSERT INTO t3 VALUES(NULL);
1 ROW created.

SQL> INSERT INTO t3 VALUES(NULL);
1 ROW created.

SQL> SELECT * FROM t3;
      ID
----------
       1
       2
       3
```

查看 10046 trace 的结果。可以看到执行计划中直接调用了 SEQUENCE，就跟之前插入记录的时候明确指定 SEQ.NEXTVAL 一样。其实 Oracle 的实现方法非常简单,这一列其实就是 Number 类型，然后将这一列的 Default 值设置为"KAMUS"."ISEQ$$_91624".nextval，仅此而已。

```
insert into t3 values (null)

call     count    cpu     elapsed    disk    query    current    rows
-------  ------   ----    -------    -----   -----    -------    ------
Parse    1        0.00    0.00       0       0        0          0
Execute  1        0.00    0.00       0       1        3          1
Fetch    0        0.00    0.00       0       0        0          0
-------  ------   ----    -------    -----   -----    -------    ------
total    2        0.00    0.00       0       1        3          1
Misses in library cache during parse: 1
Optimizer mode: ALL_ROWS
Parsing user id: 104
Number of plan statistics captured: 1
```

```
Rows (1st)  Rows (avg)  Rows (max)  Row Source Operation
----------  ----------  ----------  ----------------------------------------
         0           0           0  LOAD TABLE CONVENTIONAL  (cr=1 pr=0 pw=0 time=90 us)
         1           1           1  SEQUENCE  ISEQ$$_91624 (cr=0 pr=0 pw=0 time=14 us)
Elapsed times include waiting on following events:
  Event waited on                         Times Waite  Max. Wait  Total Waited
  --------------------------------------  -----------  ---------  ------------
  SQL*Net message to client                         1       0.00          0.00
  SQL*Net message from client                       1       5.28          5.28
********************************************************************************
```

使用 DBMS_METADATA.GET_DDL 获取到的 DDL 信息，已经符合 12c 语法的样式了，显示出了 Sequence 的具体信息。

```
SQL> SELECT dbms_metadata.GET_DDL('TABLE','T3') FROM dual;
DBMS_METADATA.GET_DDL('TABLE','T3')
--------------------------------------------------------------------
  CREATE TABLE "KAMUS"."T3"
   ( "ID" NUMBER GENERATED BY DEFAULT ON NULL AS IDENTITY MINVALUE 1 MAXVALUE 99
9999999999999999999999 INCREMENT BY 1 START WITH 1 CACHE 20 NOORDER  NOCYCLE
NOT NULL ENABLE,
     "COMMENTS" VARCHAR2(100)
   ) SEGMENT CREATION IMMEDIATE
  PCTFREE 10 PCTUSED 40 INITRANS 1 MAXTRANS 255
 NOCOMPRESS LOGGING
  STORAGE(INITIAL 65536 NEXT 1048576 MINEXTENTS 1 MAXEXTENTS 2147483645
  PCTINCREASE 0 FREELISTS 1 FREELIST GROUPS 1
  BUFFER_POOL DEFAULT FLASH_CACHE DEFAULT CELL_FLASH_CACHE DEFAULT)
  TABLESPACE "USERS"
```

系统自动产生的序列无法手工修改属性。

```
SQL> ALTER SEQUENCE "ISEQ$$_91624" INCREMENT BY 10;
ERROR at line 1:
ORA-32793: cannot ALTER a system-generated SEQUENCE


SQL> host oerr ora 32793
32793,0000, "cannot alter a system-generated sequence"
// *Cause:  An attempt was made TO ALTER a system-generated SEQUENCE.
// *Action: A system-generated SEQUENCE, such AS one created FOR an
//          IDENTITY COLUMN, cannot be altered.
```

系统自动产生的序列也不允许删除。

```
SQL> DROP SEQUENCE "ISEQ$$_91624";
ERROR at line 1:
ORA-32794: cannot DROP a system-generated SEQUENCE


SQL> host oerr ora 32794
32794,0000, "cannot drop a system-generated sequence"
// *Cause:  An attempt was made TO DROP a system-generated SEQUENCE.
// *Action: A system-generated SEQUENCE, such AS one created FOR an
//          IDENTITY COLUMN, cannot be dropped.
```

在 11gR2 中，错误信息编号在 ORA-32790 和 ORA-32800 之间是空白，而 12c 使用了这其间的 8 个错误号作为新特性的报错。

ORA-32791: prebuilt table managed column cannot have a default on null.

Cause: An attempt was made to create a materialized view on a prebuilt table that has a managed column with a default on null expression.

Action: Either remove the default on null property, or do not include the column in the materialized view definition.

ORA-32792: prebuilt table managed column cannot be an identity column

Cause: An attempt was made to create a materialized view on a prebuilt table that has a managed column that is an identity column.

Action: Either remove the identity property, or do not include the column in the materialized view definition.

ORA-32793: cannot alter a system-generated sequence

Cause: An attempt was made to alter a system-generated sequence.

Action: A system-generated sequence, such as one created for an identity column, cannot be altered.

ORA-32794: cannot drop a system-generated sequence

Cause: An attempt was made to drop a system-generated sequence.

Action: A system-generated sequence, such as one created for an identity column, cannot be dropped.

ORA-32795: cannot insert into a generated always identity column

Cause: An attempt was made to insert a value into an identity column created with GENERATED ALWAYS keywords.

Action: A generated always identity column cannot be directly inserted. Instead， the associated sequence generator must provide the value.

ORA-32796: cannot update a generated always identity column.

Cause: An attempt was made to update an identity column created with GENERATED ALWAYS keywords.

Action: A generated always identity column cannot be directly updated.

ORA-32797: identity column sequence mismatch in ALTER TABLE EXCHANGE PARTITION.

Cause: The two tables specified in the EXCHANGE have identity columns with sequences that are neither both increasing nor decreasing.

Action: Ensure that the identity columns have sequences with INCREMENT BY having the same sign.

ORA-32798: cannot use ANSI RIGHT or FULL outer join with a left correlation.

Cause: An attempt was made to use a lateral view with a left correlation to the first operand of an ANSI RIGHT or FULL outer join.

Action: Rewrite the query without the left correlation.

到此为止可以休息一下了,从 ORA-65096 开始大概花费了 1 个多小时的时间,我学习到了:

(1)什么是 common user,什么是 local user?

(2)如何查询现在的环境是 CDB 还是某个 PDB?

(3)如何登录 PDB?

(4)如何启动 PDB?

(5)PDB 和 CDB 中视图看到的内容有怎样的不同?

(6)如何创建 Identity 类型的列?

(7)删除表以后,对应的 Sequence 如何处理?

(8)Oracle 后台对于 Identity 列是如何处理的?

你是不是也是这样学习的呢?

5.3 Dump Block 是否会引起 Block 读入内存

中国有一句古话说"熟视无睹,常见不疑",指的是我们可能忽视那些随时可见的事物,并且对常见之处深信不疑,这期间可能存在巨大的误区。能够基于常见问题提出辨析和思考,并通过实践验证,是最为考验一个人知识体系的。

在一次讨论中,以下问题被提出:当我们使用 Dump Block 方式进行数据块转储时,是否需要将数据读入内存呢?

这是个常用的操作,可是很少有人思考过这个问题,有了思考还要有方法去验证,这整个过程代表着一个工程师在技术上的成熟。

我们立即动手,通过实例来求解一下这个问题的答案。

(1)重启一下数据库,这样 buffer cache 中几乎就没什么用户数据了,方便测试。

(2)找一张表看看是在哪个 file 哪个 block 里面(测试表,一行数据)。

```
SQL> select dbms_rowid.rowid_relative_fno(rowid)  fno,dbms_rowid.rowid_block_number(rowid) block# from t1;
```

```
FNO      BLOCK#
----     --------
1        103001
```

（3）T1 表在数据文件 1 中，第一个 block 是 103001，检查 v$bh，看看这个 block 有没有在 buffer cache 中。

```
SQL> select count(*)
  2  from v$bh
  3  where file# = 1 and block# = 103001;

  COUNT(*)
----------
         0
```

v$bh 视图保存着 buffer cache 中每一个 block 的信息，是一个重要视图。

（4）目前 buffer cache 中没有这个 block，做一次 Dump 再看看有没有。

```
SQL> alter system Dump datafile 1 block 103001;
System altered
SQL> select count(*)
  2  from v$bh where file# = 1 and block# = 103001;
  COUNT(*)
----------
         0
```

（5）这就验证了做 block Dump 不会把数据块先读入 buffer cache。

（6）继续做一次 select 看看，这次一定是读进 buffer cache 了。

```
SQL> select * from ops$kamus.t1;
SQL> select count(*)
  2  from v$bh
  3  where file# =1 and block# =103001;
  COUNT(*)
----------
         1
```

这就证明了我们的结论：Dump Block 操作不会引发 Block 读入 Buffer Cache。

5.4 Dump Block 是否会引起脏数据写入磁盘

伴随着上一个问题，随之而来的问题是：Dump Block 会否触发脏数据写入磁盘？

这一次我们尝试一个不同的工具 **BBED**。BBED（Oracle Block Browser and Editor）工具是 Oracle 内部提供的数据块级别查看和修改工具，借助这个工具，我们可以方便的查看 Oracle 块 block 级别的存储细节信息，更好地了解 Oracle Internal 结构的技术细节，DBA 们应当了解这个工具的简单使用方法。

掌握尽量多的工具，会让我们具备选择的基础，这也是 DBA 的基本技能要求。

首先亮出结论：**Dump Block 不会引起 Buffer cache 中的脏数据回写入磁盘**。然后我们使用 bbed 工具来验证一下。

（1）创建一个测试表。

```
SQL> CREATE TABLE t (n NUMBER);
TABLE created
```

（2）插入一条数据，提交，然后强制 checkpoint。

```
SQL> INSERT INTO t VALUES(1);
1 ROW inserted
SQL> commit;
Commit complete
SQL> ALTER system checkpoint;
System altered
```

（3）此时这条数据一定已经写回磁盘，这个无需验证，继续插入另外一条数据，提交，但是不 checkpoint。

```
SQL> INSERT INTO t VALUES(2);
1 ROW inserted
SQL> commit;
Commit complete
```

（4）此时这条脏数据在 buffer cache 中，我们可以通过 Dump block 来验证。

```
block_row_Dump:
tab 0, row 0, @0x1f9a
tl: 6 fb: --H-FL-- lb: 0x1  cc: 1
col  0: [ 2]  c1 02
tab 0, row 1, @0x1f94
tl: 6 fb: --H-FL-- lb: 0x2  cc: 1
col  0: [ 2]  c1 03
end_of_block_Dump
```

（5）通过 dbms_rowid 包取得 T 表中记录的文件号和 block 号，本例中取得是 file#=58，block#=570。

（6）关键步骤到了，现在我们要用 bbed 来获取磁盘上的数据块内容，然后跟 Dump block 的结果比较一下。

创建一个 filelist 文件，命名为 files.lst。

```
$ cat files.lst
58 /fin/u06/cnctest2data/system12.dbf 1048576000
```

5.4 Dump Block 是否会引起脏数据写入磁盘

创建一个参数文件 par.bbd，用以被 bbed 调用。

```
$ cat par.bbd
blocksize=8192
listfile=/home/oraaux/files.lst
mode=browse
```

执行 bbed。

```
$ bbed parfile=par.bbd
Password:

BBED: Release 2.0.0.0.0 - Limited Production on Mon Mar 13 17:35:32 2006
Copyright (c) 1982, 2002, Oracle Corporation.  All rights reserved.
************ !!! For Oracle Internal Use only !!! ***************

BBED> set dba 58,570
        DBA             0x0e80023a (243270202 58,570)

BBED> x /*rn rowdata
rowdata[0]                              @8182
----------
flag@8182: 0x2c (KDRHFL, KDRHFF, KDRHFH)
lock@8183: 0x01
cols@8184:    1

col    0[2] @8185:  1   --只有一条记录，值是 1

tailchk                                 @8188
-------
BBED-00210: no row at this offset
```

到目前为止我们已经验证了 **Dump block** 并不会把脏数据写回磁盘，为了看一下 **checkpoint** 的效果，我们继续往下。

（7）做 checkpoint

```
SQL> ALTER system checkpoint;
System altered
```

（8）再次运行 bbed。

```
$ bbed parfile=par.bbd
Password:

BBED: Release 2.0.0.0.0 - Limited Production on Mon Mar 13 17:35:32 2006
Copyright (c) 1982, 2002, Oracle Corporation.  All rights reserved.
************ !!! For Oracle Internal Use only !!! ***************

BBED> set dba 58,570
        DBA             0x0e80023a (243270202 58,570)
BBED> x /*rn rowdata
```

```
rowdata[0]                              @8176
----------
flag@8176: 0x2c (KDRHFL, KDRHFF, KDRHFH)
lock@8177: 0x02
cols@8178:    1
col    0[2] @8179: 2   --这是后来插入的记录，值是 2
rowdata[6]                              @8182
----------
flag@8182: 0x2c (KDRHFL, KDRHFF, KDRHFH)
lock@8183: 0x01
cols@8184:    1
col    0[2] @8185: 1   --这是第一条记录，值是 1
tailchk                                 @8188
-------
BBED-00210: no row at this offset
```

checkpoint 将 buffer cache 中的脏数据写回数据文件了。

5.5　如何验证 ASM 的块头备份块的位置

大家都知道，在 Oracle 10.2.0.5 之前，ASM 磁盘的头块并没有自己的备份，因此一旦头块损坏，如果没有以前 kfed read 备份出来的信息，也就没有办法使用 kfed merge 来作头块恢复，特别是如果一个磁盘组中所有的磁盘头块都出现问题（比如被人为地创建了 PV），恢复 ASM 磁盘头块的操作就会非常麻烦。

但是从 Oracle 10.2.0.5 之后，ASM 磁盘的头块会自动备份在另外一个块中，这实际上是 Oracle 11g 出现的功能，不过经过测试，在 Oracle 10.2.0.5 版本中，这个备份也是存在的。正是因为存在这个备份，所以 Oracle 10.2.0.5 之后的 kfed 程序才有了新的 repair 命令，该命令将备份块直接覆盖到磁盘头块，完成修复工作。

在 Oracle 10.2.0.4 中，如果尝试执行 kfed repair，则会报错说命令行参数不正确，此报错说明并不存在 repair 命令。

```
$ kfed repair
KFED-00101: LRM error [102] while parsing command line arguments
```

但是在 Oracle 10.2.0.5 中，执行 kfed repair，则会说无法打开文件，而这正说明 repair 命令是存在的，报错是因为还需要明确指定要修复哪块磁盘。

```
$ kfed repair
KFED-00303: unable to open file ''
```

那么这个备份块具体存在哪里呢？在学习 Oracle 技术的过程中，好奇心是驱使我们进步的强大动力，设问、思考、解答，这是获得自我提升的根本。养成动手的习惯，通过动手找出真相，这是成长的必经之路。

5.5 如何验证 ASM 的块头备份块的位置

在 Solaris 下的测试，我们使用 truss 来进行跟踪。

```
$ truss -o tracedisk2.out kfed repair /asmdisks/vdisk2
```

在 trace 文件中，找到下面这段，可以明确地看到 kfed 程序从第 510 个块中读出 4096 字节，然后再写回到第 0 个块中。

```
stat("/asmdisks/vdisk2", 0xFFFFFD7FFFDFDB20)    = 0
open("/asmdisks/vdisk2", O_RDWR)                = 7
lseek(7, 0x001FE000, SEEK_SET)                  = 2088960  <-- 1FE is 510
read(7, "01820101FE\0\0\0\0\0\080".., 4096)     = 4096     <-- read 4096 bytes
lseek(7, 0, SEEK_SET)                           = 0
read(7, "01820101\0\0\0\0\0\0\080".., 4096)     = 4096
lseek(7, 0, SEEK_SET)                           = 0        <-- 0 is 0
write(7, "01820101\0\0\0\0\0\0\080".., 4096)    = 4096     <-- write 4096 bytes
close(7)
```

同样如果是在 Linux 下用裸设备作为 ASM 磁盘，并且用 strace 进行 repair 命令的跟踪，也可以得到类似结果。

```
open("/dev/raw/raw3", O_RDWR)                   = 5
lseek(5, 2088960, SEEK_SET)                     = 2088960  <-- 2088960/4096=510
read(5, "\1\202\1\1\0\0\0\0\0\0\200evx\257\0\0\0\0\0\0\0\0\0\0\0\0\0\0\0\0\0\0"..., 4096) = 4096
lseek(5, 0, SEEK_SET)                           = 0
read(5, "\1\202\1\1\0\0\0\0\0\0\200evx\257\0\0\0\0\0\0\0\0\0\0\0\0\0\0\0\0\0\0"..., 4096) = 4096
lseek(5, 0, SEEK_SET)                           = 0
write(5, "\1\202\1\1\0\0\0\0\0\0\200evx\257\0\0\0\0\0\0\0\0\0\0\0\0\0\0\0\0\0\0"..., 4096) = 4096
close(5)                                        = 0
```

那么通过 kfed 命令再来验证一下这两个块是否都标志为头块。验证结果表示块类型都为 DISKHEAD。

```
$ kfed read /asmdisks/vdisk2 blkn=0 | grep KFBTYP
kfbh.type:                     1 ; 0x002: KFBTYP_DISKHEAD
$ kfed read /asmdisks/vdisk2 blkn=510 | grep KFBTYP
kfbh.type:                     1 ; 0x002: KFBTYP_DISKHEAD  <-- this is the backup!!
```

那么下一个疑问是，在 11gR2 以后，ASM 磁盘组的 AU Size 可以指定不同的大小，是不是不同的 AU Size 下的磁盘头块备份都是在第 510 个块呢？还是用 truss 来跟踪一下，这里的 vdisk3 属于一个 AU Size = 8M 的磁盘组，此时 repair 命令需要明确指定 aus，否则会报 KFED-00320 错误。

```
truss -o tracedisk3.out kfed repair /asmdisks/vdisk3 aus=8388608
```

在 trace 文件中，可以发现已经不是读第 510 个块，而是改为读第 4094 个块。

```
stat("vdisk3", 0xFFFFFD7FFFDFDB10)              = 0
open("vdisk3", O_RDWR)                          = 7
lseek(7, 0x00FFE000, SEEK_SET)                  = 16769024 <--FFE is 4094
read(7, "01820101FE07\0\0\0\080".., 4096)       = 4096
lseek(7, 0, SEEK_SET)                           = 0
read(7, "01820101\0\0\0\0\0\080".., 4096)       = 4096
```

```
lseek(7, 0, SEEK_SET)                             = 0
write(7, "01820101\0\0\0\0\0\080".., 4096)        = 4096
close(7)
```

用 kfed 验证第 4094 个块，确实标志为 DISKHEAD。

```
$ kfed read /asmdisks/vdisk3 blkn=4094 | grep KFBTYP
kfbh.type:                    1 ; 0x002: KFBTYP_DISKHEAD
```

那么也就是 AU 1M 的磁盘组头块备份在第 510 个块上，而 AU 8M 的磁盘组头块备份在第 4094 个块上，备份块的存储位置有规律吗？有的，始终保存在第 2 个 AU 的倒数第 2 个块上。下面来验证这个观点。

对于默认的磁盘组，AU Size = 1M，每个 AU 中可以存储 256 个块，块号为 0~255。第 1 个 AU 存储 256 个块，第 2 个 AU 最后 1 个块号为 255，倒数第 2 个块号是 254，也就是整体的第 510 个块（从第 1 个 AU 的第 1 个块往后算起）。

```
$ kfed read /asmdisks/vdisk2 blkn=0 | grep ausize
kfdhdb.ausize:        1048576 ; 0x0bc: 0x00100000
$ kfed read /asmdisks/vdisk2 blkn=0 | grep blksize
kfdhdb.blksize:          4096 ; 0x0ba: 0x1000
$ let r=1048576/4096;echo $r
256
$ let r=256+255-1;echo $r
510
```

对于 AU Size = 8M 的磁盘组，每个 AU 可以存储 2048 个块，块号为 0~2047。第 1 个 AU 存储 2048 个块，第 2 个 AU 最后 1 个块号为 2047，倒数第 2 个块号是 2046，也就是整体的第 4094 个块（从第 1 个 AU 的第 1 个块往后算起）。

```
$ kfed read /asmdisks/vdisk3 blkn=0 | grep ausize
kfdhdb.ausize:        8388608 ; 0x0bc: 0x00800000
$ kfed read /asmdisks/vdisk3 blkn=0 | grep blksize
kfdhdb.blksize:          4096 ; 0x0ba: 0x1000
$ let r=8388608/4096;echo $r
2048
$ let r=2048+2047-1;echo $r
4094
```

对于其他 AU Size 磁盘组的验证，看到文章的朋友有兴趣可以自己做一下。

> 结论：从 Oracle 10.2.0.5 开始，ASM 磁盘已经开始自动将头块进行备份，备份块的位置在第 2 个 AU 的倒数第 2 个块上（对于默认 1M 的 AU 来说，是第 510 个块），如果头块损坏，可以用 kfed repair 命令来修复。

5.6 如何利用文件句柄恢复误删除的文件

动手、动手，还是动手，看到有兴趣的案例、方法，就坐言起行，通过实践将这些知识变

5.6 如何利用文件句柄恢复误删除的文件

成自己的知识储备。

这一次是客户的数据库意外被删除了整个目录中的数据文件,操作系统级别的删除,然而幸运的是这个数据库没有崩溃。仍然处于 open 状态的时候,客户就发现了问题,并求助到我们,最终完整地恢复了所有数据文件。

在 Linux 下大致重新演示一下恢复的过程,恢复的步骤与数据库版本没有太大关系,但是会因操作系统的不同有所改变。

(1)在数据库 open 的时候,直接删除 users 表空间中的数据文件。

```
SQL> select name from v$datafile;
NAME
--------------------------------------------------------------
/datafile/o1_mf_system_555wqbnk_.dbf
/datafile/o1_mf_undotbs1_555wqxgl_.dbf
/datafile/o1_mf_sysaux_555wr5p6_.dbf
/datafile/o1_mf_users_555wrj4o_.dbf
SQL> host rm /app/oracle/oradata/ORCL/datafile/o1_mf_users_555wrj4o_.dbf
```

(2)尝试在 users 表空间中创建表,开始报错。

```
SQL> create table t tablespace users as select * from dual;
ERROR at line 1:
ORA-01116: error in opening database file 4
ORA-01110: data file 4:
'/datafile/o1_mf_users_555wrj4o_.dbf'
ORA-27041: unable to open file
Linux Error: 2: No such file or directory
Additional information: 3
```

在警告日志中,同样也可以看到类似信息。

```
Mon Dec 19 21:48:17 CST 2011
Errors in file /bDump/orcl_m000_3897.trc:
ORA-01116: error in opening database file 4
ORA-01110: data file 4: '/datafile/o1_mf_users_555wrj4o_.dbf'
ORA-27041: unable to open file
Linux Error: 2: No such file or directory
Additional information: 3
```

(3)检查 dbwr 的进程 PID。

```
$ ps -ef|grep dbw0|grep -v grep
oracle    2879     1  0 21:38 ?        00:00:00 ora_dbw0_orcl
```

(4) dbwr 会打开所有数据文件的句柄。在 proc 目录中可以查到,目录名是进程 PID,fd 表示文件描述符。

```
$ cd /proc/2879/fd
$ ls -l
total 0
```

```
lr-x------ 1 oracle dba 64 Dec 19 21:50 0 -> /dev/null
lr-x------ 1 oracle dba 64 Dec 19 21:50 1 -> /dev/null
lr-x------ 1 oracle dba 64 Dec 19 21:50 10 -> /dev/zero
lr-x------ 1 oracle dba 64 Dec 19 21:50 11 -> /dev/zero
lr-x------ 1 oracle dba 64 Dec 19 21:50 12 -> /product/10.2.0/db_1/rdbms/mesg/oraus.msb
lrwx------ 1 oracle dba 64 Dec 19 21:50 13 -> /product/10.2.0/db_1/dbs/ hc_orcl.dat
lrwx------ 1 oracle dba 64 Dec 19 21:50 14 -> /product/10.2.0/db_1/dbs/lkORCL
lrwx------ 1 oracle dba 64 Dec 19 21:50 15 -> /controlfile/o1_mf_ 555wq3ng_.ctl
lrwx------ 1 oracle dba 64 Dec 19 21:50 16 -> /datafile/o1_mf_ system_555wqbnk_.dbf
lrwx------ 1 oracle dba 64 Dec 19 21:50 17 -> /datafile/o1_mf_ undotbs1_555wqxgl_.dbf
lrwx------ 1 oracle dba 64 Dec 19 21:50 18 -> /datafile/o1_mf_ sysaux_555wr5p6_.dbf
lrwx------ 1 oracle dba 64 Dec 19 21:50 19 -> /app/oracle/oradata/ORCL/datafile/o1_mf_users_555wrj4o_.dbf (deleted)
lr-x------ 1 oracle dba 64 Dec 19 21:50 2 -> /dev/null
lrwx------ 1 oracle dba 64 Dec 19 21:50 20 -> /datafile/o1_mf_temp_ 555wrbnz_.tmp
lr-x------ 1 oracle dba 64 Dec 19 21:50 21 -> /product/10.2.0/db_1/rdbms/mesg/ oraus.msb
lr-x------ 1 oracle dba 64 Dec 19 21:50 3 -> /dev/null
lr-x------ 1 oracle dba 64 Dec 19 21:50 4 -> /dev/null
l-wx------ 1 oracle dba 64 Dec 19 21:50 5 -> /uDump/orcl_ora_2871.trc
l-wx------ 1 oracle dba 64 Dec 19 21:50 6 -> /bDump/alert_orcl.log
lrwx------ 1 oracle dba 64 Dec 19 21:50 7 -> /product/10.2.0/db_1/dbs/lkinstorcl (deleted)
l-wx------ 1 oracle dba 64 Dec 19 21:50 8 -> /app/oracle/admin/orcl/bdump/alert_orcl.log
lrwx------ 1 oracle dba 64 Dec 19 21:50 9 -> /app/oracle/product/10.2.0/db_1/dbs/ hc_orcl.dat
```

注意其中 "/datafile/o1_mf_users_555wrj4o_.dbf（deleted）"字样，表示该文件已经被删除，如果是 Solaris 操作系统，ls 命令不会有如此清晰地显示，为了在 Solaris 系统中确认哪个句柄对应哪个文件，则需要使用 lsof 程序。

（5）直接 cp 该句柄文件名回原位置。

```
cp 19 /datafile/o1_mf_users_555wrj4o_.dbf
```

（6）进行数据文件 recover。

```
SQL> alter database datafile 4 offline;
Database altered.
SQL> recover datafile 4;
Media recovery complete.
SQL> alter database datafile 4 online;
Database altered.
```

完成数据文件恢复。

恢复的原理是：在 Linux 操作系统中，如果文件从操作系统级别被 rm 掉，之前打开该文件的进程仍然持有相应的文件句柄，所指向的文件仍然可以读写，并且该文件的文件描述符可以从 /proc 目录中获得。但是要注意的是，此时如果关闭数据库，则此句柄会消失，那么除了扫描磁盘进行文件恢复之外就没有其他方法了，因此在数据库出现问题的时候，如果不确认情况的复杂程度，千万不要随便关闭数据库。重启数据库往往是没有意义的，甚至是致命的。

5.7 从一道面试题看分析问题的思路

我们有一道面试题，原以为很简单，但是发现面试者能够完美解出的几乎没有，一部分人有思路，但是可能是因为面试紧张，很难在指定时间内完成解题，而更大一部分人连思路也不清晰。

题目是：**请将 emp.empno=7369 的记录 ename 字段修改为"ENMOTECH"并提交，你可能会遇到各种故障，请尝试解决。**

其实题目设计得非常简单，一个 RAC 双节点的实例环境，面试人员使用的是实例 2，而我们在实例 1 中使用 select for update 将 EMP 表加锁。

```
SQL> SELECT * FROM emp FOR UPDATE;
```

此时在实例 2 中，如果执行以下 SQL 语句尝试更新 ename 字段，必然会被行锁堵塞。

```
SQL> UPDATE emp SET ename='ENMOTECH' WHERE empno=7369;
```

这道面试题中包含的知识点有以下几点。

（1）如何在另外一个 session 中查找被堵塞的 session 信息。

（2）如何找到产生行锁的 blocker。

（3）在杀掉 blocker 进程之前会不会向面试监考人员询问："我已经找到了产生堵塞的会话，是不是可以 kill 掉？"

（4）在获得可以 kill 掉进程的确认回复后，正确杀掉另一个实例上的进程。

这道题我们期待可以在 5 分钟之内获得解决，实际上大部分应试者在 15 分钟以后都完全没有头绪。

正确的思路和解法应该如下。

5.7.1 检查被阻塞会话的等待事件

更新语句以后没有响应，明显是被锁住了，那么现在这个会话经历的是什么等待事件呢？

```
SQL> SELECT sid,event,username,SQL.sql_text
  2  FROM v$session s,v$sql SQL
  3  WHERE s.sql_id=SQL.sql_id
  4  AND SQL.sql_text LIKE 'update emp set ename%';
```

```
SID EVENT                            USERNAME  SQL_TEXT
--- ------------------------         ---       ------------------------------------------
 79 enq: TX - ROW LOCK contention    ENMOTECH  UPDATE emp SET ename='ENMOTECH' WHERE empno=7369
```

以上使用的是关联 v$SQL 的 SQL 语句，实际上通过登录用户名等也可以快速定位被锁住的会话。

5.7.2 查找 blocker

得知等待事件是 enq: TX – row lock contention，行锁，接下来就是要找到谁锁住了这个会话。在 10gR2 以后，只需要 gv$session 视图就可以迅速定位 blocker，通过 BLOCKING_INSTANCE 和 BLOCKING_SESSION 字段即可。

```
SQL> SELECT SID,INST_ID,BLOCKING_INSTANCE,BLOCKING_SESSION FROM gv$session WHERE
INST_ID=2 AND SID=79;
       SID    INST_ID BLOCKING_INSTANCE BLOCKING_SESSION
---------- ---------- ----------------- ----------------
        79          2                 1               73
```

上述方法是最简单的，如果是使用更传统的方法，实际上也并不难，从 gv$lock 视图中去查询即可。

```
SQL> SELECT TYPE,ID1,ID2,LMODE,REQUEST FROM v$lock WHERE sid=79;
TY        ID1        ID2      LMODE    REQUEST
-- ---------- ---------- ---------- ----------
TX     589854      26267          0          6
AE        100          0          4          0
TM      79621          0          3          0
SQL> SELECT INST_ID,SID,TYPE,LMODE,REQUEST FROM gv$Lock WHERE ID1=589854 AND ID2=26267;
   INST_ID        SID TY      LMODE    REQUEST
---------- ---------- -- ---------- ----------
         2         79 TX          0          6
         1         73 TX          6          0
```

5.7.3 乙方 DBA 需谨慎

第三个知识点是考核作为乙方的谨慎，即使你查到了 blocker，是不是应该直接 kill 掉？必须要先征询客户的意见，确认之后才可以杀掉。

5.7.4 清除 blocker

已经确认了可以 kill 掉 session 之后，需要再找到相应 session 的 serail#，这是 kill session 时必须输入的参数。

```
SQL> SELECT SID,SERIAL# FROM gv$session WHERE INST_ID=1 AND SID=73;
SID    SERIAL#
-----  -------
73     15625
```

如果是 11gR2 数据库,那么直接在实例 2 中加入 @1 参数就可以杀掉实例 1 中的会话,如果是 10g,那么登入实例 1 再执行 kill session 的操作。

```
SQL> ALTER system KILL SESSION '73,15625,@1';
System altered.
```

再检查之前被阻塞的更新会话,可以看到已经更新成功了。

```
SQL> UPDATE emp SET ename='ENMOTECH' WHERE empno=7369;
1 ROW updated.
```

对于熟悉整个故障解决过程的人,5 分钟之内就可以解决问题。

5.7.5 深入一步

对于 TX 锁,在 v$lock 视图中显示的 ID1 和 ID2 是什么意思? 解释可以从 v$lock_type 视图中获取。

```
SQL> SELECT ID1_TAG,ID2_TAG FROM V$LOCK_TYPE WHERE TYPE='TX';
ID1_TAG            ID2_TAG
---------------    ----------
usn<<16 | slot     SEQUENCE
```

所以 ID1 是事务的 USN+SLOT,而 ID2 则是事务的 SQN。这些可以从 v$transaction 视图中获得验证。

```
SQL> SELECT taddr FROM v$session WHERE sid=73;
TADDR
----------------
000000008E3B65C0
SQL> SELECT XIDUSN,XIDSLOT,XIDSQN FROM v$transaction WHERE
 addr='000000008E3B65C0';
XIDUSN     XIDSLOT    XIDSQN
---------- ---------- ----------
9          30         26267
```

如何和 ID1=589854 and ID2=26267 对应呢? XIDSQN=26267 和 ID2=26267 直接就对应了,没有问题。 那么 ID1=589854 是如何对应的?将之转换为 16 进制,是 0x9001E,然后分高位和低位分别再转换为 10 进制,高位的 16 进制 9 就是十进制的 9,也就是 XIDUSN=9,而低位的 16 进制 1E 转换为 10 进制是 30,也就是 XIDSLOT=30。

文章写到这里,忽然感觉网上那些一气呵成的故障诊断脚本其实挺误人的,只需要给一个参数,运行一下脚本就列出故障原因。所以很少有人愿意再去研究这个脚本为什么这么写,各个视图之间的联系是如何环环相扣的。所以当你不再使用自己的笔记本,不再能迅速找到你赖以生存的那些脚本，你还能一步一步地解决故障吗？

5.8 涓涓细流终聚海

整理这一章最主要的目的是想把我多年的经验和方法介绍给大家，让大家有所借鉴，并阐释我们欣赏的优秀技术人员的一些特质，最后我在这里简单总结一下。

（1）合适的学习方法。就像我在前面说过的，对于一个学习技术的人而言，掌握正确的学习方法很重要，虽然方法可能因人而异因事而异，但正确的方法总能让你在技术的道路上走得更快些。

（2）有用的学习工具。学会学习就是要充分利用一切可用的资源。Oracle 内部提供了很多有用的工具，能够很好地帮助我们，除此而外，还有很多有用的外部工具也能够有效提升我们的效率。

（3）相关的各方面知识。所有的知识都不是单一的存在，要掌握一门技术或者意向能力，要求学习者掌握与其相关的领域和行业知识，扩展知识面，这样才能以全局的思维去认识事情的本质。

（4）解决问题的思路。在掌握了一定知识后，学会在实际问题中去应用很重要。灵活分析，善于思考，才能学以致用，让我们学的东西的价值最大化。

（5）解决问题的经验。无论是作为一个运维者还是其他工作的人，积累经验很重要。我们总会遇到各种各样的 case，这些 case 之间有相同的地方也有不同的地方，当我们处理完一件事，要善于总结，吸取教训和经验，让每一次故障都成为我们成功路上的铺路石。

这就是我想要说的话。

第 6 章　使用 XTTS 技术进行 U2L 跨平台数据迁移

——杨俊

在国内"去 IOE"浪潮的席卷之下，整个行业快速向前演进，小型机转眼已成昨日黄花。IDC 的《刀片服务器推动企业基础架构走向新 IT 时代》白皮书中指出：10 年间，刀片服务器出货量由 2005 年的 49.1 万台增长到 99.4 万台，年复合增长率达到 7.4%；中国 x86 刀片服务器市场出货量由 2005 年的 0.9 万台快速增长到 2015 年的 12.2 万台，10 年间翻了 10 倍之多，年复合增长率达到 29.2%，在 2014 年和 2015 年均占据销售额市场份额第一的位置。

随着 x86 架构服务器的迅速发展，企业从 RISC 架构服务器向 x86 架构服务器的转型也成为了潮流，即 UNIX to Linux，在行业里被简称为 U2L。

那么，在 U2L 如火如荼的今天，如何快捷、高效、平稳、安全地将 Oracle 数据库"小型机+集中式存储"环境迁移至"x86 架构平台+分布式存储"，如何将这些单库容量达到 10TB 级，甚至 30TB、50TB 的数据库在有限的停机时间内平稳、安全地迁移至 x86 服务器中，如何利用好每一个数据迁移工具的优缺点来达到方便、快捷、高效地完成数据库管理工作，是每个数据库管理员（DBA）需要认真思考的问题。

6.1　XTTS 概述

XTTS（Cross Platform Transportable Tablespaces，跨平台传输表空间），是 Oracle 自 10g 推出的用来移动单个表空间数据，以及创建一个完整的数据库从一个平台到另一个平台的迁移备份方法。

- XTTS 源自 Oracle 8i 引入的一种基于表空间传输的物理迁移方法——TTS，不过 8i 的表空间迁移仅支持相同平台、相同块大小之间的表空间传输,然而在那个年代还未像今天的技术日新月异,TTS 的光芒一直被埋没在历史的尘埃里。图 6-1 所示为 TTS 和 XTTS 迁移对比图。

- 从 Oracle 9i 开始,TTS 开始支持同平台中,不同块大小的表空间传输,这个时候很多数据库管理员就注意到了 TTS 在实际工作中的应用,不过由于每次移动表空间都需要停机、停业务,而 9i 的 TTS 只能在相同平台之间进行数据移动,相比 Oracle RMAN 本身的快捷、方便,更多人更愿意选择使用 RMAN 进行数据备份和数据移动。

- 基于这些原因,Oracle 10g 时代引入了跨平台的表空间传输方案 XTTS,标志着第一代 XTTS 的诞生。

图 6-1

在 Oracle 10.1 中第一代 XTTS 是基于表空间的传输,到 Oracle 11gR1 后,跨平台数据的迁移可以支持传输表空间中的某个特定分区,不过在数据移动过程中,仍然需要将主库设置为只读状态、停机、停业务下才能进行数据迁移,对于业务不可间断的系统仍旧需要花费大量的停机时间才能达到跨平台物理迁移的效果,所以把 11gR2 以前的 XTTS 技术称作第一代 XTTS 技术。

从 11gR2 开始,为了应对越来越大的数据量,相对停机时间要求日益减少的情况,Oracle 推出了新的解决方案——加强版 XTTS,使用增量备份方式实现跨平台的数据迁移。从真正意义上讲,能够减少停机时间、进行增量备份的 XTTS,才真正是今天所说的 XTTS。

XTTS 各版本的功能对比如表 6-1 所示。

表 6-1

版本说明	跨平台	不同块	增量备份
Oracle 8i			
Oracle 9i		√	
Oracle 10g	√	√	
Oracle 10gR2	√	√	
Oracle 11gR1	√	√	
Oracle 11gR2	√	√	√
Oracle 12c	√	√	√

6.2 XTTS 技术迁移应用场景

6.2.1 应用场景一：全国"去 IOE"战略实施

所谓"去 IOE"，是对"去 IBM、Oracle、EMC"的简称，其中，IBM 代表硬件以及整体解决方案服务商，Oracle 代表数据库，EMC 代表数据存储。我国很多领域的核心业务系统都运行在"IOE"的软硬件架构之上。

另外随着 x86 平台市场份额的不断扩大，以英特尔至强系列处理器为主导的服务器，在传统行业数据中心中的应用越来越广泛，同时通过降低硬件采购及功耗和散热成本，采购性价比与同等 RISC 架构相比总体拥有成本也大幅降低。至此，U2L 数据库迁移，即把数据库服务器从小型机迁移到 x86 平台，从 UNIX 操作系统迁移到 Linux 操作系统，成为当前最热门的趋势。那么如何在保证有限的停机时间内，完成大量数据的迁移成为各行各业在去"IOE"的路上必须要迈过的一道坎。

而 Oracle 在 11.2.0.4 中提出了 XTTS 增量备份表空间传输的 U2L 迁移方式，成为当前为止在面对大批量数据中最高效、最安全的解决方案之一。U2L 迁移示意图如图 6-2 所示。

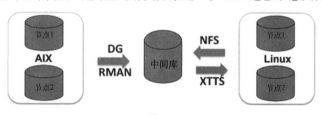

图 6-2

6.2.2 应用场景二："云平台"数据中心建设

随着技术的发展，传统数据中心在多方面都面临挑战，加上大数据时代的来临，对传统数据中心提出了新的挑战，为了解决这一问题，随之出现了云数据中心。同时随着 x86 虚拟化技术的成熟，以 x86 架构为基础的服务器计算资源通过虚拟化配置资源池，形成云化平台，完成统一调度、集中计算，而传统小型机数据库迁移到 x86 架构的云平台使用 XTTS 迁移也成为最直接、有效的方式之一，如图 6-3 所示。

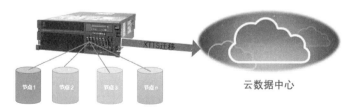

图 6-3

6.2.3 应用场景三：老旧环境淘汰改造

随着我国信息化建设程度的深入，一批又一批老旧物理硬件设备已经达到使用年限，甚至一些核心系统的服务器处于脱保状态。从配置层面来看，更是呈现服务器配置低下、内存不足、资源过度利用等情况。而随着应用模块的增加，数据量也呈几何增长，业务需求变化快过市场需求的今天，业务逻辑的复杂度也呈现倍数增长。在当前这种传统系统架构落后、设备陈旧、数据安全无法得到保障的情况下，购买新的设备进行更新和升级无疑是大家从物理上解决性能瓶颈的办法之一。可面对昂贵的 UNIX 服务器成本，是选择买还是不买同样成为摆在决策者面前的两难选择，而 x86 平台的强势崛起，低成本、高性能的表现无疑更符合用户长远利益的选择。使用 XTTS 跨平台迁移 UNIX 到新架构 x86 同样也更符合停机时间短、迁移周期短、回退简单等要求。

6.2.4 应用场景四：数据库分布式存储重构

随着越来越多业务功能的上线，在今天的数据中心环境下，Oracle RAC（Real Application Clusters）等企业数据库系统正在发展成为高度集成的中央系统。而 RAC 也面临数据库性能优化这一巨大挑战，性能瓶颈可能出现在网络和处理器等多个领域中，但最常见的来自于缓慢的硬盘驱动器。随着应用对更快的随机输入/输出需求不断地增加，这些机械硬盘驱动器更难满足这些需求。构造一套灵活的纯软件解决方案，充分利用基于 PC 服务器的内部直连式存储来创建一个虚拟的、可扩展的 SAN，性能或优于外部传统光纤通道 SAN，而成本和复杂性大幅度降低的分布式存储架构成为当今数据库重构的新方向，在节省物理硬件成本的开支外还能提高性能，其中 XTTS 在进行分布式存储重构实施中也能起到良好的作用，分布式存储重构解决方案如图 6-4 所示。XTTS 利用 RMAN 直接路径复制数据文件的方式，速度更快、性能更好地完成传统集中存储向分布式存储数据迁移。

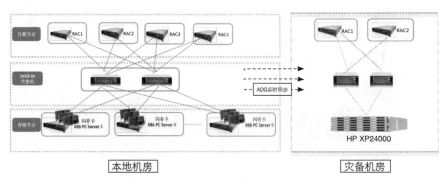

图 6-4

6.2.5 应用场景五：其他应用场景

可能的其他应用场景还包括跨平台恢复数据永久保存、数据仓库搭建、历史数据跨平台归档、非结构化数据迁移、单独表空间跨平台备份恢复等。

6.3 XTTS 迁移步骤

说了那么多 XTTS 在 U2L 之中的迁移应用场景，下面就从 XTTS 的迁移步骤出发，一步一步来了解使用 XTTS 进行表空间传输的整个迁移步骤。

首先通过图 6-5 来看一下 Oracle 11gR2 以前版本的 XTTS 典型的迁移步骤。

通过图 6-5 可以看出，在 Oracle 11gR2 以前使用 XTTS 迁移数据中是没有**增量前滚**这一步，必须在源库表空间完全只读状态下才能进行数据迁移。由于数据传输又必须在数据库启动中只读状态下来读取，另外整个迁移数据的过程是一个串行的步骤，所以以此过程所需要的时间同数据库数据量的大小成正比。如果数据量很大，数据传输和转换的时间可能会很长。因此在 Oracle 11gR2 以后，Oracle 推出了通过前滚数据文件，复制数据后再进行多次增量备份的 XTTS 来完成迁移过程。在这个过程中通过**开启块跟踪特性**，根据 SCN 号来执行一系列的增量备份，并且通过对块跟踪文件的扫描，来完成增量数据的增量备份应用。最后再通过一定的停机时间，在源库 Read Only 的状态下进行最后一次增量备份转换应用，使得整个迁移过程的停机时间同源库数据块的变化率成正比，这样大大缩短了停机时间。

如图 6-6 所示，可以把 Oracle 11gR2 以后版本的整个迁移过程分为 7 个阶段。

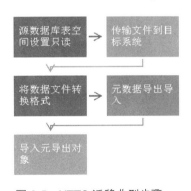

图 6-5 XTTS 迁移典型步骤

图 6-6 XTTS 迁移步骤图

6.4 XTTS 迁移方式

XTTS 基于一组 rman-xttconvert 的脚本文件包（参考 MOS 1389592.1）来实现跨平台的数

据迁移，主要文件是 Perl 脚本 xttdriver。xttdriver.pl 是备份、转换、应用的执行脚本，xtt.properties 则是属性文件，其中包含 XTTS 配置的路径、参数。

采用 XTTS 迁移方式，具备跨平台字序转换和全量初始化加增量 Merge 的功能，非常适用于异构 OS 跨平台迁移，成为数据库实施人员中公认的大数据量跨平台迁移的最佳选择。

传统的 XTTS 要求数据由源端到目标端传输的整个过程中，表空间必须置于只读模式，严重影响业务可用性。XTTS 方式可以在业务正常运行的情况下，进行物理全量初始化、增量数据块备份、数据高低字节序转换、增量数据块应用、保持目标端与源端数据的同步，整个过程不影响源端数据库使用。在最后的增量数据块应用完毕后，利用停机窗口进行数据库切换，显著地减少了停机时间。

rman-xttconvert 包参数说明如表 6-2 所示。

表 6-2

参数范例	意义
tablespaces=TS1,TS2	需要传输的表空间
platformid=2	源库的 platform_id，v$database 中得到
srcdir=src1,src2	当使用 dbms_file_transfer 时使用，表示源库存放数据文件的路径
dstdir=dst1,dst2	当使用 dbms_file_transfer 时使用，目标库存放数据文件的路径
srclink=ttslink	从目标端指向源端的 dblink，当使用 dbms_file_transfer 时使用
dfcopydir=/storage	源端用于存放数据文件的 copy，使用 rman 时使用
backupformat=/storage	源端用于存放增量备份的目录，无论哪种方式都需要设置
stageondest=/storage	目标端存放数据文件 copy 目录和存放增量备份的目录
storageondest=/oradata/prod/%U	数据文件的最终存放点
backupondest=/storage	增量备份格式转换后的输出目录
cnvinst_home=	不同的增量转换目录使用时设置该参数
cnvinst_sid=	不同中转 sid 使用时使用
parallel=3	默认为 3
rollparallel=2	
Getfileparallel=8	默认为 8，使用 rman 时的并行设置

6.4.1 方式一：dbms_file_transfer

DBMS_FILE_TRANSFER 包是 Oracle 提供的一个用于复制二进制数据库文件或在数据库之间传输二进制文件的程序包。在 XTTS 迁移中，利用不同的参数进行数据文件传输转换完成迁移。dbms_file_transfer 参数示例如表 6-3 所示。

表 6-3

Subprogram 参数	Description 描述
COPY_FILE	从源目录中读取一个文件并创建一个在目标目录的副本,源和目标目录都可以是本地文件系统,或是自动存储管理(ASM)磁盘组,或本地文件系统和 ASM 之间复制
GET_FILE	从远程数据库读取远程文件,然后创建一个本地文件系统中的文件副本或 ASM
PUT_FILE	读取本地文件或 ASM 和从远程数据库创建一个远程文件系统中文件的副本

dbms_file_transfer 语法示例如表 6-4 所示。

表 6-4

参数	示例

COPY_FILE	BEGIN
	DBMS_FILE_TRANSFER.COPY_FILE('SOURCEDIR','t_xdbtmp.f', 'DGROUP', 't_xdbtmp.f');
	END;

GET_FILE	BEGIN
	  DBMS_FILE_TRANSFER.GET_FILE ('df' , 'a1' , 'dbs1', 'dsk_files' , 'oa5.dat');
	  DBMS_FILE_TRANSFER.GET_FILE ('dsk_files' , 'a2.dat', 'dbs1' , 'dsk_files' , 'a2back.dat');
	END ;

PUT_FILE	BEGIN
	DBMS_FILE_TRANSFER.PUT_FILE ('ft1_1' , 'a2.dat', 'df' , 'a4.dat' , 'dbs2') ;
	END ;

注:如果数据库存在大量的表空间文件,那么使用 dbms_file_transfer 包来进行迁移是速度最快的。

6.4.2 方式二:RMAN Backup

RMAN Backup 方式是基于 RMAN 备份原理,通过使用 rman-xttconvert 包提供的参数,对数据库进行基于表空间的备份,将备份生产的备份集写到本地或者 NFS 盘上,然后再通过 rman-xttconvert 包中包含的不同平台之间数据文件格式转换的包对数据文件格式进行转换,最后通过记录的表空间 FILE_ID 号生成元数据的导入脚本,通过 db_link 执行完成。rman 方式 XTTS 执行步骤如表 6-5 所示。

表 6-5

使用阶段	脚本名字	用途
准备	perl xttdriver.pl -p (源端)	为了生成下述文件
	rmanconvert.cmd	rman 数据文件转化格式的命令
	xttplan.txt	记录 scn 号
	perl xttdriver.pl -c (目标端)	转换文件格式,直接到最终目录

续表

使用阶段	脚本名字	用途
增量前滚	perl xttdriver.pl -i （源端）	
	tsbkupmap.txt	表空间对应增备
	incrbackups.txt	增备的备份集的路径
	xttplan.txt.new	记录新的 scn 号
	perl xttdriver.pl -s （源端）	将 xttplan.txt.new 替换 xttplan.txt
	perl xttdriver.pl －r （目标端）	应用增备
传输阶段	perl xttdriver.pl -e	生成导入元数据的脚本

6.4.3 方式三：手工 XTTS 迁移

在实践中，可以总结提炼割接的步骤，创建手工脚本执行 XTTS 迁移，以加强自动化和迁移速度。以下总结了迁移的过程，可以把需要进行迁移的工作根据任务以及子任务的方式固化，形成一套可重复执行的迁移技术方案。

如表 6-6 所示是手工 XTTS 迁移说明（以实际生产为例），实践中可以参考这些步骤组织适合自己的流程和方法。

表 6-6

操作对象	操作说明
目标数据库安装配置	根据标准数据库安装手册进行数据库软件，数据库实例创建，配置参数
源库新监听创建	srvctl add listener -l listener_1523 -p 1523 源端数据库创建监听用于远程访问数据比对的监听，此处用于正式切换时，源库生产监听停止，启用此新建的监听，使得其他无业不能连接
源库 Dblink 创建	create user enmoXTTStest identified by "xxxxxxx"; grant dba to enmoXTTStest; 创建迁移切换数据一致性比对用户、DBLINK 连接用户，权限为 DBA 权限
目标库 Dblink 创建	创建基于 DBA 权限用户的 DB_LINK 连接到源端，使用 listener_1523
目标库其他配置	安全规范、性能参数配置，删除 USERS 表空间，新建一个默认的表空间
NFS 存储准备	一期 NFS 存储挂载用于数据初始化工作，二期 NFS 存储挂载用于增量备份
源库开启块跟踪	源端开启块跟踪，需关注块跟踪大小 ALTER DATABASE ENABLE BLOCK CHANGE TRACKING USING FILE '+SATADATA/changetracking.chg';
源库表空间自包含	SELECT * from sys.transport_set_violations;
源库获取用户相关环境信息	用户，DBLINK，PROFILE，PRIV
源库临时表信息	select o.owner, object_name, created 　from dba_objects o, dba_tables t 　where o.object_name = t.table_name 　and o.owner = t.owner and t.TEMPORARY = 'Y' order by 1, 3;

6.4 XTTS 迁移方式

续表

操作对象	操作说明
排除系统表空间	select to_char(wm_concat(TS#)) from v$tablespace where NAME not in ('UNDOTBS1','SYSTEM','SYSAUX','TEMP','UNDOTBS2','UNDOTBS3','UNDOTBS4','XDB','BASE') and name not in (select distinct TABLESPACE_NAME from DBA_TEMP_FILES);
源库准备 rman 备份脚本	根据生成的 TS# 号制订出 RMAN COPY 脚本
执行 rman copy 脚本	分别在源库多个 RAC 节点发起 RMAN 数据文件到 NFS 存储
准备数据文件转换脚本	根据 NFS 磁盘上复制数据文件完成后的结果，通过脚本拉取 convert 转换时所需要的执行脚本
执行数据文件格式转换	在目标库多个节点执行：RMAN CONVERT 将数据文件转成 Linux 平台，并存放在 ASM 里面
准备增量备份脚本	根据第一次 RMAN COPY 确认过的 SCN 进行增量同步备份
获取增量转换清单	增量备份文件转换成 Linux 平台生成转换清单，增量转换文件到/dump/enmo/incr 目录，生成增量备份转换的结果集存放到执行脚本中
应用增量备份	先将 asmcmd 中转换好的文件名作为列表存放在列表文件中，使用 for 循环读取，list 文件名为 apply_list.txt

XTTS 迁移割接步骤，如表 6-7 所示。

表 6-7

操作对象	操作说明
源库	在业务停业务时，清理回收站数据 purge dba_recyclebin 在业务停业务时，创建验证用户 create user enmo_test identified by xxxxxxx; conn enmo_test/xxxxxxxx 在业务停止后，创建数据验证表并插入验收数据 create table enmo_test(enmo_a number) insert into select lenvel from dual connect by lenve <1000;
修改 JOB 参数	停服务与修改 JOB 参数 alter system set job_queue_processes=0;
监听	源库和目标库监听停止 srvctl stop listener -l listener srvctl stop listener -l listener_scan1
查杀活动进程	查杀进程，确认无活动事务与死事务，包括分布式事务 ps -ef|grep LOCAL=NO|grep -v grep|awk '{print $2}'|xargs kill -9 select local_tran_id,state from dba_2pc_pending ; KILL JOB 进程
表空间只读状态	源库表空间置为 read only 状态
增量备份转换应用	根据上一次增量备份 SCN 号进行最后一次增量备份转换并应用
源库元数据、元对象导出	在做增量备份的同时，做元数据、元对象导出 exp、expdp
创建用户	根据 XTTS 执行前收集到的用户信息创建用户
创建 database link	根据 XTTS 执行前收集到的用户信息创建 dblink
目标库全备	备份数据库，用于回退

续表

操 作 对 象	操 作 说 明
目标库表空间读写模式	修改目标库表空间 read write
元数据导入	目标库导入传输表空间的元数据（imp、impdp）
数据库可用性、验证数据验证	查询测试表是否有数据
业务核心数据验证	验证核心表的数据是否一致，核心表验证由开发、应用进行
元对象导入	二次导入（存储过程、试图等信息）导入数据库的元对象
设置默认表空间	将用户的默认表空间修改成原来的值
用户权限	给角色及对象授权
对象对比	迁移对象比对

编译无效对象	开启并行进程，编译无效对象
	DECLARE
	    threads pls_integer := 150;
	BEGIN
	    utl_recomp.recomp_parallel(threads);
	END;
	/

验证数据清理	删除测试表、数据验证用户、迁移使用的 DBLINK
统计信息导入	导入表与索引的统计信息，可以按 schema 或者对象来导入
数据量比对	对象数、表记录数
迁移 contab 内容	提前将 crontab 里面的脚本迁移到目标数据库
开启服务	业务确认没有问题后开启监听与启动服务 srvctl start service -d orcl -s
收集字典表统计信息 不影响业务	EXEC DBMS_STATS.GATHER_DICTIONARY_STATS; EXEC DBMS_STATS.GATHER_FIXED_OBJECTS_STATS;
新库：打开 JOB	alter system set job_queue_processes=1000;

6.5 XTTS 前置条件检查

如本章前文概述所示，传输表空间技术从 Oracle 8i 诞生，经过多个版本的改进完善，时至今日已经发展成为跨平台大数据量迁移的利器，尤其从 Oracle 11.2.0.3 以后 XTTS 推出使用跨平台增量备份的方式，通过迁移不同字节序格式系统之间的数据，大大地减少了停机的时间。在方便的同时，来看一看使用 XTTS 进行数据迁移需要具备的前置条件，具体如下所示。

- 源库的操作系统不能是 Windows。
- 源库的 COMPATIBLE 参数必须设置为 11.1.0 或更高。
- 源库的 COMPATIBLE 参数值不能大于目标库的 COMPATIBLE 参数值。

- 源和目标库时区、字符集保持一致。
- 目标端 db_files 参数大于源库。
- 源库必须处于 ARCHIVELOG 模式。
- 源库的 RMAN 配置里 DEVICE TYPE DISK 不能设置为 COMPRESSED。
- 要迁移的表空间的数据文件必须都是 online 或者不包含 offline 的数据文件。
- 排除系统表空间，避免冲突，并且检查业务表空间是否存在自包含。

当然以上所说的目标端数据库版本均为 11.2.0.4 版本或者以上，如果在使用过程中，目标库的版本是 11.2.0.3 或者更低，那么需要创建一个单独的 11.2.0.4 版本数据库作为中间库来在目标端进行数据文件的格式转换，而使用 DBMS_FILE_TRANSFER 包目标端的数据库版本必须是 11.2.0.4。

迁移检查

基于以上这些条件，在开始 XTTS 迁移之前，需对整个实施环境进行一次全面的检查，保证在实施之前环境满足迁移要求。

（1）数据库时区检查，保持源库和目标库时区一致性。

```
select dbtimezone from dual;
DBTIME
------
+08:00
```

（2）数据库字符集检查，保持源库和目标库字符集一致性。

```
select * from nls_database_parameters where parameter like '%CHARACTERSET%';
PARAMETER                   VALUE
--------------------        ----------------
NLS_CHARACTERSET            AL32UTF8
NLS_NCHAR_CHARACTERSET      UTF8
```

（3）数据库目标端补丁情况检查。

目标端 psu 根据数据库安装配置最佳实践规范配置安装最新 PSU。

```
select 'opatch',comments from dba_registry_history
'OPATC COMMENTS
------ ----------------
opatch PSU 11.2.0.4.4
```

使用 dbms_file_transfer 方法，在 11g 中目标端建议安装的补丁如下：

- Patch 19023822，修复目标端使用 dbms_file_transfer.get_file 包获取源端数据文件出现 ORA-03106 的情况。
- Patch 22171097: MERGE REQUEST ON TOP OF DATABASE PSU 11.2.0.4.6 FOR BugS 17534365 19023822。

如果准备阶段使用 rman 方法，目标端没有小补丁安装需求。

（4）检查目标库端数据库组件安装情况，需包含且多于源库组件。

源端组件：

```
COMP_NAME
-------------------------------------
Oracle Application Express
Oracle Multimedia
Oracle XML Database
Oracle Expression Filter
Oracle Rules Manager
Oracle Workspace Manager
Oracle Database Catalog Views
Oracle Database Packages and Types
JServer JAVA Virtual Machine
Oracle XDK
Oracle Database Java Packages
```

目标端组件：

```
COMP_NAME
-------------------------
Oracle Enterprise Manager
Oracle Workspace Manager
Oracle Database Catalog Views
Oracle Database Packages and Types
```

注：组件不同，可能会导致一些特殊的业务受到影响，需同开发、应用进行确认。

（5）检查是否使用了 key compression 的索引组织表。

```
Select index_name,table_name from dba_indexes where compression='ENABLED';
Select owner,table_name from dba_tables where iot_type is not null;
```

如果存在 key compression 的索引组织表，目标端需要安装 patch 14835322，否则索引组织表无法导入到目标端，需要手动重建，或者通过手工方式导入。

（6）检查源端数据文件、表空间异常情况。

```
Select name from v$datafile where instr(name,' ') > 0;
```

如果存在空格或者换行符数据文件，需要将该数据文件 rename，否则在传输过程中会报错终止。如果存在同名表空间，而且该表空间中的对象需要传输，建议将目标端中的表空间 rename，以避免冲突。

（7）检查相同表空间下是否存在不同目录下的同名数据文件。

使用 dbms_file_transfer 方式进行 XTTS 数据迁移，需要通过修改准备阶段的 getfile.sql 和 xttnewdatafile 文件来调整数据文件生成的路径。

```
select substr(file_name,-6,2) from dba_data_files where tablespace_name='TBS_NAME' order by 1;
```

如果目标端有多个目录，指定数据文件生成在不同的目录，可以避免冲突。

如果准备阶段使用 rman 方式进行同步，则所有的数据文件会转换为 xtf 结尾的文件，文件名会重新命名，所以不需要提前修改。

（8）检查表空间自包含。

在传输阶段，可能因为目标端数据文件目录所限制，需要将各个表空间拆分进行传送。在导入元数据阶段，考虑到字包含特性，需要将所有的表空间汇总进行传送。检查表空间时，只检查业务表空间的自包含情况，系统表空间、临时表空间、undo 表空间不在检查列。因为在 XTTS 数据迁移时，无需对系统表空间、temp、undo 进行传输，需要在目标端重建。

```
SQL>execute dbms_tts.transport_set_check(TBS_NAME ,TRUE);
SQL> SELECT * from sys.transport_set_violations;
no rows selected
```

（9）检查源端 compatible 参数。

源端不可以是 Windows，源端的 compatible.rdbms 必须大于 10.2.0，且不大于目标端 compatible.rdbms。

```
show parameter compatible
```

如果目标端数据库版本是 11.2.0.3 或更低。那么需要在目标端安装 11.2.0.4 并创建实例，然后用来进行备份集转换。如果 11.2.0.4 中转实例使用 ASM，那么 ASM 版本也必须是 11.2.0.4，否则报错 ORA-15295。

（10）启用 block change tracking（块跟踪）功能。

块跟踪功能在源端数据量较大，或者数据改变较大时启用，需要在源库安装补丁 Bug 16850197。该补丁在 11.2.0.3.9 和 11.2.0.4 版本 psu 中提供。如果源库是在上述版本前，需要安装个别补丁，注意块跟踪设置的目录大小，避免因块跟踪目录满而导致源数据库 hang。

（11）检查目标端的 db_files 参数。

在元数据导入阶段，如果目标端的 db_files 参数小于源端的 db_files 参数，会导致元数据因无法关联创建数据文件而导入出错，所以要确保目标端参数大于或者等于源端。

```
Show parameter db_files
```

6.6 XTTS 最佳实践方案论证

6.6.1 技术方案概况

在不影响现有生产环境的前提下，缩短业务停机时间，搭建一套同平台的 data guard 数据库为中间库，通过 XTTS 迁移技术来完成生产环境跨平台到新环境的数据迁移。

6.6.2　技术方案实施步骤

（1）目标环境准备。

（2）搭建中间 data guard 环境。

（3）挂载 NFS 存储到老生产环境减少传输时间。

（4）通过 XTTS RMAN COPY 中间 DG 数据库的数据文件。

（5）通过中间环境 DG 数据库进行增量备份。

（6）新环境进行增量转换、应用。

（7）正式切割时生产环境表空间 read only。

（8）最后一次增量备份。

（9）导入导出元数据。

（10）数据校验。

6.6.3　技术方案模型

技术方案模型如图 6-7 所示。

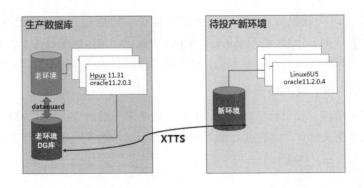

图 6-7

6.6.4　方案可行性说明

（1）采用 Data guard 中间环境作为 XTTS 前期准备阶段的源库，因 DG 数据库作为生产环境的备库同生产环境数据库版本一致、数据实时同步一致、操作系统环境一致，从环境要求方面符合 XTTS 迁移条件。

（2）采用中间环境来作为数据文件复制的源库，中间环境因是备库，可以在不影响业务的情况下进行备库停机。

（3）待投产新环境为 11.2.0.4 版本满足增强版 XTTS 数据库版本要求。

6.6.5 方案优缺点论述

使用这一方案的优缺点简要概括如下要点。

优点：备库操作不影响生产环境业务；备库复制数据文件不影响生产环境 IO 性能；增量备份可以在中间库进行多次，缩短停机时间；数据一致性能得到保障。

缺点：需准备中间库环境资源。

6.6.6 技术方案论证结论

通过搭建 Data guard 数据库作为中间库环境来进行 XTTS 迁移的方案，经过实际测试效果良好，不影响生产环境业务，且实施安全可靠，对缩短停机时间进行大批量数据迁移具备最佳实践效果，方案可行。

6.7 XTTS RMAN Backup 步骤

（1）环境检查参照 XTTS 前置条件检查第二项迁移检查。

（2）创建 db_link。

```
create public database link ttslink connect to system identified by ******
using '(DESCRIPTION =
   (ADDRESS = (PROTOCOL = TCP)(HOST = 10.111.118.103)(PORT = 1521))
   (CONNECT_DATA =
     (SERVER = DEDICATED)
     (SERVICE_NAME = orcl)
   )
 )';
```

（3）XTTS 执行脚本配置。

```
存放转换脚本
/home/Oracle/enmotest/xtt
存放执行脚本
/home/Oracle/enmotest/xtt-scripts
配置必要脚本
SCZXB[/home/Oracle/enmotest/xtt-scripts]$more xtt.properties
```

```
tablespaces=ENMOTEST_TABS---表空间名称
platformid=4 ---平台代号
dfcopydir=/dump2/enmotest/rman_copy---数据文件复制存放的路径
backupformat=/dump2/enmotest/rman_incr---增量备份集存放的路径
stageondest=/dump2/enmotest/rman_incr----目端存放数据文件复制目录和存放增量备份的目录
storageondest=+FLASHDATA/orcl/DATAFILE---目标端数据文件转换后的最终目录
backupondest=/dump2/enmotest/rman_incr-----备份文件转换后生成的目录
```

（4）源库发起 rman copy。

```
--源端进行第一次 rman copy 数据文件
perl /home/Oracle/enmotest/xtt-scripts/xttdriver.pl -p
```

（5）目标库复制生成脚本。

在目标端，通过 Oracle 用户复制源端/home/Oracle/enmotest/xtt/rmanconvert.cmd 到目标端的/home/Oracle/enmotest/xtt 目录下。

```
scp source:/home/Oracle/enmotest/xtt/rmanconvert.cmd dest:/home/Oracle/enmotest/xtt
```

（6）目标端进行转换。

```
----目标端进行转换
perl /home/Oracle/enmotest/xtt-scripts/xttdriver.pl -c
```

（7）源端增量备份。

```
cd /home/Oracle/enmotest/xtt-scipts
perl xttdriver.pl -i
```

结束会生成一个新的记录 scn 的文件：xttplan.txt.new。

（8）复制生产文件后转换应用。

将 tsbkupmap.txt xttplan.txt 传到目标端/home/Oracle/enmotest/xtt，将 incrbackups.txt 传到目标端/dump2/enmotest/rman_incr。

```
cd /home/Oracle/xtt
cp incrbackups.txt /dump2/enmotest/rman_incr
cp xttplan.txt tsbkupmap.txt /dump2
```

目标端开始应用增量备份。

```
cd /home/Oracle/enmotest/xtt-scritps
perl xttdriver.pl -r
```

（9）确定新的 scn。

```
perl xttdriver.pl -s
```

该步骤会将-i 时生成的 xttplan.txt.new 改名为 xttplan.txt，并将原来的 xttplan.txt 备份。

（10）源库 read only 进行最后一次增量备份。

```
--源库表空间 read only
alter tablespace ENMOTEST_TABS  read only;
```

```
--最后一次增量备份
perl xttdriver.pl -i
将 tsbkupmap.txt xttplan.txt 传到目标端/home/Oracle/enmotest/xtt
将 incrbackups.txt 传到目标端/dump2/enmotest/rman_incr
cd /home/Oracle/xtt
cp incrbackups.txt /dump2/enmotest/rman_incr
cp xttplan.txt tsbkupmap.txt /dump2
cp xttplan.txt tsbkupmap.txt /home/Oracle/enmotest/xtt
```

（11）应用增量备份，完成最后一次增量恢复。

```
cd /home/Oracle/xtt-scripts
perl xttdriver.pl -r
```

（12）元数据导入脚本生成。

```
--目标端运行
cd /home/Oracle/xtt-scripts
perl xttdriver.pl -e
```

（13）目标端创建用户。

```
--目标库先创建用户，并且权限同源库一样
create user enmoXTTStest identified by ********* default tablespace USERS;
```

（14）导入元数据。

```
impdp \'/ as sysdba\' directory=DATA_PUMP_DIR logfile=tts_imp.log network_link=ttslink
transport_full_check=no transport_tablespaces=ENMOTEST_TABS
transport_datafiles='+ FLASHDATA/orcl/datafile/enmotest_tabs_1024.xtf'
```

（15）数据验证。

应当根据数据特点，定制脚本、程序，进行基本的数据验证。

6.8 XTTS 实战案例分享

实践是检验真理的唯一标志，下面通过一个跨平台迁移 32TB 数据库的 XTTS 实战案例，来解析在大数据量迁移过程中的手工脚本应用情况。以下案例从 XTTS 原理出发，涉及操作系统、NFS 存储、RMAN 备份、系统字节序转换、数据验证以及网络知识。

6.8.1 案例现状介绍

某省交管核心系统自上线运行 6 年以来，从最初 GB 为单位的数量级上升到今天 32TB 的业务数据量，其中照片信息的 LOB 字段占有 27TB。随着近几年信息化行业深化改革发展，信息化、互联网+、大数据已经成为交管业务支撑不可或缺的组成元素，但该交管局核心系统却存在严重问题，已然不能满足现有业务的发展。为解决这套老旧的核心业务系统，通过调研，最终确定为客户采用 zDATA 分布式存储方案。组建一个高速、安全、稳定的高性能分布式存

储数据库架构，通过去除老旧 HP 机器，采用高配置的 X86 PC 服务器作为计算和存储节点，不仅提供强大的 CPU、I/O、Memory 支持能力，还为后续的横向扩容存储提供不停机服务，可如何进行这大批量的数据迁移成为本项目的一个难点。

6.8.2 系统现状评估

系统资源配置如表 6-8 所示。系统现状评估如图 6-8 所示。

表 6-8

主　机	生产库主库为 3 节点集群，其中 2 台为 HP 8640
存储	HP XP24000
应用	Websphere（约 20 台主机，14 台为 HP 小机，其他为 Windows 环境）
连接数	单节点连接数为 260～300
容灾备份	无容灾环境
数据容量	约 32TB，集中存储 HP XP24000 已经满了，无法扩容

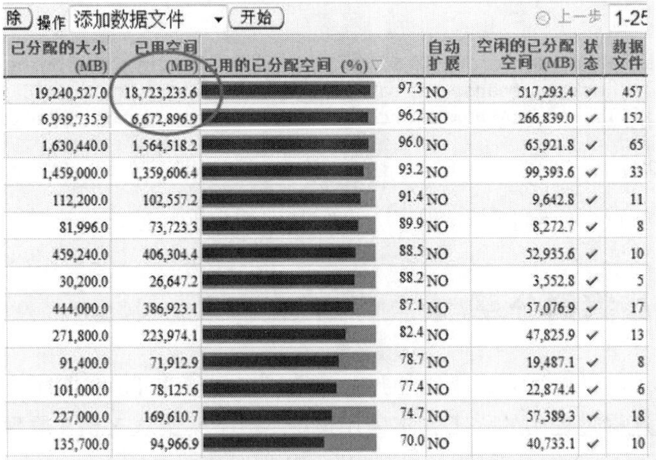

已分配的大小(MB)	已用空间(MB)	已用的已分配空间(%)	自动扩展	空闲的已分配空间(MB)	状态	数据文件
19,240,527.0	18,723,233.6	97.3	NO	517,293.4	✓	457
6,939,735.9	6,672,896.9	96.2	NO	266,839.0	✓	152
1,630,440.0	1,564,518.2	96.0	NO	65,921.8	✓	65
1,459,000.0	1,359,606.4	93.2	NO	99,393.6	✓	33
112,200.0	102,557.2	91.4	NO	9,642.8	✓	11
81,996.0	73,723.3	89.9	NO	8,272.7	✓	8
459,240.0	406,304.4	88.5	NO	52,935.6	✓	10
30,200.0	26,647.2	88.2	NO	3,552.8	✓	5
444,000.0	386,923.1	87.1	NO	57,076.9	✓	17
271,800.0	223,974.1	82.4	NO	47,825.9	✓	13
91,400.0	71,912.9	78.7	NO	19,487.1	✓	8
101,000.0	78,125.6	77.4	NO	22,874.4	✓	6
227,000.0	169,610.7	74.7	NO	57,389.3	✓	18
135,700.0	94,966.9	70.0	NO	40,733.1	✓	10

图 6-8

该系统存在的问题还包括以下情况：3 节点 RAC 架构不合理；集中存储使用 10 年以上、计算节点服务器设备老旧；资源配置低，物理扩容到达瓶颈；数据爆发式增长；业务应用模块增多、数据库表存放 LOB 字段；基层业务人员反馈系统各种性能问题。

6.8.3 迁移需求分析

系统迁移面临的问题主要有如下情况：HP-UX 跨平台迁移到 Linux；数据库总量为 32TB，

LOB 数据大小为 27TB；单个数据库表空间为 17TB；数据库版本为 11.2.0.3（无任何补丁）；计划内停机切换时间为 8 小时；计划内完成时间为 15 天；数据库账号密码不能改变；无应用测试；无资源提供测试环境（存储、资源）。

如何设计方案，解决这些挑战，是项目实施中的重要内容。

6.8.4 迁移方案选型

如图 6-9 所示，通过需求调研分析后，因系统涉及 30 多 TB 数据量，并且业务停机时间只有 8 个小时，另外需要跨平台进行数据迁移。我方经过几次测试论证后，排除如下方案，在此也请各位思考一下，如果遇到此类需求作为 DBA 应该如何应对。

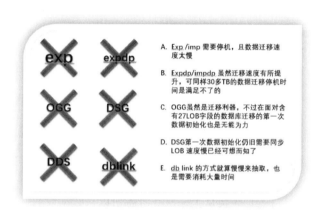

图 6-9

最终选择了最具挑战的 XTTS 来完成这次 32TB 的跨平台迁移挑战。迁移前后的资源配置情况如表 6-9 所示。

表 6-9

配置类型	源库	目标库
数据库版本	11.2.0.3	11.2.0.4.160419
数据库名称	orcl	orcl
数据库字符集	AMERICAN_AMERICA.ZHS16GBK	AMERICAN_AMERICA.ZHS16GBK
数据库节点	RAC 3 节点	RAC 4 节点
操作系统版本	HPUX11.31	Linux6.5
磁盘组大小	35TB	80TB
数据库大小	32TB	
块大小	16384	16384

6.8.5 迁移的具体实施

在以下的描述中，我们再现了生产过程中的实战步骤，供读者在学习和实践中参考。

（1）XTTS 环境检查，如表 6-10 所示。

表 6-10

检 查 项	源 库	目 标 库
时区是否一致	时区为东八区	东八区
字符集是否一致	16GBK	16GBK
检查目标端补丁情况	无	需打最新 PSU
组件检查	—	包含源库组件
key compression 索引组织表	存在	需手工重建
表空间规范检查	不同磁盘组下数据文件名称命名相同	—
TEMP 表空检查	存在	需手导入
检查目标端的 db_files 参数	1 024	4 096
检查源端 compatible 参数	不可以是 windows 且大于 10.2.0	11.2.0.4
检查表空间自包含	存在自包含 需手工 MOVE	—
用户	DBLINK、PROFILE、PRIV	需手工创建

（2）开启块跟踪。

Block change tracking 进程记录自从上一次 0 级备份以来数据块的变化，并把这些信息记录在跟踪文件中。RMAN 使用这个文件判断增量备份中需要备份的变更数据。这极大地提高了备份性能和速度，RMAN 可以不再扫描整个文件以查找变更数据。

```
SQL> ALTER DATABASE ENABLE BLOCK CHANGE TRACKING USING FILE '+SATADATA/changetracking.chg';
Database altered.
```

（3）挂载 NFS 存储。

NFS 存储挂载在源库和目标库之间，用于传输数据文件和增量备份节省数据文件的传输时间。

```
mount -o llock,rw,bg,vers=3,proto=tcp,noac,forcedirectio,hard,nointr,timeo=600,rsize=32768,wsize=32768,suid 10.160.118.236:/dump1 /dump1
```

源端、目标端需要挂载 35TB 存储用于存放所有数据文件的镜像文件，建议将存储远程从源端挂载到目标端，减少备份传送时间。图 6-10 所示为 XTTS 迁移工作示意图-NFS 存储初始化挂载。

6.8 XTTS 实战案例分享

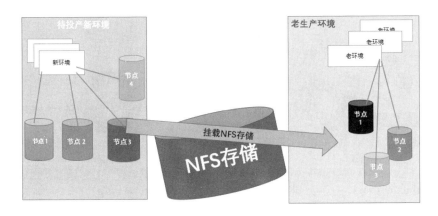

图 6-10

（4）SCN 确认记录。

SCN（System Chang Number）作为 Oracle 中的一个重要机制，在数据恢复、Data Guard、Streams 复制、RAC 节点间的同步等各个功能中起着重要作用，在此需确认 SCN，且该 SCN 号用于后续增量备份的起始点。

```
alter system checkpoint;
select current_scn from v$database;
```

（5）开始 RMAN Copy。

基于数据文件的 RMAN COPY 生成的文件存放于挂载的 NFS 目录下。

```
rman target / <<EOF
run{
allocate channel c1 type disk;
allocate channel c2 type disk;
backup as copy datafile 18,19,20,21,22........ format '/dump1/enmo/copy/enmo_%U';
release channel c1;
release channel c2;
}
EOF
```

rman backup 工作示意图如图 6-11 所示。

（6）数据文件格式转换。

Convert 用于转换数据文件的字节序，转换后的新数据文件直接写入新环境的磁盘组，转换过程消耗目标端的 CPU 资源，此处需要关注目标端磁盘组的大小，避免造成磁盘组满引起转换失败，如图 6-12 所示。

```
convert from platform 'HP-UX IA (64-bit)' datafile '/dump1/ccm/vvstart_tabs.dbf' format '+FLASHDATA/ORCL/DATAFILE/vvstart_new_01.dbf';
```

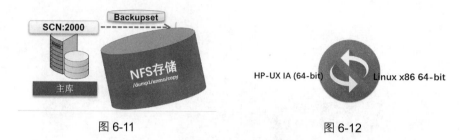

图 6-11　　　　　　　　　　　图 6-12

（7）增量备份阶段。

开启块跟踪后基于块进行快速增量，增量备份前先查询并记录当前的 SCN，根据全备时记录的 SCN 进行增备（假定记录的 SCN 是 1000），增量备份文件存放于 NFS 存储上，增量备份后生成的字节序是 HP-UX 的需进一步转换。

```
set until scn=1850
backup incremental from scn 1000 datafile 18,19,20,21,22...... format '/dump1/enmo/incr/copy_%d_%T_%U';3;
```

（8）增量转换应用。

增量备份的转换和应用是两个过程，首先是增量备份集从 HP-UX 平台转换成 Linux 平台格式，转换完毕后的备份集在 Linux 平台数据库才能识别。

```
sys.dbms_backup_restore.backupBackupPiece(bpname => '/dump1/enmo/incr/copy_ORCL_20160707_78ra40o7_1_1',
    fname => '/dump1/enmo/incr/copy_ORCL_20160707_78ra40o7_1_1_conv',handle => handle,media=>media,
    comment=> comment, concur=> concur,recid=> recid,stamp => stamp, check_logical => FALSE,copyno=> 1,
    deffmt=> 0, copy_recid=> 0,copy_stamp => 0,npieces=> 1,dest=> 0,pltfrmfr=> 4);
```

其次是增量备份集的应用，这个过程和 rman 恢复的原理是一样的。

```
sys.dbms_backup_restore.restoreBackupPiece(done => done, params => null, outhandle => outhandle,outtag => outtag, failover => failover);
```

（9）循环进行增量备份。

循环进行增量备份操作在正式环境的切割之前进行，其目的是为了减少最后一次数据库表空间 read only 时生产环境的停机时间，需要特别注意的是，操作之前务必查询并记录当前 SCN 号。这个 SCN 号是下一次开始增量备份的起点，如图 6-13 所示。

（10）正式割接准备。

正式切换的准备阶段是整个 XTTS 迁移过程中最重要的一步。为保证数据的一致性，需要在源库停止业务后，对活动的数据库会话进行查杀处理，并且在 read only 表空间之前需要进行几次检查点，确保检查点成功完成才能进行下一步操作。另外，针对计划窗口任务的 JOB 需要提前关闭 JOB 避免因 JOB 执行、批处理等导致数据不一致，如图 6-14 所示。

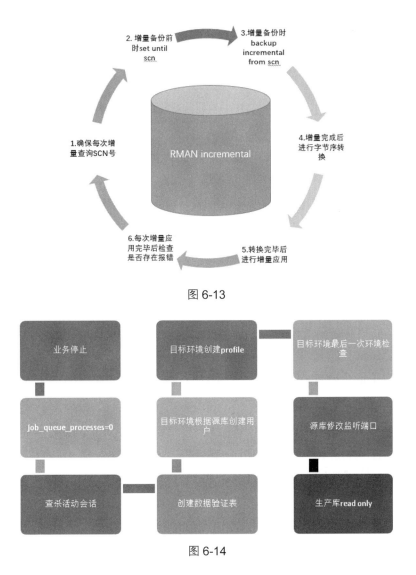

图 6-13

图 6-14

（11）最后一次增量备份。

最后一次增量备份是在生产源库表空间全部 read only 的情况下进行的，需要根据前一次记录的 SCN 号进行最后一次增量备份、转换、应用。在转换应用之后建议把新环境做一个闪回点或者进行一次全备，作为下一步导入元数据失败的回退方案，如图 6-15 所示。

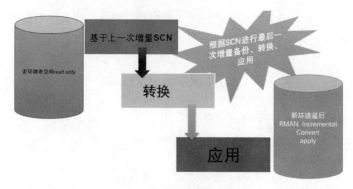

图 6-15

注：转换应用之前建议先把新环境进行一次全备。

（12）元数据的导入和导出。

导出需在表空间 read only 下才能进行，应排除系统表空间。导入时可能会遇到 type 不存在的情况，建议使用数据泵进行。

如下所示，在源库导出表空间的元数据：

```
exp \'/ as sysdba\'
transport_tablespace=y
tablespaces='TBS_NAME'
STATISTICS=none
file=/dump1/enmo/exp/orcl_XTTS_0715.dmp
```

根据导出的 dmp 包导入元数据：

```
imp \'/ as sysdba\'
transport_tablespace=y
TABLESPACES='TBS_NAME'
file=/dump1/enmo/exp/orcl_XTTS0715.dmp
log=/dump1/enmo/exp/orcl_imp_XTTS0715.log
datafiles=('+FLASHDATA/orcl/datafile/VIO_DATA_u01',
'+FLASHDATA/orcl/datafile/base_image_fno65')
```

（13）元对象的导入和导出。

开始导入之前先把表空间 read write，可以使用 dblink 进行远程不落地导入，指定需要的 schemas，导入存在权限不足可进行手工授权，导入完毕后即可开始验证。如下所示：

1）迁移列表 schema 对象导入。

```
impdp "'/ as sysdba'"  metrics=yes network_link=ENMO_TEST schemas=VIO EXCLUDE=table,index content=metadata_only  directory=enmo_exp logfile=imp_full_metadata_`date +"%d%H%M"`.log
```

2）临时表导入。

3）组织索引表创建。

4）其他手工导入用户。

（14）数据校验收尾阶段。

1）统计信息收集或者导入。

```
EXEC DBMS_STATS.gather_schema_stats('DRV_ZW', estimate_percent => 10,degree => 64);
```

2）无效对象重新检查编译。

```
SQLplus / as sysdba <<EOF
DECLARE
   threads pls_integer := 150;
BEGIN
   utl_recomp.recomp_parallel(threads);
END;
/
```

3）对象数量比对

```
select owner,object_type,count (*) from dba_objects where owner ='用户名称'
group by owner,object_type order by owner,object_type;
```

4）主键索引核对。

```
select count(1),a.status from dba_constraints a where a.owner='用户名称'
and a.constraint_type='P' GROUP BY A.status;
```

5）大表数据校验。

```
--num_rows 行数验证
select table_name,num_rows from all_tables where owner='用户名称'
group by table_name,num_rows
having num_rows>500 order by table_name;
--大小验证
 select owner, segment_name, bytes / 1024 / 1024
 from dba_segments where segment_type = 'TABLE' and owner = '用户名称';
```

6）账号权限、同义词验证。

```
Set lines 180
Col object_name for a40
select object_name,object_type,status from dba_objects
where owner in ('账号名称') and status<>'VALID';
```

7）数据文件头状态。

```
select STATUS,ERROR,TABLESPACE_NAME from V$DATAFILE_HEADER;
```

8）表空间校验。

确认 owner 用户的 DEFAULT_TABLESPACE、TEMPORARY_TABLESPACE 以及所有用户在相关表空间上的配额情况,将之前创建的 owner 用户的默认表空间改为正确的默认表空间。

9）启动数据库。

启动监听,应用开始连接测试:

```
srvctl start scan_listener
```

删除测试用户信息:

```
drop user enmoXTTStest cascade;
drop public database link enmo_test;
```

验证结果比对表如表 6-11 所示。

表 6-11

数据校对项	结　果
测试数据比对	True
迁移列表的表空间数量比对	True
Schema 数量比对	True
对象数比对	True
表记录数比对	True
包、函数、存储过程、索引等状态比对	True
权限比对	True
同义词比对	True
临时表数量比对	True

(15)回退方案。

一个完整的方案还需要包括回退方案,迁移切换失败进行回退。

1)前提条件:在不影响现有生产环境后续的可用性的情况下进行切换。

2)回退条件:仅需把源生产数据库表空间置换为 read write、源库 JOB 进程调整为 1000、源库监听启动。

3)回退时间:执行回退方案可保证在 5 分钟之内完成。

4)回退影响:本次切换失败。

6.9　XTTS 风险预估

在执行迁移时需要充分估计项目的风险点,在关键的环节要加强测试、把关,规避可能的风险,以下是在各个阶段可能出现的问题点,需要特别关注。

(1)第一次全量备份消耗源生产库资源需关注。

(2)全量备份挂载 NFS 会占用网络流量。

(3)筛选排除系统表空间需认真仔细。

(4)自包含检查需排除系统表空间。

(5)使用 exp 导元数据可能会遇到 Bug。

(6)最后一次增量备份只读源库表空间可能会因活动会话占用,表空间 read only 过慢。

(7)切换之前一定要先对目标库做一个 rman 全备,避免失败无法回退。

(8)每次增量备份之前都需记录 SCN。

6.10 XTTS 总结

对于数据库的跨平台迁移,大家所熟悉的方法有很多,每种方法都各有利弊,关键还要看实际需求再来决定使用哪一种方式更能切合业务,提高工作效率。物理迁移的方式具备很多优势,这样减少了很多繁琐的数据校验过程,尤其是面对超过 10TB 的数据库时,物理迁移能在一定程度上提高迁移速度,节省大量的时间。

本章节向大家推荐的 XTTS 在面对 U2L 大数量迁移中更能发挥其优势,不过也有很多不为人知的陷阱。我们建议如果想使用好这个方法,需要对 XTTS 的原理非常熟悉,尽量采用手工脚本的方式来进行数据迁移。官方推出的 DBMS_FILE_TRANSFER 包由于 Bug 太多,在同步过程中经常会遇到很多莫名其妙的错误而中断,所以并不建议大家使用。对于 RMAN 备份的方式,因为本身 rman_xttconvert 包是通过执行不同参数来自动进行数据文件的复制、转换、应用以及增量等,需要大家对它的执行过程非常熟悉才不容易造成混乱。如果当您面对需要传输的表空间非常多时,建议还是采用手工的方式会比较保险。

第 7 章 Oracle 的参数和参数文件

——盖国强

在 Oracle 数据库的世界里，充满了挑战和意外，大到容灾备份，小到安装配置，一个小小的参数就可能引发无穷的麻烦。所以在资深 DBA 的职业生涯中，总是路走的越多，胆子反而变得更小。如果说回顾一下，我最愿意和大家分享的就是：谨慎、谨慎、再谨慎，哪怕是在最微小的环节，也容不得半点疏忽。在本章中，我就从小小的、最不重要的参数文件入手，和大家探讨一下，这其中可能存在的点点滴滴你可能疏忽的细节，而这样的疏忽又可能在紧要关头对数据库运维造成巨大的威胁。

在 Oracle 数据库中，有一系列的初始化参数用来进行数据库约束和资源限制，这些参数通常存储在一个参数文件中，在数据库实例启动时读取并加载。

初始化参数对数据库来说非常重要，很多参数通过合理的调整可以极大地提高数据库性能。而几乎所有的 DBA 都经历过参数调整错误时，对于参数文件手忙脚乱的修改过程，所以本书的读者一定要掌握对于参数文件的修改备份及恢复过程。

本章撰写的内容，详细描述了参数文件在各个数据库版本中的变化，以帮助读者了解 Oracle 数据库的演变过程，请读者在阅读的过程中注意相应的版本描述。（注意，通常旧版本中的命令在新版本中是完全支持的，这就是向后兼容）

7.1 参数的分类

参数的分类方式有很多，一般按照得出方式不同，可以将其分为三类：推导参数、系统依赖参数和可变参数。另外随着 Oracle 版本的升级，对参数做出了相应的调整，因此不推荐参数和废弃参数也占了很大一部分。了解这些能够帮助我们更好地理解 Oracle 管理模式。

7.1.1 推导参数

推导参数通常来自于其他参数的运算，依赖其他参数得出，所以这类参数通常不需要修改。如果强制修改，那么修改值会覆盖推导值。

常见的此类参数有很多，例如 SESSIONS 参数，在 Oracle 12c 文档中，该参数按以下公式运算得出：

```
(1.5 * PROCESSES) + 22
```

PROCESSES 参数代表操作系统上能够并发向 Oracle 数据库发起的连接进程数量。当 PROCESSES 被修改时，此参数会自动计算并生效。以下是一个示范数据库中这两个参数的设置：

```
SQL> select name,value from v$parameter where name in ('processes','sessions');
NAME            VALUE
------------    -----------
processes       200
sessions        322
```

如果该参数设置过低，则在应用并非高时，超过 PROCESSES 数量的进程将无法连接到数据库。所以在规划数据库时，合理设置 PROCESSES 参数是十分重要的。但是很多时候由于应用的异常可能导致业务环境的进程数量激增，所以在生产环境中对进程数量进行必要监控是必需的。

7.1.2 操作系统依赖参数

某些参数的有效值或者取值范围依赖或者受限于操作系统，比如 MEMORY_TARGET 参数，设置 Oracle 使用的内存大小，该参数的最大值就要受限于物理内存，而且同时受限于系统设置。

下面的错误是经常会遇到的。

```
SQL*Plus: Release 11.2.0.3.0 Production on Sat Apr 9 03:42:40 2016
Copyright (c) 1982, 2011, Oracle. All rights reserved.
Connected to an idle instance.
SQL> startup
ORA-00845: MEMORY_TARGET not supported on this system
```

在 Linux 平台，这种情况一般情况下意味着/dev/shm 设置过小。

```
SQL> create pfile='/tmp/pfile.ora11g' from spfile;
File created.
SQL> !grep -i 'memory_target' /tmp/pfile.ora11g
*.memory_target=511705088
SQL> !df -h /dev/shm
Filesystem      Size    Used    Avail   Use%    Mounted on
tmpfs           400M    0       400M    0%      /dev/shm
```

用 root 用户增大并 remount 之后，就可以正常启动实例。

```
SQL> !df -h /dev/shm
Filesystem      Size  Used Avail Use% Mounted on
tmpfs           600M     0  600M   0% /dev/shm
SQL> startup
ORACLE instance started.
Database mounted.
Database opened.
```

需要特别注意的是，如果有多个实例，那么其大小要大于所有实例所需要内存的总和。

这一类参数通常被称为操作系统依赖参数。

7.1.3 可变参数

可变参数包含绝大多数潜在影响系统性能的可调整参数，某些可变参数设置的是限制条件，如 OPEN_CURSORS；有的参数是设置容量，如 DB_CACHE_SIZE 等。这类参数通常可以为 DBA 或最终用户调整，从而产生限制或性能变化，对 Oracle 至关重要。

7.1.4 不推荐参数

随着版本的更新和新特性的发布，部分参数已经不再推荐使用。对于这类参数，一般情况下还可以设置并生效，但也有部分参数尽管可以设置，但不会产生效果。

可以通过如下查询获得这些参数（Oracle 版本 11.2.0.3）。

```
SQL> select name from v$parameter where ISDEPRECATED='TRUE';

NAME
--------------------------------
lock_name_space
instance_groups
resource_manager_cpu_allocation
active_instance_count
buffer_pool_keep
buffer_pool_recycle
log_archive_start
standby_archive_dest
log_archive_local_first
parallel_server
parallel_server_instances
```

在生产环境下使用这类参数需要非常谨慎，因为随着后续版本的发布，这类参数可能变成废弃参数从而根本无法使用。

7.1.5 废弃参数

由于 Oracle 数据库的参数众多，在新版本中可能废弃很多旧的参数，了解这些废弃参数，明确废弃原因，是 DBA 需要关注的内容之一。

在 Oracle Database 11gR2 中，有大约 130 个参数被废弃。

```
SQL> select * from v$version where rownum <2;
BANNER
--------------------------------------------------------------------------------
Oracle Database 11g Enterprise Edition Release 11.2.0.3.0 - 64bit Production

SQL> select count(*) from V$OBSOLETE_PARAMETER;
  COUNT(*)
----------
       131
SQL> select * from V$OBSOLETE_PARAMETER;
NAME                     ISSPE
------------------       ------------
spin_count               FALSE
use_ism                  FALSE
lock_sga_areas           FALSE
instance_nodeset         FALSE
……
```

这个视图的创建语句如下：

```
SQL> select view_definition from v$fixed_view_definition
  2  where view_name='GV$OBSOLETE_PARAMETER';
VIEW_DEFINITION
--------------------------------------------------------------------------------
select inst_id,kspponm,decode(ksppoval,0,'FALSE','TRUE')  from x$ksppo
```

底层的 X$KSPPO 是这些废弃参数的来源。

除了以上参数类别，初始化参数通常还有一些其他分类方式：

● 按照修改方式划分，初始化参数又可以分为静态参数和动态参数。其中静态参数只能在参数文件中修改，在重新启动后方能生效；动态参数可以动态调整，调整后通常可以立即生效。

● 按照获取方式不同，初始化参数又可以分为显式参数和隐含参数。显式参数可以通过 V$PARAMETER 查询得到；而隐含参数通常以 '_' 开头，必须通过查询系统表方能获得。

比较常用的几个隐含参数有以下几个。

```
NAME                            VALUE   PDESC
--------------------------      ------- --------------------------------------------
_allow_resetlogs_corruption     FALSE   allow resetlogs even if it will cause corruption
_offline_rollback_segments              offline undo segment list
_corrupted_rollback_segments            corrupted undo segment list
```

总之，虽然分类方式不同，但是参数都是这些，我们更多需要了解的是这些参数的用途。

7.2 参数文件管理和使用

参数文件是一个包含一系列参数及参数对应值的操作系统文件。参数文件有两种类型。

● 初始化参数文件（Initialization Parameters Files）——Oracle 9i 之前 Oracle 一直采用 PFILE 方式存储初始化参数，该文件为文本文件，可以通过文本编辑器手工修改，但无法通过数据库修改。

● 服务器参数文件（Server Parameter Files）——从 Oracle 9i 开始，Oracle 引入的 SPFILE 文件，该文件为数据格式，不能通过手工修改，但可以通过数据库直接修改。

从操作系统上我们也可以看到这两者的区别，PFILE 为 ASCII 文本文件，SPFILE 为数据文件。

```
[oracle@jumper oracle]$ cd $ORACLE_HOME/dbs
[oracle@jumper dbs]$ file initconner.ora
initconner.ora: ASCII text
[oracle@jumper dbs]$ file spfileconner.ora
spfileconner.ora: data
```

SPFILE 的引入使得对于参数的修改都可以在命令行完成，我们可以彻底告别手工修改初始化参数文件的历史，这就大大减少了人为错误的发生。

使用 startup 命令启动数据库时，Oracle 将会按照以下顺序在默认目录（Windows 默认目录为%ORACLE_HOME%\database；Linux/UNIX 下默认目录为$ORACLE_HOME/dbs）中搜索参数文件：spfile<ORACLE_SID>.ora，spfile.ora 和 init<ORACLE_SID>.ora。

当在默认位置创建了 SPFILE 并重新启动数据库，Oracle 会按顺序搜索以上目录的 SPFILE。

如果想使用 PFILE 启动数据库，可以在启动时指定 PFILE 或者删除 SPFILE。通过指定 PFILE 启动数据库的命令格式类似如下：

```
SQL> startup pfile='E:\Oracle\admin\eyglen\pfile\init.ora';
```

不能以同样的方式指定 SPFILE，但是可以创建一个包含 SPFILE 参数的 PFILE 文件，指向 SPFILE。SPFILE 是一个自 Oracle 9i 引入的初始化参数，类似于 IFILE 参数，用于定义非默认路径下的 SPFILE 文件。使用时修改 PFILE 文件内容如下：

```
#Pfile link to SPFILE
SPFILE= 'E:\Oracle\Ora9iR2\database\SPFILEEYGLEN.ORA'
```

7.2.1 参数文件的创建

SPFILE 必须由 PFILE 创建，其语法如图 7-1 所示。

7.2 参数文件管理和使用

图 7-1

命令 CREATE SPFILE 需要 SYSDBA 或者 SYSOPER 的权限，注意其中的 MEMORY 选项是从 Oracle 11g 引入的。例如：

```
SQL> create spfile from pfile;
```

默认的，SPFILE 会创建到系统默认目录（$ORACLE_HOME/dbs）。如果 SPFILE 已经存在，那么创建会返回以下错误，这也可以用来判断当前是否使用了 SPFILE 文件。

```
SQL> create spfile from pfile;
ERROR 位于第 1 行:
  ORA-32002: 无法创建已由例程使用的 SPFILE
```

从 Oracle 11g 开始，为了增强参数文件的恢复，一个新的命令被引入用于从当前运行实例创建参数文件。这个命令如下：

```
create <spfile|pfile> from memory;
```

这个命令可以使用当前的参数设置在默认位置创建一个 SPFILE 文件，当然也可以指定一个不同的位置：

```
SQL> create spfile='/tmp/spfile.ora' from memory;
File created.
```

这一命令简化了我们在某些条件下的参数文件恢复。

当使用 DBCA 自定义（不使用模板）创建数据库时，在最后一个步骤，选择生成数据库创建脚本，可以将创建数据库所需要执行的脚本保存下来。通过这些脚本，可以进一步研究 Oracle 数据库的创建过程（当然也可以通过手工执行这些脚本，手工创建数据库），如图 7-2 所示。

图 7-2

以 Windows 为例，在 scripts 目录下，通常可以看到这样一些脚本（根据安装选项不同，脚本可能不同）：

```
C:\oracle\admin\eygle\scripts>dir
2005-01-06  13:23              918 CreateDB.sql
2005-01-06  13:23              631 CreateDBCatalog.sql
2005-01-06  13:23              134 CreateDBFiles.sql
2005-01-06  13:23              781 eygle.bat
2005-01-06  13:23            2,847 init.ora
2005-01-06  13:24              409 postDBCreation.sql
```

手工创建过程通常可以通过 eygle.bat 批处理文件执行开始，系统会根据脚本自动执行创建过程。在前面的章节已经描述过这个创建过程，和本章内容有关的是，这里存在一个 init.ora 文件（或 init.ora.<时间戳> 文件）。这个文件是根据创建数据库之前定义的参数自动生成的。该参数文件被用来在创建过程中启动数据库，通过 CreateDB.sql 可以看到这个引用：

```
startup nomount pfile="C:\oracle\admin\eygle\scripts\init.ora";
CREATE DATABASE eygle
MAXINSTANCES 1 MAXLOGHISTORY 1 MAXLOGFILES 5 MAXLOGMEMBERS 3 MAXDATAFILES 100
DATAFILE 'd:\oradata\eygle\system01.dbf' SIZE 250M REUSE AUTOEXTEND ON NEXT  10240K
MAXSIZE UNLIMITED EXTENT MANAGEMENT LOCAL
```

在数据库创建完成之后，Oracle 调用 postDBCreation.sql 脚本来进行一系列的后续处理，最后 Oracle 通过 init.ora 文件创建了 SPFILE 文件。该脚本的内容大致如下：

```
set echo on
create  spfile='D:\Oracle\11.2.0\database\spfileeygle.ora'  FROM  pfile='D:\Oracle\admin\eygle\scripts\init.ora';
shutdown immediate;
```

这就是数据库创建过程中 PFILE 和 SPFILE 的交接。建议每个试图深入学习 Oracle 的人都仔细研究一下自动建库的脚本，深入了解该过程非常有助于学习与领悟 Oracle。

除了第一次启动数据库需要 PFILE（然后可以根据 PFILE 创建 SPFILE），数据库可以不再需要 PFILE，Oracle 强烈推荐使用 SPFILE，应用其新特性来存储和维护初始化参数设置。

7.2.2　12c create spfile 的警示

在 12c 中，create spfile 命令又有了新的参数变更，引入了 as copy 选项，这个变化是由于一个 Bug 引入的。通过以下的测试和验证过程，大家会发现新版本中的这个变化，避免在新版本中遭遇陷阱。以下验证环境为 Oracle RAC 12.1.0.2.0，先记录当前 DB 的资源配置。

```
[oracle@rac12-node1 ~]$ srvctl config database -db rac12
Database unique name:rac12
Database name:rac12
Oracle home:/u01/app/oracle/product/12.1.0
Oracle user:oracle
Spfile:+DATA/rac12/spfilerac12.ora
```

```
Password file:+DATA/RAC12/PASSWORD/pwdrac12.276.902472499
Domain:
Start option:open
Stop option:immediate
Database role:PRIMARY
Management policy:AUTOMATIC
Server pools:racpool
Disk Groups:DATA
Mount point paths:
Services:racdb
Type:RAC
Start concurrency:
Stop concurrency:
OSDBA group:dba
OSOPER group:dba
Database instances:
Configured nodes:
Database is policy managed
```

对于 RAC 环境，一般都推荐使用共享的 SPFILE，方便维护初始化参数。下面的连续测试主要观察命令执行后对这个参数的影响。

首先测试生成 PFILE 或者 SPFILE，同时指定生成文件的位置，请注意后者直接导致了集群参数文件指向的变更。

```
SQL>create pfile='/tmp/ffile.ora' from spfile='+DATA/rac12/spfilerac12.ora';
File created. SQL>! srvctl config database -db rac12|grep -i 'spfile'
Spfile:+DATA/rac12/spfilerac12.ora
SQL>create spfile='/tmp/ffile.spfile' from pfile='/tmp/ffile.ora';
File created.
SQL>! srvctl config database -db rac12|grep -i 'spfile'
Spfile:/tmp/ffile.spfile
```

从内存生成 PFILE 或者 SPFILE，同时指定生成文件的位置，后者对于 SPFILE 同样更新了集群配置。

```
SQL> create pfile='/tmp/fmem.ora' from memory;
File created.
SQL>! srvctl config database -db rac12|grep -i 'spfile'
Spfile:/tmp/ffile.spfile
SQL>create spfile='/tmp/fmem.spfile' from memory;
File created.
SQL>! srvctl config database -db rac12|grep -i 'spfile'
Spfile:/tmp/fmem.spfile
```

从文件生成 PFILE 或者 SPFILE，不指定生成文件的位置。

```
 SQL> create pfile from spfile='DATA/rac12/spfilerac12.ora';
File created.
SQL>! srvctl config database -db rac12|grep -i 'spfile'
Spfile:/tmp/fmem.spfile
SQL>create spfile from pfile='/tmp/ffile.ora';
```

```
File created
SQL>! srvctl config database -db rac12|grep -i 'spfile'
Spfile:+DATA/spfilerac12_1.ora
```

指定生成文件位置，但源文件默认使用默认位置。

```
SQL> create pfile='/tmp/ffile2.ora' from spfile;
File created.
SQL>!srvctl config database -db rac12|grep -i 'spfile'
Spfile:+DATA/spfilerac12_1.ora
SQL>create spfile='/tmp/ffile2.spfile' from pfile;
File created.
SQL>! srvctl config database -db rac12|grep -i 'spfile'
Spfile:/tmp/ffile2.spfile
```

生成文件和源文件均使用默认位置。

```
SQL>create pfile from spfile;
File created.
SQL>! srvctl config database -db rac12|grep -i 'spfile'
Spfile:/tmp/ffile2.spfile
SQL>create spfile from pfile;
File created.
SQL>! srvctl config database -db rac12|grep -i 'spfile'
Spfile:+DATA/spfilerac12_1.ora
```

通过测试可见每一次生成 SPFILE，都同时更新了 **Database** 资源配置里面的 **SPFILE** 设定！

由于这个命令执行时没有任何提示会更新 Database 资源设定，所以很容易导致 SPFILE 的设定被更改到某个节点的本地文件系统，这样有可能会导致其他节点在重启动之后找不到指定的 SPFILE，从而启动失败。

幸运的是，通常 RAC 在安装完成后，在初始化参数的默认位置（$ORACLE_HOME/dbs）一般会创建一个 PFILE，里面用 SPFILE 参数指向了共享的 SPFILE。

```
[oracle@rac12-node3 ~] $ cd $ORACLE_HOME/dbs
[oracle@rac12-node3 dbs] $ ls
hc_rac12_3.dat id_rac12_3.dat init.ora initrac12_3.ora
[oracle@rac12-node3 dbs] $ cat initrac12_3.ora
SPFILE='+DATA/rac12/spfilerac12.ora'
[oracle@rac12-node3 dbs] $
```

这会如果不知情地执行了之前的创建操作，这会导致部分节点使用不同的 SPFILE：

```
SQL>select INSTANCE_NAME,NAME,VALUE from gv$parameter gp,gv$instance gi
  2  where gp.INST_ID=gi.INST_ID and gp.name='spfile';

INSTANCE_NAME   NAME       VALUE
-------------   --------   ---------------------------------------------
rac12_3         spfile     +DATA/rac12/spfilerac12.ora
rac12_1         spfile     /tmp/ffile.spfile
rac12_4         spfile     +DATA/rac12/spfilerac12.ora
```

在 MOS 网站上搜索，确认如下 Bug，Oracle 提供了补丁修正：

Bug 18799993 - CREATE SPFILE updates the DB resource by default as of 12.1（Doc ID 18799993.8）。

在以下 Bug 描述中，Oracle 详细阐述了这个问题，这是一个仅在 RAC 环境中出现的问题，并在补丁中提供了 AS COPY 选项。

```
As of 12c creating an spfile also updates the spfile location in the cluster. This is different
to 11.2 behaviour and can affect scripts that create a  local SPFILE that is not accessible
to other RAC nodes.
Rediscovery Notes  After an spfile is created, the spfile location is updated in the cluster.
Other nodes may then be unable to access the new spfile.
Workaround  None other than be sure to create SPFILE on a shared disk accessible to  all
nodes.
Note:  This fix extends the CREATE SPFILE syntax to add an "AS COPY" option.  If 'AS COPY'
is specified the cluster wide spfile location is not updated.
```

安装后进行简单测试。

```
SQL>create spfile='/tmp/aferpatch_ffile.spfile' from pfile='/tmp/ffile.ora';
File created.
SQL>! srvctl config database -db rac12|grep -i 'spfile'
Spfile:/tmp/aferpatch_ffile.spfile
SQL>create spfile='/tmp/aferpatch_fmen.spfile' from memory;
File created.
SQL>! srvctl config database -db rac12|grep -i 'spfile'
Spfile:/tmp/aferpatch_ffile.spfile
SQL>create spfile='/tmp/aferpatch_ffile2.spfile' from pfile;
File created.
SQL>! srvctl config database -db rac12|grep -i 'spfile'
Spfile:/tmp/aferpatch_ffile2.spfile
SQL>create spfile from pfile='/tmp/ffile.ora';
File created.
SQL>! srvctl config database -db rac12|grep -i 'spfile'
Spfile:/tmp/aferpatch_ffile2.spfile
SQL>create spfile from memory;
File created.
SQL>! srvctl config database -db rac12|grep -i 'spfile'
Spfile:/tmp/aferpatch_ffile2.spfile
SQL>create spfile from pfile;
File created.
SQL>! srvctl config database -db rac12|grep -i 'spfile'
Spfile:/tmp/aferpatch_ffile2.spfile
```

可以看到有一些改变，现在 create spfile from pfile 命令只有在指定生成文件路径才会更新 Database 资源配置，create spfile from memory 不再更新 Database 资源配置。

继续来检查 as copy 的使用情况。

```
SQL>create spfile='/tmp/aferpatch_ffile.spfile' from pfile='/tmp/ffile.ora' as copy;
File created.
SQL>! srvctl config database -db rac12|grep -i 'spfile'
```

```
Spfile:/tmp/aferpatch_ffile2.spfile
SQL>create spfile='/tmp/aferpatch_ffile.spfile' from pfile as copy;
File created.
SQL>! srvctl config database -db rac12|grep -i 'spfile'
Spfile:/tmp/aferpatch_ffile2.spfile
SQL>create spfile='/tmp/aferpatch_ffile.spfile' from memory as copy;
ERROR at line 1:
ORA-009333:SQL command not properly ended

SQL>create spfile from pfile='/tmp/ffile.ora' as copy;
File created.
SQL>! srvctl config database -db rac12|grep -i 'spfile'
Spfile:/tmp/aferpatch_ffile2.spfile
SQL>create spfile from pfile as copy;
File created.
SQL>! srvctl config database -db rac12|grep -i 'spfile'
Spfile:/tmp/aferpatch_ffile2.spfile
SQL>create spfile from memory as copy;
ERROR at line 1:
ORA-00933:SQL command not properly ended
```

可以看到 from memory 不支持 as copy 选项，同时加了 as copy 选项后，即使指定了 spfile 生成文件的路径，也不再更新 Database 资源配置。

通过以上测试和验证过程，得出以下结论。

- create spfile from memory：不支持 as copy 选项，但是也不再更新 Database 资源配置。
- create spfile from pfile：在指定生成文件路径而且不加 as copy 选项时，仍然会更新 Database 资源配置。

通过这个案例可以看出，一个新的版本变化，会改变很多数据库细节上的行为。如果不关注这些细节，就有可能在运维时遭遇困境。所以当我们使用一个新版本时，需要尽可能关注新特性，并保持对于数据库修正的持续跟踪。

7.3 12c 参数与参数文件新特性

在 Oracle Database 12c 中，由于 PDB 的引入，对 PDB 进行独立的参数管理成为一个现实的需求。CDB 中的 PDB 必须允许进行独立的参数配置与管理。

最基本的，每个独立的 PDB 都可以有独立的参数设置，所以在底层的 X$对象中记录的参数也按照 CON_ID 进行多条独立存储。通过如下的查询，可以看到针对每个独立的 PDB 都存在独立的参数设置（针对 PDB 的查询需要对以前的常用 SQL 进行适当改写）。

```
SQL> col ksppinm for a30
SQL> select con_id,ksppinm from x$ksppi where ksppinm='_allow_resetlogs_corruption';

   CON_ID KSPPINM
```

```
         1 _allow_resetlogs_corruption
         2 _allow_resetlogs_corruption
         3 _allow_resetlogs_corruption
```

7.3.1 参数表的引入

12c 中的参数文件是针对 CDB 的设置,对于 PDB 的参数设置不记录在 SPFILE 文件,以下测试过程可以验证。

首先修改一个参数设置。

```
SQL> select banner from v$version where rownum < 2;
BANNER
--------------------------------------------------------------------------------
Oracle Database 12c Enterprise Edition Release 12.1.0.1.0 - 64bit Production

SQL> show con_name
CON_NAME
------------------------------
CDB$ROOT

SQL> alter system set open_cursors=500 container=all;
System altered.

SQL> show parameter spfile
NAME                                 TYPE        VALUE
------------------------------------ ----------- ------------------------------
spfile                               string      /u01/app/oracle/product/ga/db_2/dbs/spfilemomo.ora
```

检查此时的参数文件,可以看到其中记录的修改参数。

```
SQL> ! strings /u01/app/oracle/product/ga/db_2/dbs/spfilemomo.ora
momo.__data_transfer_cache_size=0
momo.__db_cache_size=721420288
momo.__java_pool_size=16777216
momo.__large_pool_size=33554432
momo.__oracle_base='/u01/app/oracle'#ORACLE_BASE set from environment
momo.__pga_aggregate_target=570425344
momo.__sga_target=1090519040
momo.__shared_io_pool_size=50331648
momo.__shared_pool_size=251658240
momo.__streams_pool_size=0
*.compatible='12.1.0.0.0'
*.control_files='/u01/app/oracle/oradata/MOMO/controlfile/o1_mf_8qs9wxhs_.ctl'
*.db_block_size=8192
*.db_create_file_dest='/u01/app/oracle/oradata'
*.db_domain=''
*.db_name='momo'
*.db_recovery_file_dest='/u01/app/oracle/fast_recovery_area'
*.db_recovery_file_dest_size=10737418240
*.diagnostic_dest='/u01/app/oracle'
```

```
*.dispatchers='(PROTOCOL=TCP) (SERVICE=momoXDB)'
*.enable_pluggable_database=true
*.memory_target=1584m
*.open_cursors=500
.processes=300
*.remote_login_passwordfile='EXCLUSIVE'
*.undo_tablespace='UNDOTBS1'
```

如果修改 PDB 的参数，其修改并不会记录在参数文件中。

```
SQL> alter session set container=enmo;
Session altered.

SQL> alter system set open_cursors=2000;
System altered.

SQL> ! strings /u01/app/oracle/product/ga/db_2/dbs/spfilemomo.ora |grep open
*.open_cursors=500
```

那么对于 PDB 的参数的修改记录在何处呢？通过跟踪修改过程，可以解析 12c 中的这个变化。以下是跟踪过程。

```
SQL> show con_name

CON_NAME
------------------------------
ENMO
SQL> oradeBug setmypid
Statement processed.
SQL> oradeBug EVENT 10046 TRACE NAME CONTEXT FOREVER, LEVEL 12
Statement processed.
SQL> oradeBug TRACEFILE_NAME
/u01/app/oracle/diag/rdbms/momo/momo/trace/momo_ora_18582.trc
SQL> alter system set open_cursors=888;
System altered.

SQL> oradeBug EVENT 10046 trace name context off
Statement processed.
```

检查跟踪文件可以发现，修改参数最终被转化为对于底层参数表 PDB_SPFILE$的插入和更新过程。

```
=====================
PARSING IN CURSOR #140326162677680 len=33 dep=0 uid=0 oct=49 lid=0 tim=174451122363
hv=2037700194 ad='0' SQLid='94vufttwr9pm2'
    alter system set open_cur
END OF STMT
PARSE #140326162677680:c=0,e=197,p=0,cr=0,cu=0,mis=0,r=0,dep=0,og=0,plh=0,tim=174451122362
WAIT #140326162677680: nam='reliable message' ela= 92 channel context=3265888440 channel handle=3265759608 broadcast message=3268513048 obj#=-1 tim=174451122576
=====================
PARSING IN CURSOR #140326162675256 len=102 dep=1 uid=0 oct=2 lid=0 tim=174451123080
hv=4009187694 ad='bbbcf678' SQLid='0h3wm2vrgfqbf'
```

7.3 12c 参数与参数文件新特性

```
    insert into pdb_spfile$(db_uniq_name, pdb_uid, sid, name, value$, comment$)  values
(:1,:2,:3,:4,:5,:6)
    END OF STMT
    PARSE #140326162675256:c=0,e=422,p=0,cr=0,cu=0,mis=1,r=0,dep=1,og=4,plh=0,tim=174451123079
    BINDS #140326162675256:
     Bind#0
      oacdty=01 mxl=32(04) mxlc=00 mal=00 scl=00 pre=00
      oacflg=10 fl2=0001 frm=01 csi=178 siz=32 off=0
      kxsbbbfp=7fffc5764742  bln=32  avl=04  flg=09
      value="momo"
     Bind#1
      oacdty=02 mxl=22(22) mxlc=00 mal=00 scl=00 pre=00
      oacflg=00 fl2=1000001 frm=00 csi=00 siz=24 off=0
      kxsbbbfp=7fa03b070170  bln=22  avl=06  flg=05
      value=655250352
     Bind#2
      oacdty=01 mxl=32(01) mxlc=00 mal=00 scl=00 pre=00
      oacflg=10 fl2=0001 frm=01 csi=178 siz=32 off=0
      kxsbbbfp=7fffc57635f8  bln=32  avl=01  flg=09
      value="*"
     Bind#3
      oacdty=01 mxl=32(12) mxlc=00 mal=00 scl=00 pre=00
      oacflg=10 fl2=0001 frm=01 csi=178 siz=32 off=0
      kxsbbbfp=0bb8bad4  bln=32  avl=12  flg=09
      value="open_cursors"
     Bind#4
      oacdty=01 mxl=32(03) mxlc=00 mal=00 scl=00 pre=00
      oacflg=10 fl2=0001 frm=01 csi=178 siz=32 off=0
      kxsbbbfp=7fffc576364c  bln=32  avl=03  flg=09
      value="888"
     Bind#5
      oacdty=01 mxl=32(00) mxlc=00 mal=00 scl=00 pre=00
      oacflg=10 fl2=0001 frm=01 csi=178 siz=32 off=0
      kxsbbbfp=00000000  bln=32  avl=00  flg=09
    PARSING IN CURSOR #140326162672080 len=106 dep=1 uid=0 oct=6 lid=0 tim=174451133817
hv=502170820 ad='b8a9e008' SQLid='d8cf9hcfyx164'
    update pdb_spfile$ set value$=:5, comment$=:6  where  name=:1 and pdb_uid=:2 and
db_uniq_name=:3 and sid=:4
    END OF STMT
    PARSE #140326162672080:c=0,e=305,p=0,cr=0,cu=0,mis=1,r=0,dep=1,og=4,plh=0,tim=174451133816
    BINDS #140326162672080:
     Bind#0
      oacdty=01 mxl=32(03) mxlc=00 mal=00 scl=00 pre=00
      oacflg=10 fl2=0001 frm=01 csi=178 siz=32 off=0
      kxsbbbfp=7fffc576364c  bln=32  avl=03  flg=09
      value="888"
     Bind#1
      oacdty=01 mxl=32(00) mxlc=00 mal=00 scl=00 pre=00
      oacflg=10 fl2=0001 frm=01 csi=178 siz=32 off=0
      kxsbbbfp=00000000  bln=32  avl=00  flg=09
     Bind#2
      oacdty=01 mxl=32(12) mxlc=00 mal=00 scl=00 pre=00
```

```
 oacflg=10 fl2=0001 frm=01 csi=178 siz=32 off=0
 kxsbbbfp=0bb8bad4  bln=32  avl=12  flg=09
 value="open_cursors"
Bind#3
 oacdty=02 mxl=22(22) mxlc=00 mal=00 scl=00 pre=00
 oacflg=00 fl2=1000001 frm=00 csi=00 siz=24 off=0
 kxsbbbfp=7fa03b109338  bln=22  avl=06  flg=05
 value=655250352
Bind#4
 oacdty=01 mxl=32(04) mxlc=00 mal=00 scl=00 pre=00
 oacflg=10 fl2=0001 frm=01 csi=178 siz=32 off=0
 kxsbbbfp=7fffc5764742  bln=32  avl=04  flg=09
 value="momo"
Bind#5
 oacdty=01 mxl=32(01) mxlc=00 mal=00 scl=00 pre=00
 oacflg=10 fl2=0001 frm=01 csi=178 siz=32 off=0
 kxsbbbfp=7fffc57635f8  bln=32  avl=01  flg=09
 value="*"
```

也就是说，在 Oracle 12c 中，通过增加参数表 PDB_SPFILE$，Oracle 实现了对于不同 PDB 的参数设置区分。这些非默认的参数设置被记录在 CDB 中。

```
SQL> alter session set container=cdb$root;

Session altered.

SQL> select db_uniq_name,name,value$ from pdb_spfile$;
DB_UNIQ_NAME         NAME             VALUE$
--------------       --------------   -----------------
momo                 open_cursors     888
```

7.3.2 参数表在 PDB 启动中的作用

初始化参数在数据库实例启动时发挥作用，而对于 PDB 来说，参数表中的参数在 PDB 打开时发挥作用，跟踪 PDB 的 Open 过程，可以从后台看到 PDB_SPFILE$的调用过程。

如下步骤实现 PDB 的打开过程跟踪。

```
SQL> startup
ORACLE instance started.
Database mounted.
Database opened.
SQL> oradeBug setmypid
Statement processed.
SQL> oradeBug EVENT 10046 trace name context forever,level 12
Statement processed.
SQL> oradeBug tracefile_name
/u01/app/oracle/diag/rdbms/momo/momo/trace/momo_ora_18461.trc
SQL> alter pluggable database all open;
Pluggable database altered.

SQL> oradeBug event 10046 trace name context off;
Statement processed.
```

7.4 参数修改及重置

检查跟踪文件,可以看到该过程首先执行的即是参数表信息的读取,由此可见,参数表和参数文件发挥了完全类似的功能。

```
=====================
PARSING IN CURSOR #140418418360024 len=89 dep=1 uid=0 oct=3 lid=0 tim=173430521346
hv=1466406983 ad='ba282d08' SQLid='4sbjpw1bqg627'
select name, value$, comment$, sid from pdb_spfile$ where db_uniq_name=:1 and pdb_uid=:2
END OF STMT
PARSE #140418418360024:c=0,e=255,p=0,cr=0,cu=0,mis=1,r=0,dep=1,og=4,plh=0,tim=173430521346
BINDS #140418418360024:
 Bind#0
  oacdty=01 mxl=32(04) mxlc=00 mal=00 scl=00 pre=00
  oacflg=10 fl2=0001 frm=01 csi=178 siz=56 off=0
  kxsbbbfp=7fb5b6124d60  bln=32  avl=04  flg=05
  value="momo"
 Bind#1
  oacdty=02 mxl=22(22) mxlc=00 mal=00 scl=00 pre=00
  oacflg=00 fl2=1000001 frm=00 csi=00 siz=0 off=32
  kxsbbbfp=7fb5b6124d80  bln=22  avl=06  flg=01
```

7.4 参数修改及重置

用户可以通过 ALTER SYSTEM 或者导入、导出来更改 SPFILE 的内容。修改参数的完整命令如下:

```
alter system set <parameter_name> =<value> scope = memory|spfile|both [sid=<sid_name>]
```

SCOPE 参数有三个可选值:MEMORY、SPFILE 和 BOTH。

- MEMORY:只改变当前实例运行(对于动态参数而言),重新启动数据库后失效。
- SPFILE:只改变 SPFILE 的设置,不改变当前实例运行,重新启动数据库后生效。
- BOTH:同时改变实例及 SPFILE,当前更改立即生效,重新启动数据库后仍然有效,对于静态参数不能使用这个选项修改。

关于 SID 还需要注意的一点是:如果命令中没有指定 SID,那么意味着实例名为'*'。

针对 RAC 环境,ALTER SYSTEM 还可以指定 SID 参数,对不同实例进行不同设置。但是需要注意的是,即使指定的实例名不存在,系统也不会报任何的错误,这点对单实例同样适用。

```
[oracle@rac12-node1 ~]$ export ORACLE_SID=rac12_1
[oracle@rac12-node1 ~]$ SQLplus "/as sysdba"
SQL*Plus: Release 12.1.0.2.0 Production on Sun Apr 10 08:30:12 2016
Copyright (c) 1982, 2014, Oracle. All rights reserved.

SQL> alter system set open_cursors=500 sid='noexist';
System altered.
```

以下通过简单的例子来看一下 SCOPE 参数的几个用法。

1. SCOPE=MEMORY

修改当前实例的 DB_CACHE_ADVICE 参数为 OFF：

```
SQL> show parameter db_cache_ad
NAME                    TYPE        VALUE
----------------------- ----------- --------
db_cache_advice         string      ON
SQL> alter system set db_cache_advice=off scope=memory;
System altered.
SQL> show parameter db_cache_ad
NAME                    TYPE        VALUE
----------------------- ----------- --------
db_cache_advice         string      OFF
```

如果观察 alert_<sid>.log 文件，我们可以发现其中记录了如下一行。

```
Wed Apr 26 21:18:57 2006
ALTER SYSTEM SET db_cache_advice='OFF' SCOPE=MEMORY;
```

由于修改的参数没有写入参数文件，如果重新启动数据库，这个更改将会丢失。

```
SQL> startup force;
SQL> show parameter db_cache_ad
NAME                    TYPE       VALUE
----------------------- ---------- -------
db_cache_advice         string     ON
```

也就是说 SCOPE=MEMORY 的修改影响，不会跨越一次数据库的重新启动。

当使用 MEMORY 选项时，要明确地知道这样的修改仅对当前实例有效，忽略这一点则可能遭遇后续的故障。以下是实际生产系统中关于 SPFILE 的一个案例问题，数据库在重新启动时无法正常启动，检查发现 UNDO 表空间丢失。

此次故障诊断，首先检查告警日志文件，发现其中包含如下错误信息。

```
Thu Apr  1 11:11:28 2004
Errors in file /oracle/admin/gzhs/udump/gzhs_ora_27781.trc:
ORA-30012: undo tablespace 'UNDOTBS1' does not exist or of wrong type
Thu Apr  1 11:11:28 2004
Error 30012 happened during db open, shutting down database
USER: terminating instance due to error 30012
Instance terminated by USER, pid = 27781
ORA-1092 signalled during: alter database open...
```

在告警日志末尾显示了数据库在 OPEN 状态因为错误而异常终止。最后出错的错误号是 ORA-30012。该错误的含义如下。

```
[oracle@jumper oracle]$ oerr ora 30012
30012, 00000, "undo tablespace '%s' does not exist or of wrong type"
// *Cause:  the specified undo tablespace does not exist or of the
```

7.4 参数修改及重置

```
//            wrong type.
// *Action:  Correct the tablespace name and reissue the statement.
```

这说明是 UNDO 表空间不存在导致出现问题。既然是 UNDO 表空间丢失，接下来需要确认相关数据文件，看 UNDO 表空间数据文件是否存在。

```
bash-2.03$ cd /u01/ oradata/gzhs
bash-2.03$ ls -l *UNDO*
total 55702458
-rw-r-----   1 oracle    dba       1073750016 Apr  1 11:44 UNDOTBS2.dbf
```

通过检查发现存在文件 UNDOTBS2.dbf，大小约为 1GB。

既然存在一个 UNDO 表空间文件，用户又没有主动执行删除操作，那么极其可能是参数设置出了问题，将数据库启动到 MOUNT 状态，检查当前参数设置。

```
SQL> startup mount;
ORACLE 例程已经启动。
数据库装载完毕。
SQL> show parameter undo
NAME                                 TYPE         VALUE
------------------------------------ ----------- ------------
undo_management                      string       AUTO
undo_retention                       integer      10800
undo_suppress_errors                 boolean      FALSE
undo_tablespace                      string       UNDOTBS1
SQL> show parameter spfile
NAME                                 TYPE         VALUE
------------------------------------ ----------- ------------
spfile                               string
```

检查发现系统没有使用 SPFILE，初始化参数设置的 UNDO 表空间为 UNDOTBS1，数据库中存在的 UNDO 文件为 UNDOTBS2.dbf。由此可以判定，参数设置可能和数据库的实际情况不符。

告警日志文件中记录了对于数据库重要操作的信息，可以从中查找对于 UNDO 表空间的操作。除了数据库创建时建立的 UNDOTBS1，我们发现了创建 UNDOTBS2 的记录信息。

```
Wed Mar 24 20:20:58 2004
/* OracleOEM */ CREATE UNDO TABLESPACE "UNDOTBS2" DATAFILE '/u01/oradata/gzhs/UNDOTBS2.dbf'
SIZE 1024M AUTOEXTEND ON NEXT  100M MAXSIZE UNLIMITED
```

通过修改参数指定了新的 UNDO 表空间。

```
Wed Mar 24 20:24:25 2004
ALTER SYSTEM SET undo_tablespace='UNDOTBS2' SCOPE=MEMORY;
```

问题就在这里，创建了新的 UNDO 表空间以后，因为使用的是 PFILE 文件，切换表空间的修改只对当前实例生效，操作人员忘记了修改 PFILE 文件。

如果使用 SPFILE，默认的修改范围是 BOTH，会同时修改 SPFILE 文件，就可以避免以上问题的出现。

随后发现删除 UNDOTBS1 的信息。

```
Wed Mar 24 20:25:01 2004
/* OracleOEM */ DROP TABLESPACE "UNDOTBS1" INCLUDING CONTENTS  AND DATAFILES  CASCADE CONSTRAINTS
Wed Mar 24 20:25:03 2004
Deleted file /u01/oradata/gzhs/undotbs01.dbf
Completed: /* OracleOEM */ DROP TABLESPACE "UNDOTBS1" INCLUDI
```

这样再次重新启动数据库时，问题出现了，PFILE 中定义的 UNDOTBS1 找不到了，而且操作时间较久，没人能回忆起来。对于这个案例，找到了问题的根源，解决起来就简单了，修改 PFILE 参数文件，就可以启动数据库解决问题。在这里我们可以看到，使用 SPFILE 可以免去手工修改 PFILE 文件的麻烦，减少犯错的可能。

2. SCOPE=SPFILE

当指定 SCOPE=SPFILE 时，当前实例运行不受影响：

```
SQL> alter system set db_cache_advice=off scope=spfile;
System altered.
SQL> show parameter db_cache_ad
NAME                    TYPE        VALUE
------------------- ---------- --------
db_cache_advice         string      ON
```

同样可以从告警日志文件中看到这个修改：

```
Wed Apr 26 21:24:02 2006
ALTER SYSTEM SET db_cache_advice='OFF' SCOPE=SPFILE;
```

这个修改将在下次数据库启动后生效：

```
SQL> startup force;
SQL> show parameter db_cache_ad
NAME                 TYPE       VALUE
--------------- --------- --------
db_cache_advice     string      OFF
```

但需要知道的是，对于静态参数，只能指定 SCOPE=SPFILE 进行修改。通过 SCOPE=SPFILE 修改的参数，虽然对当前实例无效，但是其参数值可以从 V$SPPARAMETER 视图中查询得到：

```
SQL> show parameter  db_cache_advice
NAME                 TYPE       VALUE
--------------- --------- --------
db_cache_advice    string     OFF
SQL> alter system set db_cache_advice=on scope=spfile;
System altered.
SQL> select name,value from v$spparameter where name='db_cache_advice';
NAME                    VALUE
--------------- -----------
db_cache_advice         ON
```

```
SQL> show parameter db_cache_ad
NAME                 TYPE        VALUE
-------------------- ----------- ----------
db_cache_advice      string      OFF
```

3. SCOPE = BOTH

使用 BOTH 选项实际上等同于不带参数的 ALTER SYSTEM 语句。

```
SQL> alter system set db_cache_advice=off scope=both;
System altered.
SQL> alter system set db_cache_advice=off;
System altered.
SQL> show parameter db_cache_ad
NAME                 TYPE        VALUE
-------------------- ----------- ----------
db_cache_advice      string      OFF
```

在告警日志文件中可以看到如下信息:

```
Wed Apr 26 21:28:21 2006
ALTER SYSTEM SET db_cache_advice='OFF' SCOPE=BOTH;
Wed Apr 26 21:28:28 2006
ALTER SYSTEM SET db_cache_advice='OFF' SCOPE=BOTH;
```

注意到不带 SCOPE 参数和 SCOPE=BOTH 实际上是等价的。但是如果修改静态参数，那么需要指定 SPFILE 参数，不能指定 BOTH 参数，否则数据库将会报错。需要注意的一种情况是，如果数据库以 PFILE 启动，那么因为 SPFILE 无法使用，这个时候 SCOPE=MEMORY 是唯一选项，也意味着静态参数无法通过命令进行修改。

当我们想恢复某个参数为默认值时，可以使用如下命令:

```
alter system reset parameter <scope=memory|spfile|both> sid='sid|*'
```

该命令通常用于 RAC 环境中。在单实例环境中，在 Oracle10g 之前，需要指定 sid='*' 来 reset 一个参数，Oracle 将从 SPFILE 文件中去除该参数。

```
[oracle@jumper dbs]$ strings spfileconner.ora |grep open
*.open_cursors=150
[oracle@jumper dbs]$ SQLplus "/ as sysdba"
SQL> alter system reset open_cursors scope=spfile sid='*';
System altered.
[oracle@jumper dbs]$ strings spfileconner.ora |grep open
```

可以看到，reset 之后 SPFILE 文件中不再存在 OPEN_CURSORS 参数。

对于修改 SPFILE，需要清楚地认识到一点: 仅仅修改当前实例 SPFILE 里面对应实例的对应参数 (如果不指定实例名则实例名为'*')，如果找不到对应项，则报错。

```
SQL> alter system reset open_cursors;
System altered.
SQL> alter system reset open_cursors;
```

```
ERROR at line 1:
ORA-32010: cannot find entry to delete in SPFILE
```

7.4.1 解决参数文件的修改错误

在使用 SPFILE 之后,可能会遇到一些不同以往的错误。比如修改了错误的参数导致数据库无法启动,手工修改 SPFILE 导致参数文件损坏。

了解参数的优先级和生效原则,就可以就此修改和绕过错误。

- 实例级别的参数级别高于数据库层面的参数。即只有当实例级别未做设置时,在数据库级别的同一参数才会生效。
- 级别相同,那么后面出现的参数生效。

下面介绍参数修改错误常见问题的处理办法。比如修改 SGA_MAX_SIZE 超过系统最大内存数量。

```
SQL> alter system set sga_max_size=5G scope=spfile;
System altered.
```

那么下次启动,内存不足,数据库是无法启动的,数据库出现 ORA-27102 号错误。

```
SQL> shutdown immediate;
Database closed.
Database dismounted.
ORACLE instance shut down.
SQL> startup
ORA-27102: out of memory
```

如果是在 Linux/UNIX 上,可以在实例未启动时连接,创建 PFILE,然后手工修改 PFILE,用 PFILE 启动数据库即可。

```
[oracle@jumper oracle]$ SQLplus "/ as sysdba"
Connected to an idle instance.
SQL> create pfile from spfile;
File created.
```

也可以编辑一个参数文件,位置在$ORACLE_HOM/dbs(Windows 为 database 目录)下,包含如下两行。

```
[oracle@test126 dbs]$ cat initeygle.ora
SPFILE='/opt/oracle/product/10.2.0/dbs/spfileeygle.ora'
sga_max_size=1073741824
```

第一行指向 SPFILE,第二行写上出错的参数,给一个正确的值。这个值在实例启动时会覆盖之前错误的设置,然后就可以使用这个文件启动数据库实例了。

```
SQL> startup pfile=$ORACLE_HOME/dbs/initeygle.ora
ORACLE instance started.
Database mounted.
Database opened.
```

7.4.2 通过 event 事件来跟踪对参数文件的修改

Events 事件是 Oracle 的重要诊断工具及问题解决办法，通常需要通过 Events 设置来屏蔽或者更改 Oracle 的行为。下面我们来看一下怎样修改 SPFILE，增加 Events 事件设置。

```
SQL> alter system set event='10841 trace name context forever' scope=spfile;
System altered.
SQL> startup force;
SQL> show parameter event
NAME              TYPE      VALUE
----------------- --------- ----------
event             string    10841 trace name context forever
```

10841 事件是用于解决 Oracle 9i 中 JDBC Thin Driver 问题的一个方法，如果你的 alert.log 文件中出现以下错误提示。

```
Wed Jan  7 17:17:08 2004
Errors in file /opt/oracle/admin/phsdb/udump/phsdb_ora_1775.trc:
ORA-00600: internal error code, arguments: [ttcgcshnd-1], [0], [], [], [], [], [], []
Wed Jan  7 17:17:18 2004
Errors in file /opt/oracle/admin/phsdb/udump/phsdb_ora_1777.trc:
ORA-00600: internal error code, arguments: [ttcgcshnd-1], [0], [], [], [], [], [], []
```

那么你很可能是遇到了 Bug: 1725012。通过设置以上事件，可以屏蔽和解决这个 ORA-00600 错误，具体你可以参考 Metalink 相关文档。

如果想取消 event 参数设置，同样可以采用 reset 参数的方法。

```
SQL> show parameter event
NAME              TYPE      VALUE
----------------- --------- ---------
event             string    10046 trace name context forever,level 12
SQL> alter system reset event scope=spfile sid='*';
System altered.
SQL> startup force;
SQL> show parameter event
NAME              TYPE      VALUE
----------------- --------- --------
event             string
```

7.5 参数的查询

对于数据库设置的初始化参数，有很多查询和获取方法，以下由简入繁重点阐述查询和获得参数设置的方法。

7.5.1 参数查询的基本方式

Oracle 提供了多种可以查询初始化参数的方法，例如在 SQL*Plus 中通过 show parameters，show spparameters 查看，或者查询 v$parameter 等。其中，show spparameters 是 11g 中的新特性，直观地显示 v$spparameter 中参数的值，show parameters 的结果来自于 v$parameter。

```
SQL> show parameters sga
NAME                      TYPE         VALUE
------------------------- ------------ -------
lock_sga                  boolean      FALSE
pre_page_sga              boolean      FALSE
sga_max_size              big integer  900M
sga_target                big integer  900M
SQL> select name,value from v$parameter where name like '%SGA%';
NAME                      TYPE         VALUE
------------------------- ------------ -------
lock_sga                  boolean      FALSE
pre_page_sga              boolean      FALSE
sga_max_size              big integer  900M
sga_target                big integer  900M
```

使用 SQL_TRACE 跟踪当前会话，可以获得 show parameters 的内部操作，跟踪大致步骤如下。

```
alter session set SQL_trace=true;
show parameters sga
alter session set SQL_trace=false
```

找到跟踪文件，可以发现 SQL*Plus 的 show parameters 命令的本质是通过如下一条 SQL 查询得到的数据库参数。

```
SELECT NAME NAME_COL_PLUS_SHOW_PARAM,DECODE(TYPE,1,'boolean',2,'string',3,
   'integer',4,'file',5,'number',     6,'big integer', 'unknown') TYPE,
   DISPLAY_VALUE VALUE_COL_PLUS_SHOW_PARAM
FROM V$PARAMETER
WHERE UPPER(NAME) LIKE UPPER('%sga%') ORDER BY NAME_COL_PLUS_SHOW_PARAM,ROWNUM
```

show parameters 既然是从 V$PARAMETER 视图来查询参数设置，那么对应视图 GV$PARAMETER 的定义就决定了能够获得的内容输出。

通过 GV$PARAMETER 视图的创建语句，我们可以观察到，这个视图实际上是建立在两个底层数据字典表 X$KSPPI 和 X$KSPPCV 之上的。

通过以下查询我们可以从内部表直接获得所有参数及其描述信息：

```
SELECT x.ksppinm NAME, y.ksppstvl VALUE, x.KSPPDESC PDESC
FROM SYS.x$ksppi x, SYS.x$ksppcv y
WHERE x.indx = y.indx
   AND x.ksppinm LIKE '%&par%';
```

7.5 参数的查询

汇总 Oracle 获取初始化参数方法：SHOW PARAMETERS、SHOW SPPARAMETERS、CREATE PFILE、V$PARAMETER、V$PARAMETER2、V$SYSTEM_PARAMETER、V$SYSTEM_PARAMETER2、V$SPPARAMETER。

（1）SHOW PARAMETERS 是 SQL*Plus 工具提供的查询初始化参数的方法，用于查询当前会话生效的初始化参数。

（2）SHOW SPPARAMETERS 也是 SQL*Plus 工具提供的方法，用于查询当前会话生效的 SPFILE 参数包含的初始化参数，这个命令在 11g 以后 SQLplus 版本中有效。

（3）CREATE PFILE 命令可以将 SPFILE 中或当前内存中设置的初始化文件保存到 PFILE 文件中，然后可以通过文本编辑工具直观地看到 SPFILE 中或当前内存中设置了哪些初始化参数。这种方法虽然看上去比较麻烦，但是列出的参数都是用户设置的参数，所有默认值的参数并不会列出来，因此看到的结果更直观。在 11g 以后的版本允许 CREATE PFILE FROM MEMORY。

（4）V$PARAMETER 视图提供了当前会话可见的初始化参数的设置，可以查询 RAC 数据库的所有实例的设置。

● V$PARAMETER2 视图和 V$PARAMETER 相似，唯一的区别在于对于包括多值的初始化参数，从这个视图会返回多条记录，每条记录对应一个值。同样的，对于 RAC 环境可以查询 GV$PARAMETER2 视图。

（5）V$SYSTEM_PARAMETER 视图记录当前实例生效的初始化参数设置。注意这里是实例生效而不是会话生效。同样，GV$SYSTEM_PARAMETER 则包含了所有实例生效的初始化参数信息。

● V$SYSTEM_PARAMETER2 视图与 V$SYSTEM_PARAMETER 视图的关系类似于 V$PARAMETER2 视图与 V$PARAMETER 视图的关系，都是对于包含多个值的参数采用了分行处理的方式。

（6）V$SPPARAMETER 记录了来自 SPFILE 文件中初始化参数。如果 SPFILE 文件中没有设置参数，则字段 ISSPECIFIED 对应的值为 FALSE。同样可以查询 GVSPPARAMETER 参数来显示 RAC 环境所有实例的设置。

按照参数的生效级别分类

按照参数的生效级别分类，可以将参数分为当前 session 级别参数、当前 system 参数和重启生效参数。

（1）session 级别的视图：v$parameter、gv$parameter. v$parameter2 和 gv$parameter2。

- v$parameter 显示的是当前 session 级的参数。如果没有使用 alter session 单独设置当前 session 的参数值,每一个新 Session 都是从 v$system_parameter 上取得系统的当前值而产生 Session 的 v$parameter view。与 v$parameter 之间的区别则在于 v$parameter2 把 LIST 的值分开来了,一行变多行数据,用 ORDINAL 来指示相对的位置。

- session 级别的参数的优先级最高,可以通过 alter session 改变当前会话的动态初始化参数值。

(2) system 级别的视图:v$system_parameter、v$system_parameter、v$system_parameter4、gv$system_parameter、gv$system_parameter2 和 gv$system_parameter4。

- system 级别即当前实例生效的参数。修改 system/实例级别的参数一般使用 alter system set scope=memory/both 命令。同样,当只用 scope=memory 时,只能修改动态初始化参数。

- 通常情况下,v$parameter 与 v$system_parameter 查询的的值相同,但若使用了延迟参数 defer 修改参数,则 v$parameter 的查询结果为当前会话的值,而 v$system_parameter 查询的结果则针对所有新建的连接生效。另外 v$system_parameter4 记录的是当前数据库级别所有用户设置的初始化参数。当使用 create pfile/spfile from memory 生成参数文件时,此时参数来源于 v$system_parameter4。

(3) 重启生效参数视图:v$spparameter 和 gv$spparameter。

- v$spparameter 显示的就是保存在 spfile 中的参数值(scope=both 或者 spfile),记录了来自 SPFILE 文件中初始化参数。如果在 SPFILE 文件中没有设置参数,则字段 ISSPECIFIED 对应的值为 FALSE。同样可以查询 GV$SPPARAMETER 参数来显示 RAC 环境所有实例的设置。

7.5.2 参数值的可选项

Oracle 的很多参数具有多个不同的可选值,可以通过 V$PARAMETER_VALID_VALUES 来查询,例如以下查询获得 CURSOR_SHARING 参数的 3 个可选设置。

```
SQL> select * from V$PARAMETER_VALID_VALUES where name like '%cursor%';
NUM NAME              ORDINAL    VALUE      ISDEFAULT
--- ----------------- ---------- ---------- -----------
901 cursor_sharing    1          FORCE      FALSE
901 cursor_sharing    2          EXACT      TRUE
901 cursor_sharing    3          SIMILAR    FALSE
```

这个视图是基于 X$KSPVLD_VALUES 建立起来的,可以通过查询 X$视图直接获得这些设置选项。

```
SQL> SELECT
  2    INST_ID,
  3    PARNO_KSPVLD_VALUES      pvalid_par#,
  4    NAME_KSPVLD_VALUES       pvalid_name,
  5    VALUE_KSPVLD_VALUES      pvalid_value,
```

```
    6    DECODE(ISDEFAULT_KSPVLD_VALUES, 'FALSE', '', 'DEFAULT' ) pvalid_default
    7  FROM X$KSPVLD_VALUES
    8  WHERE LOWER(NAME_KSPVLD_VALUES) LIKE LOWER('%&1%')
    9  ORDER BY
   10     pvalid_par#,pvalid_default,pvalid_Value;
Enter value for 1: cursor

   INST_ID   PAR#  PARAMETER                              VALUE       DEFAULT
---------- ------ -------------------------------------- ---------- ---------
         1    901 cursor_sharing                          EXACT      DEFAULT
         1        cursor_sharing                          FORCE
         1        cursor_sharing                          SIMILAR
         1   1003 _optimizer_extended_cursor_sharing      NONE
         1        _optimizer_extended_cursor_sharing      UDO
```

当修改某些参数提供的参数值出现错误时，数据库会将可选的正确参数值抛出，也可以使用这个方法获得参数的可选值，以下测试返回了两个常见参数的设置值。

```
SQL> alter system set optimizer_features_enable='';
ERROR at line 1:
ORA-00096: invalid value  for parameter optimizer_features_enable, must be from
among 12.1.0.1.1, 12.1.0.1, 11.2.0.4, 11.2.0.3, 11.2.0.2, 11.2.0.1, 11.1.0.7,
11.1.0.6, 10.2.0.5, 10.2.0.4, 10.2.0.3, 10.2.0.2, 10.2.0.1, 10.1.0.5, 10.1.0.4,
10.1.0.3, 10.1.0, 9.2.0.8, 9.2.0, 9.0.1, 9.0.0, 8.1.7, 8.1.6, 8.1.5, 8.1.4,
8.1.3, 8.1.0, 8.0.7, 8.0.6, 8.0.5, 8.0.4, 8.0.3, 8.0.0
SQL> alter system set cursor_sharing='';
ERROR at line 1:
ORA-00096: invalid value  for parameter cursor_sharing, must be from among SIMILAR, EXACT, FORCE
```

7.6　不同查询方法之间的区别

获取参数的方法很多，但各种方法之间有很大的区别。

7.6.1　V$PARAMETER 和 V$PARAMETER2 的区别

首先看一下 V$PARAMETER 和 V$PARAMETER2 的区别，这个区别同样适用于 V$SYSTEM_PARAMETER 和 V$SYSTEM_PARAMETER2。

```
SQL> SELECT NAME, VALUE FROM V$PARAMETER
  2  MINUS
  3  SELECT NAME, VALUE FROM V$PARAMETER2;
NAME                      VALUE
------------------------- ---------------------------------------
control_files             E:\ORACLE\ORADATA\YTK102\CONTROL01.CTL,
                          E:\ORACLE\ORADATA\YTK102\CONTROL02.CTL,
                          E:\ORACLE\ORADATA\YTK102\CONTROL03.CTL
SQL> SELECT NAME, VALUE FROM V$PARAMETER2
  2  MINUS
  3  SELECT NAME, VALUE FROM V$PARAMETER;
NAME                      VALUE
------------------------- ---------------------------------------
```

```
control_files                    E:\ORACLE\ORADATA\YTK102\CONTROL01.CTL
control_files                    E:\ORACLE\ORADATA\YTK102\CONTROL02.CTL
control_files                    E:\ORACLE\ORADATA\YTK102\CONTROL03.CTL
```

7.6.2　V$PARAMETER 和 V$SYSTEM_PARAMETER 的区别

一般在查询初始化参数的时候都习惯性地使用 SHOW PARAMETER，也就是查询 V$PARAMETER 视图，但有时这样得到的结果并不准确。

```
SQL> show parameter query_rewrite_enabled
NAME                        TYPE        VALUE
--------------------------- ----------- ---------
query_rewrite_enabled       string      TRUE
SQL> select name, value
  2  from v$parameter where name = 'query_rewrite_enabled';
NAME                        VALUE
--------------------------- ----------
query_rewrite_enabled       TRUE
SQL> select name, value
  2  from v$system_parameter where name = 'query_rewrite_enabled';
NAME                        VALUE
--------------------------- --------
query_rewrite_enabled       TRUE
```

此时如果在会话级修改初始化参数 query_rewrite_enabled。

```
SQL> alter session set query_rewrite_enabled = false;
会话已更改。
SQL> show parameter query_rewrite_enabled
NAME                        TYPE        VALUE
--------------------------- ----------- ---------
query_rewrite_enabled       string      FALSE
SQL> select name, value
  2  from v$parameter where name = 'query_rewrite_enabled';
NAME                        VALUE
--------------------------- --------
query_rewrite_enabled       FALSE
SQL> select name, value
  2  from v$system_parameter where name = 'query_rewrite_enabled';
NAME                        VALUE
--------------------------- --------
query_rewrite_enabled       TRUE
```

可以看到，show parameter 和查询 v$parameter 视图的结果都是 FALSE，而刚才做的修改只是会话级，并没有修改系统的初始化参数。V$PARAMETER 视图反映的是初始化参数在当前会话中生效的值，而 V$SYSTEM_PARAMETER 反映的才是实例级上的初始化参数。

再来看延迟参数修改的情况。

```
SQL> select name, value from v$parameter where name = 'recyclebin';
NAME               VALUE
------------------ -----------
```

7.6 不同查询方法之间的区别

```
                       recyclebin          on
SQL> select name, value from v$system_parameter where name = 'recyclebin';
NAME                    VALUE
---------------         ------
recyclebin              on
SQL> alter system set recyclebin = off deferred scope = memory;
系统已更改。
SQL> select name, value from v$parameter where name = 'recyclebin';
NAME                    VALUE
---------------         --------
recyclebin              on
SQL> select name, value from v$system_parameter where name = 'recyclebin';
NAME                    VALUE
---------------         --------
recyclebin              OFF
```

结果和前面的恰好反过来,v$parameter 视图中的结果没有改变,而 v$system_parameter 视图的结果变成了 OFF。这是因为延迟修改对数据库中当前存在的会话不生效,因此反映当前会话情况的 v$parameter 视图结果不变。而对于系统而言,初始化参数已经改变,所有新建会话的参数也会改变,所以 v$system_parameter 视图的结果发生了改变。

```
SQL> CONN YANGTK/YANGTK@YTK111
已连接。
SQL> select name, value from v$parameter where name = 'recyclebin';
NAME                    VALUE
----------------        ---------
recyclebin              OFF
SQL> select name, value from v$system_parameter where name = 'recyclebin';
NAME                    VALUE
----------------        ---------
recyclebin              OFF
```

根据这两个例子,利用 V$PARAMETER 视图获取系统的启动初始化参数是不准确的,应该从 V$SYSTEM_PARAMETER 视图来获取。Oracle 在视图 V$SYSTEM_PARAMETER 中提供了一个列 ISDEFAULT,表示当前设置的值是否是数据库的默认值。

```
SQL> select name, value, isdefault
  2  from v$system_parameter where name = 'open_cursors';
NAME                    VALUE                   ISDEFAULT
----------------        ----------------------  ----------------
open_cursors            400                     FALSE
SQL> select isdefault, count(*)
  2  from v$system_parameter group by isdefault;
ISDEFAULT   COUNT(*)
---------   ----------
TRUE        267
FALSE       22
```

根据这个结果可以看到,数据库中绝大部分的初始化参数设置都是默认值。

```
SQL> select name, value, isdefault
  2  from v$system_parameter where name = 'undo_retention';
```

```
NAME                   VALUE              ISDEFAULT
-------------------    ---------------    ---------------
undo_retention         900                TRUE
SQL> select sid, name, value
  2  from v$spparameter where name = 'undo_retention';
SID    NAME                       VALUE
----   --------------------       ------------
*      undo_retention
SQL>   alter system set undo_retention = 900;
系统已更改。
SQL> select name, value, isdefault
  2  from v$system_parameter where name = 'undo_retention';
NAME                   VALUE              ISDEFAULT
-----------------      ---------------------   ------------
undo_retention         900                     TRUE
SQL> select sid, name, value
  2  from v$spparameter where name = 'undo_retention';
SID    NAME                       VALUE
----   ---------------------      ----------
*      undo_retention             900
```

对于手工设置的初始化参数与系统默认值相同的情况，通过 v$system_parameter 视图是无法区分的。实际上查询 V$SYSTEM_PARAMETER4 视图就可以获取到所有用户设置的初始化参数。

当数据库执行 CREATE PFILE FROM MEMORY 命令时，Oracle 创建 PFILE 的数据源是 V$SYSTEM_PARAMETER4 视图。

7.6.3　GV$SPPARAMETER 和 V$SPPARAMETER 的区别

这里有一个问题：GV$SPPARAMETER 是否有意义。因为 V$SPPARAMETER 参数本身包含了 SID 列，SPFILE 中本身包含了所有实例的设置，那么查询 GV$SPPARAMETER 视图是否意义不大呢？其实不然。

因为 RAC 的各个节点可以使用统一的 SPFILE 启动，同样也可以选择不同的 SPFILE 进行启动，这时 GV$SPPARAMETER 视图中获取的结果，才是各个实例 SPFILE 中设置的结果。

这样说比较难以理解，下面来看一个简单的例子。

```
SQL> select inst_id, name, value
  2  from gv$system_parameter where name = 'open_cursors';
   INST_ID    NAME                  VALUE
   --------   -----------------     ------------
         1    open_cursors          600
         2    open_cursors          400
SQL> select sid, name, value
  2  from v$spparameter where name = 'open_cursors';
SID      NAME                 VALUE
------   -----------------    ------------
```

7.6 不同查询方法之间的区别

```
*         open_cursors      300
test1     open_cursors      500
test2     open_cursors      700
SQL> select inst_id, sid, name, value
  2  from gv$spparameter where name = 'open_cursors';
 INST_ID SID       NAME              VALUE
-------- --------  ---------------- ----------
       1  *         open_cursors      300
       1  test1     open_cursors      500
       1  test2     open_cursors      700
       2  *         open_cursors      300
       2  test1     open_cursors      500
       2  test2     open_cursors      700
SQL> select inst_id, name, value
  2  from gv$system_parameter where name = 'spfile';
   INST_ID NAME             VALUE
-------- --------------- ---------------------------
       1  spfile           +DATA/test/spfiletest.ora
       2  spfile           +DATA/test/spfiletest.ora
```

由内存中参数创建 SPFILE，并利用新建的 SPFILE 启动当前实例。

```
SQL> create spfile='/export/home/oracle/spfiletest1.ora' from memory;

文件已创建。
SQL> host
$ vi /export/home/oracle/inittest1.ora
"/export/home/oracle/inittest1.ora" [New file]
spfile=/export/home/oracle/spfiletest1.ora
"/export/home/oracle/inittest1.ora" [New file] 2 lines, 44 characters
$ exit
SQL> shutdown immediate

数据库已经关闭。
已经卸载数据库。
ORACLE 例程已经关闭。
SQL> startup pfile=/export/home/oracle/inittest1.ora
ORACLE 例程已经启动。
数据库装载完毕。
数据库已经打开。
```

检查 spfile 中的设置。

```
SQL> select inst_id, name, value
  2  from gv$system_parameter where name = 'spfile';
   INST_ID NAME             VALUE
---------- ---------------  ---------------------------------------
         1 spfile           /export/home/oracle/spfiletest1.ora
         2 spfile           +DATA/test/spfiletest.ora
SQL> select inst_id, name, value
  2  from gv$system_parameter where name = 'open_cursors';
   INST_ID NAME             VALUE
---------- ---------------- ---------------------------
         1 open_cursors     600
         2 open_cursors     400
SQL> select sid, name, value
  2  from v$spparameter where name = 'open_cursors';
SID       NAME              VALUE
```

```
test1         open_cursors          600
test2         open_cursors          400
SQL> select inst_id, sid, name, value
  2  from gv$spparameter where name = 'open_cursors';
   INST_ID SID        NAME                           VALUE
---------- ---------- ------------------------------ --------
         2 *          open_cursors                   300
         2 test1      open_cursors                   500
         2 test2      open_cursors                   700
         1 test1      open_cursors                   600
         1 test2      open_cursors                   400
```

可以看到，由于两个实例采用了不同的 SPFILE，导致两个实例上设置的对方实例的初始化参数值，与对方实例上当前设置值不符。

在上面的例子中，两个实例上真正的参数设置查询方式如下：

```
SQL> select inst_id, sid, name, value
  2  from gv$spparameter
  3  where name = 'open_cursors' and substr(sid, -1) = to_char(inst_id);
   INST_ID SID        NAME                              VALUE
---------- ---------- --------------------------------- -----------------------------------
         2 test2      open_cursors                      700
         1 test1      open_cursors                      600
```

7.7　RAC 下参数的维护

在 RAC 环境下，由于多节点不同实例在启动时都需要依赖参数文件，所以其管理更加复杂。本节就 RAC 下参数文件管理进行阐述。

7.7.1　RAC 下共享 spfile

RAC 环境下数据库在启动时，首先尝试寻找 Cluster 里面 Database 资源的 Spfile 配置选项。如果找不到对应的文件，那么继续按照单实例的寻找顺序在默认位置查找。

建议在 RAC 环境下使用共享的 SPFILE，并在默认位置保留一个 PFILE，里面通过 SPFILE 参数指向共享的 SPFILE。默认在 RAC 安装配置完成，就自动生成了一个 PFILE 文件。

下面是 RAC 环境中一个参数文件的设置范例。

```
[oracle@raclinux1 ~]$ cd $ORACLE_HOME/dbs
[oracle@raclinux1 dbs]$ more initRACDB1.ora
SPFILE='+MY_DG2/RACDB/spfileRACDB.ora'
```

在此环境中，需要谨慎使用 create spfile from pfile 的命令，很多朋友因为草率地执行这样的操作而导致数据库故障。

在 ASM 或 RAC 环境中，通常的 init<sid>.ora 文件中只有如上示例的一行，如果此时执行 create spfile from pfile 命令，则新创建的 SPFILE 文件将也只有这样一行信息，数据库将无法启动。

7.7.2 使用 ASM 存储参数文件

在 ASM 环境中，参数文件可以存储在 ASM 磁盘组上，而在 Oracle RAC 环境中，默认使用存储在 ASM 上的参数文件，在维护 RAC 环境的参数文件时要格外谨慎。

以下是一个测试过程，用于指导大家如何将参数文件转移到 ASM 存储并使之生效。

首先检查参数文件的位置，并通过 SPFILE 创建一个 PFILE 文件，进而在 ASM 磁盘上创建 SPFILE 文件。

```
SQL> connect / as sysdba
SQL> show parameter spfile
NAME      TYPE     VALUE
------    -------  ------------------------------
spfile    string   /oracle/product/11.2.0/db_1/dbs/spfileracdb11.ora

SQL> create pfile from spfile
File created.

SQL> create spfile='+RACDB_DATA' from pfile='/oracle/product/11.2.0/db_1/dbs/initracdb11.ora';
File created.
```

检查 ASM 上的参数文件。

```
[grid@rac1 ~]$ asmcmd

ASMCMD> ls RACDB_DATA/racdb1/spfile*
spfileracdb1.ora
```

同步 RAC 两个节点上的参数文件，更改其内容，设置 SPFILE 参数指向 ASM 中的参数文件。

```
[oracle@rac1   dbs]$    echo    "SPFILE='+RACDB_DATA/racdb1/spfileracdb1.ora'"    >
/oracle/product/11.2.0/db_1/dbs/initracdb11.ora

[oracle@rac1 dbs]$ ssh rac2 "echo \"SPFILE='+RACDB_DATA/racdb1/spfileracdb1.ora'\" > /oracle/product/11.2.0/db_1/dbs/initracdb12.ora"
```

通过 srvctl 修改 OCR 中关于参数文件的配置。

`[oracle@rac1 dbs]$ srvctl modify database -d racdb1 -p +RACDB_DATA/racdb1/spfileracdb1.ora`

现在通过 CRS 启动数据库，将不再需要 dbs 目录下的参数文件，可以将其移除。

`[oracle@rac1 dbs]$ mv /oracle/product/11.2.0/db_1/dbs/spfileracdb11.ora /oracle/product/11.2.0/db_1/dbs/spfileracdb11.ora_bak`

`[oracle@rac1   dbs]$   ssh  rac2  "mv  /oracle/product/11.2.0/db_1/dbs/spfileracdb12.ora /oracle/product/11.2.0/db_1/dbs/spfileracdb12.ora_bak"`

在下次重新启动数据库时，新的配置将会生效。

```
[oracle@rac1 dbs]$ srvctl stop database -d racdb1
[oracle@rac1 dbs]$ srvctl start database -d racdb1

[oracle@rac1 dbs]$ srvctl status database -d racdb1
Instance racdb11 is running on node rac1
Instance racdb12 is running on node rac2
```

检查数据库，新的参数文件已经被使用和生效。

```
SQL> SHOW parameter spfile
NAME   TYPE   VALUE
------ ------ -----------------------------
spfile string +RACDB_DATA/racdb1/spfileracdb1.ora
```

7.7.3 谨慎修改 RAC 参数

在 RAC 环境中，即使按照正常的方式修改参数，也有可能因为遭遇 Bug 而导致事故，所以在进行重要的环境变更前，一定要进行测试，并详细检查与变更有关的文件，确保变更不会引起错误。

在 Oracle 10.1 的版本中，会遇到这样的问题，在 RAC 环境下修改 UNDO_RETENTION 参数，使用如下命令：

```
alter system set undo_retention=18000 sid='*';
```

这条命令直接导致了 RAC 的其他节点挂起，Oracle 记录了一个相关 Bug，Bug 号为：4220405（这个 Bug 在 Oracle 10gR2 中修正），其 Workaround 就是分别修改不同实例。

```
alter system set undo_retention=18000 sid='RAC1';
alter system set undo_retention=18000 sid='RAC2';
alter system set undo_retention=18000 sid='RAC3';
```

这个案例告诉我们，Bug 无处不在，数据库调整应当极其谨慎，最好在测试环境中测试过再应用到生产环境。

再次重申，在 RAC 环境中，每一个维护操作都要相当谨慎！

7.7.4 RAC 环境下初始化参数的查询方法

下面介绍 RAC 环境下初始化参数的查询方法。

一个简单的例子：

```
SQL> show parameter open_cursors
NAME                 TYPE        VALUE
-------------------- ----------- --------
open_cursors         integer     300
SQL> alter system set open_cursors = 500 scope = both sid = 'test1';
系统已更改。
```

7.7 RAC 下参数的维护

```
SQL> disc
从 Oracle Database 11g Enterprise Edition Release 11.1.0.6.0 - 64bit Production
With the Partitioning, Real Application Clusters, OLAP, Data Mining and Real Application
Testing options 断开
SQL> set instance test2
Oracle Database 11g Release 11.1.0.0.0 - Production
SQL> conn sys as sysdba
输入口令:
已连接。
SQL> alter system set open_cursors = 400 scope = both sid = 'test2';
系统已更改。
SQL> disc
从 Oracle Database 11g Enterprise Edition Release 11.1.0.6.0 - 64bit Production
With the Partitioning, Real Application Clusters, OLAP, Data Mining
and Real Application Testing options 断开
SQL> set instance local
Oracle Database 11g Release 11.1.0.0.0 - Production
SQL> conn / as sysdba
已连接。
```

不同的查询方法得到的结果。

```
SQL> select name, value
  2  from v$parameter where name = 'open_cursors';
NAME                    VALUE
----------------     ---------
open_cursors             500
SQL> select inst_id, name, value
  2  from gv$parameter where name = 'open_cursors';
   INST_ID NAME              VALUE
---------- ---------------   -------
         1 open_cursors        500
         2 open_cursors        400
SQL> show parameter open_cursors
NAME                 TYPE        VALUE
----------------     ---------   --------
open_cursors         integer     500
SQL> select sid, name, value
  2  from v$spparameter where name = 'open_cursors';
SID         NAME                       VALUE
---------   ------------------         ----------
*           open_cursors                300
test1       open_cursors                500
test2       open_cursors                400
SQL> show spparameter open_cursors
SID         NAME                TYPE         VALUE
--------    ----------------    -----------  -------
*           open_cursors        integer       300
test2       open_cursors        integer       400
test1       open_cursors        integer       500
```

似乎除了看不到全局设置外,GV$PARAMETER 参数和 V$SPPARAMETER 没有什么不同,其实不然,如果 alter system set 的时候只修改了 spfile 或者 memory 参数,结果就会不同。

```
SQL> alter system set open_cursors = 600 scope = memory sid = 'test1';
系统已更改。
SQL> alter system set open_cursors = 700 scope = spfile sid = 'test2';
系统已更改。
SQL> select name, value
  2  from v$parameter where name = 'open_cursors';
NAME            VALUE
--------------- -------
open_cursors    600
SQL> select inst_id, name, value
  2  from gv$parameter where name = 'open_cursors';
   INST_ID NAME              VALUE
---------- ---------------   --------
         1 open_cursors      600
         2 open_cursors      400
SQL> select sid, name, value
  2  from v$spparameter where name = 'open_cursors';
SID        NAME              VALUE
---------- ----------------  ------
*          open_cursors      300
test1      open_cursors      500
test2      open_cursors      700
```

从上面的对比可以看出，通过 GV$ 视图访问的结果和 SPFILE 中包含的信息完全不同。除了上面介绍的几种视图之外，CREATE PFILE 也是一个不错的选择。

Oracle 把 SPFILE 也纳入到 RMAN 的备份恢复策略当中，如果你配置了控制文件自动备份（AUTOBACKUP），那么 Oracle 会在数据库发生重大变化（如增减表空间）时自动进行控制文件及 SPFILE 文件的备份。

7.8 参数文件备份

（1）在 RMAN 命令行，通过 show all 检查备份的默认设置。

```
[oracle@enmoedu backup]$ rman target /
Recovery Manager: Release 12.1.0.2.0 - Production on Wed Sep 21 14:03:02 2016
Copyright (c) 1982, 2014, Oracle and/or its affiliates.  All rights reserved.
connected to target database: PROD1 (DBID=2137673358)

RMAN> SHOW ALL;

using target database control file instead of recovery catalog
RMAN configuration parameters for database with db_unique_name PROD1 are:
CONFIGURE RETENTION POLICY TO REDUNDANCY 1; # default
CONFIGURE BACKUP OPTIMIZATION OFF; # default
CONFIGURE DEFAULT DEVICE TYPE TO DISK; # default
CONFIGURE CONTROLFILE AUTOBACKUP OFF; # default
```

```
CONFIGURE CONTROLFILE AUTOBACKUP FORMAT FOR DEVICE TYPE DISK TO '%F'; # default
CONFIGURE SNAPSHOT CONTROLFILE NAME TO '/oracle/app/oracle/product/12.1.0/prod_1/dbs/
snapcf_PROD1.f'; # default
```

这个设置可以在数据库中通过以下方式查询得到。

```
SQL> select * from v$rman_configuration;

    CONF# NAME                          VALU   CON_ID
---------- ------------------------- ---- ----------
        1 CONTROLFILE AUTOBACKUP        OFF          0
```

（2）默认是没有开启自动备份，现在进行设置。

```
RMAN> CONFIGURE CONTROLFILE AUTOBACKUP ON;

new RMAN configuration parameters:
CONFIGURE CONTROLFILE AUTOBACKUP ON;
new RMAN configuration parameters are successfully stored
CONFIGURE CONTROLFILE AUTOBACKUP ON;
new RMAN configuration parameters are successfully stored
```

（3）指定自动备份的位置。

```
RMAN> CONFIGURE CONTROLFILE AUTOBACKUP FORMAT FOR DEVICE TYPE DISK TO '/backup/rman_
test/control_%F';

new RMAN configuration parameters:
CONFIGURE CONTROLFILE AUTOBACKUP FORMAT FOR DEVICE TYPE DISK TO '/backup/
rman_test/control_%F';
new RMAN configuration parameters are successfully stored
```

（4）检查自动备份。

自动备份的参数文件或控制文件，可以通过视图 v$backup_spfile 来检查。

```
SQL> select * from v$backup_spfile;

RECID   STAMP    SET_STAMP  SET_COUNT MODIFICAT  BYTES  COMPLETIO DB_UNIQUE_NAME  CON_ID
------ --------- ---------- ------    --------- -------- -------  --------------- ------
1      923149162 923149162      2     21-SEP-16   32768  21-SEP-16 PROD1               0
```

（5）列出备份集。也可以利用查看备份集的形式，查看 spfile 自动备份情况。

```
RMAN> list backup of spfile;
List of Backup Sets
===================
BS Key  Type LV Size       Device Type Elapsed Time Completion Time
------- ---- -- ---------- ----------- ------------ ---------------
2       Full    9.64M      DISK        00:00:00     21-SEP-16
        BP Key: 2   Status: AVAILABLE  Compressed: NO  Tag: TAG20160921T141922
        Piece Name: /backup/rman_test/control_c-2137673358-20160921-00
    SPFILE Included: Modification time: 21-SEP-16
    SPFILE db_unique_name: PROD1
```

（6）备份包含。如果使用 RMAN 进行备份，在提示中可以看到以下信息。

```
RMAN> backup database;

Starting backup at 21-SEP-16
allocated channel: ORA_DISK_1
channel ORA_DISK_1: SID=25 device type=DISK
channel ORA_DISK_1: starting full datafile backup set
channel ORA_DISK_1: specifying datafile(s) in backup set
input datafile file number=00005 name=/oracle/oradata/PROD1/example01.dbf
input datafile file number=00001 name=/oracle/oradata/PROD1/system01.dbf
input datafile file number=00003 name=/oracle/oradata/PROD1/sysaux01.dbf
input datafile file number=00004 name=/oracle/oradata/PROD1/undotbs01.dbf
input datafile file number=00006 name=/oracle/oradata/PROD1/users01.dbf
...
channel ORA_DISK_1: backup set complete, elapsed time: 00:00:15
Finished backup at 21-SEP-16

Starting Control File and SPFILE Autobackup at 21-SEP-16
piece handle=/backup/rman_test/control_c-2137673358-20160921-00 comment=NONE
Finished Control File and SPFILE Autobackup at 21-SEP-16
```

7.9 参数文件恢复

使用自动备份恢复 spfile 文件。

```
RMAN> restore spfile to '/tmp/spfile_bak.ora' from autobackup;

Starting restore at 21-SEP-16
using channel ORA_DISK_1

recovery area destination: /oracle/fast_recovery_area
database name (or database unique name) used for search: PROD1
channel ORA_DISK_1: no AUTOBACKUPS found in the recovery area
channel ORA_DISK_1: looking for AUTOBACKUP on day: 20160921
channel ORA_DISK_1: AUTOBACKUP found: /backup/rman_test/control_c-2137673358-20160921-00
channel ORA_DISK_1: restoring spfile from AUTOBACKUP /backup/rman_test/control_c-2137673358-20160921-00
channel ORA_DISK_1: SPFILE restore from AUTOBACKUP complete
Finished restore at 21-SEP-16
```

检查一下，恢复的 SPFILE 文件自动生成到了指定位置。

```
[oracle@enmoedu tmp]$ ls -l /tmp/spfile_bak.ora
-rw-r----- 1 oracle oinstall 2560 Sep 21 14:30 /tmp/spfile_bak.ora
```

同样可以通过这种方法重建控制文件，如下所示。

7.9 参数文件恢复

```
RMAN> restore controlfile to '/tmp/controfile_bak.ctl' from autobackup;

Starting restore at 21-SEP-16
using channel ORA_DISK_1

recovery area destination: /oracle/fast_recovery_area
database name (or database unique name) used for search: PROD1
channel ORA_DISK_1: no AUTOBACKUPS found in the recovery area
channel ORA_DISK_1: looking for AUTOBACKUP on day: 20160921
channel ORA_DISK_1: AUTOBACKUP found: /backup/rman_test/control_c-2137673358-20160921-01
channel ORA_DISK_1: restoring control file from AUTOBACKUP /backup/rman_test/control_c-2137673358-20160921-01
channel ORA_DISK_1: control file restore from AUTOBACKUP complete
Finished restore at 21-SEP-16

[oracle@enmoedu tmp]$ ls -l /tmp/controfile_bak.ctl
-rw-r----- 1 oracle oinstall 10043392 Sep 21 14:32 /tmp/controfile_bak.ctl
```

自动备份控制文件的功能给我们带来了极大的收益。通过自动备份，在数据库出现紧急状况的时候，用户可能从这个自动备份中获得更为有效、及时的控制文件。

随着 ASM 的增强，Oracle 也提供了基于 ASM 的参数文件备份和恢复命令。在 ASMCMD 中，可以通过 spbackup 命令对参数文件进行备份。

```
ASMCMD> spbackup +DATA/asm/asmparameterfile/registry.253.721810181 +DATA/spfileBackASM.bak
ASMCMD> spbackup +DATA/asm/asmparameterfile/registry.253.721810181 +FRA/spfileBackASM.bak
```

新增的命令还包括 spcopy 和 spmove 等。

第三篇

SQL 之美

- 第 8 章　学习与分享
- 第 9 章　诊断 Cache buffers chains 案例一则
- 第 10 章　戒骄戒躁、细致入微

第 8 章　学习与分享

——我的成长历程（杨廷琨）

> **题记**
>
> 随着科技的飞速发展和软硬件技术的更新换代，数据库领域也在不断发生着变化，例如传统数据库与新技术的结合、数据库的云化等。改变是为了更好地生存发展，Oracle 产品本身发生了很大变化，这在一定程度上影响了 Oracle 数据库运维者的工作模式。杨廷琨将在本章阐释近年来 Oracle 在技术层面和环境的变化，以及如何才能成为一名优秀的 DBA。

8.1　对数据库开发和运维的认识

我是从数据库开发做起，经历了开发 DBA 的阶段，最终成为一个管理 DBA 的。

数据库开发主要涉及建模、表的逻辑和物理结构设计、PL/SQL 代码编写、性能优化等方面。而管理和维护 DBA 的任务包括逻辑备份、物理备份、恢复、迁移和升级等。虽然从工作范围上，二者可以明确区分，但是很多时候很难界定一个工作到底属于开发范畴还是管理范围。如果想要数据库的后期维护成本低，前期的设计非常重要。也就是说，如果 DBA 在设计物理逻辑结构时多考虑后期维护工作，就可以把很多复杂且烦琐的维护工作消除在设计阶段。

Oracle 目前有很大的市场占有率（根据 IDC 的分析报告，2015 年中国市场 Oracle 占有率高达 56%），只要是具有一定数据规模的行业，都会存在 Oracle 的身影。Oracle 数据库的优势主要体现在大数据量 OLTP 环境中，Oracle 特有的锁机制和多版本读一致性，最大限度提升了系统所能承载的并发用户数。而对于小数据量的情况或 OLAP 环境中，Oracle 并没有绝对的优势。

对于 Oracle 数据库的发展方向，我认为 Oracle 数据库会越来越智能、所集成的功能会越来越强大，而数据库和存储的一体化（Exadata）很可能是今后几年 Oracle 发展的主要方向，并且这个方向会持续扩展到云上。

8.2 行业发展给 DBA 带来的挑战

"DBA 行业将要逐渐消亡"这一说法显然是杞人忧天。Oracle 的性能很稳定，而且随着新版本新功能的不断增加，原本很多需要人工设置的工作都可以自动完成了。

例如创建一个表时，在数据字典管理的表空间中要考虑到非常多的表存储参数，PCTFREE、PCTUSED、INITRANS、MAXTRANS、INITIAL、NEXT、MINEXTENTS、MAXEXTENTS、PCTINCREASE、BUFFER_POOL，所有这些参数都要仔细规划。随着 Oracle 越来越智能化以及本地管理表空间 ASSM 的出现，绝大部分情况不需要再对表的存储参数进行额外修改，默认值的设置就能够满足绝大部分要求。这样的例子比比皆是，显然 Oracle 入门的门槛变低了。

但门槛变低并不意味着成长为高级 DBA 的难度下降。事实上，Oracle 在每个版本中都会引入大量的自动化功能。以广为人知的内存管理为例，9i Oracle 推出了自动 PGA 管理，10g 推出了自动 SGA 管理，11g 推出了自动内存管理。但是在客户的关键应用上，我并不建议客户使用 11g 的新特性自动内存管理，甚至在一些非常繁忙的 OLTP 数据库中，为了避免 Oracle 在 SGA 各个组件之间自动调整内存，建议客户手工配置各个内存组件的大小。

一个高水平的 DBA 不但要了解 Oracle 推出的新特性的功能，还要了解这个新特性的限制条件以及可能导致的问题。如果出现问题，要能对这个特性产生的问题进行深入分析和诊断。所以，从这个角度来看，DBA 需要掌握的知识会越来越多。

随着 Oracle 越来越自动化，初级 DBA 的职位可能会有一定危险。如果 Oracle 升级到了高版本，而 DBA 的知识框架却不更新，那么这个 DBA 很可能会被淘汰。因此，一个上进的 DBA 要能够随着数据架构的变化和行业的发展不断提升自己的能力，合理规划，不断进步。

对于现在仍在 Oracle 运维岗位的人来说，既然上了 Oracle 这个快行道，就义无反顾地走下去吧！

8.3 个人学习经验分享

如何成为一个优秀的 DBA 呢？

有关 Oracle 技术学习的文章非常多，方法并不是最重要的，持之以恒的学习才是成功的关键。我的学习方法是从阅读 Oracle 官方文档起步，首先看 Concept，然后看 Administrator，最后是 Backup、Performance Tunning、RAC、Data Guard、Upgrade、Utilities、Network 等。研究技术没有太多的窍门可言，对我来说，一个十分有帮助的方法是：对 Oracle 技术有一定了解后，自己多做一些总结性的工作，包括做测试、写 BLOG。这样可以把 Oracle 技术的发展梳理出来，了解了一个技术的发展过程，再去看它的改变和发展，就会更加得心应手。

1．抬头看路

除了潜心研究外，多关注新的技术发展趋势也十分关键。低头做事、抬头看路，了解最新的技术发展和行业趋势可以避免走弯路，有助于更好地理解技术的演进过程。

大部分的技术人员都喜欢埋头研究技术，而技术大会正是一个让技术人员抬头看路的机会。所以，我鼓励运维者多参加各类技术会议，了解更多的业内最新动态，结识圈内更多的技术专家。

2．坚持记录，分享知识

博客在我的技术成长中占有很重要的地位，个人这些年也一直坚持更新博客，为大家分享了很多有价值的技术和经验。

其实我开始写博客时，博客已经流行一段时间了。开始没有想过要申请一个博客，直到有一天自己通过查询文档解决了一个比较复杂的问题后，才发现如果不把这个解决问题的步骤记录下来，那么过不了多长时间，很可能就会忘记一些关键的步骤，下次即使碰到同样的问题，仍然可能需要从头开始。如果将其保存在文本文件中，时间久了，查找起来就会很不方便，而当时的博客恰好可以满足需要，于是我的博客也就这样诞生了。

坚持做一件事情并不容易，我也有一段时间没有坚持更新博客，慢慢发现如果不去更新博客，自己就会变得比较懒，工作之外花在学习 Oracle 上的时间也会随之减少。意识到这一点后，我恢复了博客的更新，并给自己定下了每天更新一篇的目标，让博客成为督促自己每天学习 Oracle 知识、总结 Oracle 经验的工具。

我建议大家多写博客，将自己有价值的技术和经验记录下来，分享给大家，这样后来的人就能够从前人的经验中学习，避免重复低级的错误。当然运维本身现在也面临着繁冗重复的问题，博客并不能解决这个问题，但无论如何，博客都是记录和分享知识的一个很好的途径。

8.4　Oracle 中的 NULL 剖析

经常看到很多人提出和 NULL 有关的问题。NULL 是数据库中特有的类型，Oracle 中很多容易出现的错误都和 NULL 有关。下面简单总结一些 NULL 的相关知识。

8.4.1 NULL 的基础概念和由来

NULL 是数据库中特有的数据类型,当一条记录的某个列为 NULL,则表示这个列的值是未知的、不确定的。既然是未知的,就有无数种的可能性。因此,NULL 并不是一个确定的值。这是 NULL 的由来,也是 NULL 的基础,所有和 NULL 相关的操作的结果都可以从 NULL 的概念推导出来。

判断一个字段是否为 NULL,应该用 IS NULL 或 IS NOT NULL,而不能用'='。对 NULL 的判断只能定性,而不能定值。简单地说,由于 NULL 存在着无数的可能,因此两个 NULL 既不是相等的关系,也不是不相等的关系,同样不能比较两个 NULL 的大小,这些操作都是没有意义的,得不到一个确切的答案的。因此,对 NULL 的=、!=、>、<、>=、<=等操作的结果都是未知的,这些操作的结果仍然是 NULL。

同理,对 NULL 进行+、-、*、/等操作的结果也是未知的,结果也是 NULL。

所以通常会这样总结 NULL,除了 IS NULL、IS NOT NULL 以外,对 NULL 的任何操作的结果还是 NULL。

上面这句话总结得很精辟,而且很好记,所以通常人们只记得这句话,而忘了这句话是如何得到的。只要清楚 NULL 的真正含义,在处理 NULL 的时候就不会出错。

首先看一个经典的例子。

```
SQL> CREATE OR REPLACE PROCEDURE P1 (P_IN IN NUMBER) AS
  2  BEGIN
  3    IF P_IN >= 0 THEN
  4      DBMS_OUTPUT.PUT_LINE('TRUE');
  5    ELSE
  6      DBMS_OUTPUT.PUT_LINE('FALSE');
  7    END IF;
  8  END;
  9  /
过程已创建。
SQL> CREATE OR REPLACE PROCEDURE P2 (P_IN IN NUMBER) AS
  2  BEGIN
  3    IF P_IN < 0 THEN
  4      DBMS_OUTPUT.PUT_LINE('FALSE');
  5    ELSE
  6      DBMS_OUTPUT.PUT_LINE('TRUE');
  7    END IF;
  8  END;
  9  /
过程已创建。
```

上面两个过程是否等价?如果你输入一个大于等于 0 的值,第一个过程会输出 TRUE,而第二个过程也会输出 TRUE。同理,如果输入一个小于 0 的值,则两个过程都会输出 FALSE。

因此对于熟悉 C 或 JAVA 的开发人员来说，可能认为这两个过程是等价的，但是在数据库中，输入的值除了大于等于 0 和小于 0 之外，还有第 3 种可能性：NULL。

当输入为 NULL 时，可以看到上面两个过程不同的输出。

```
SQL> SET SERVEROUT ON
SQL> EXEC P1(NULL)
FALSE
PL/SQL 过程已成功完成。
SQL> EXEC P2(NULL)
TRUE
PL/SQL 过程已成功完成。
```

输入为 NULL 时，无论是第一个过程还是第二个过程，在流程中的第一个判断语句的结果是一样的：不管是 NULL >= 0 还是 NULL < 0 结果都是未知，所以两个判断的结果都是 NULL。最终，在屏幕上输出的分别是两个过程中定义的判断 ELSE 分支中的输出值。

8.4.2　NULL 的布尔运算的特点

由于引入了 NULL，在处理逻辑过程中一定要考虑 NULL 的情况。同样的，布尔值的处理也需要考虑 NULL 的情况，这使得布尔值从原来的 TRUE 和 FALSE 两个值变成了 TRUE、FALSE 和 NULL 3 个值。

下面是 TRUE 和 FALSE 两种情况进行布尔运算的结果。

AND 操作如表 8-1 所示。

表 8-1

AND	TRUE	FALSE
TRUE	TRUE	FALSE
FALSE	FALSE	FALSE

OR 操作如表 8-2 所示。

表 8-2

OR	TRUE	FALSE
TRUE	TRUE	TRUE
FALSE	TRUE	FALSE

上面是熟悉的 TRUE 和 FALSE 两个值进行布尔运算的结果，如果加上一个 NULL 的情况会怎样？NULL 的布尔运算是否会像 NULL 的算术运算那样，结果都是 NULL 呢？下面通过一个过程来说明。

```
SQL> SET SERVEROUT ON SIZE 100000
SQL> DECLARE
```

```
2    TYPE T_BOOLEAN IS TABLE OF BOOLEAN INDEX BY BINARY_INTEGER;
3    V_BOOL1 T_BOOLEAN;
4    V_BOOL2 T_BOOLEAN;
5
6    PROCEDURE P(P_IN1 BOOLEAN, P_IN2 BOOLEAN, P_OPERATOR IN VARCHAR2) AS
7     V_RESULT BOOLEAN;
8    BEGIN
9     IF P_IN1 IS NULL THEN
10      DBMS_OUTPUT.PUT('NULL ');
11     ELSIF P_IN1 THEN
12      DBMS_OUTPUT.PUT('TRUE ');
13     ELSE
14      DBMS_OUTPUT.PUT('FALSE ');
15     END IF;
16
17     IF P_OPERATOR = 'AND' THEN
18      DBMS_OUTPUT.PUT('AND ');
19      V_RESULT := P_IN1 AND P_IN2;
20     ELSIF P_OPERATOR = 'OR' THEN
21      DBMS_OUTPUT.PUT('OR ');
22      V_RESULT := P_IN1 OR P_IN2;
23     ELSE
24      RAISE_APPLICATION_ERROR('-20000', 'INPUT PARAMETER P_OPERATOR ERROR');
25     END IF;
26
27     IF P_IN2 IS NULL THEN
28      DBMS_OUTPUT.PUT('NULL');
29     ELSIF P_IN2 THEN
30      DBMS_OUTPUT.PUT('TRUE');
31     ELSE
32      DBMS_OUTPUT.PUT('FALSE');
33     END IF;
34
35     IF V_RESULT IS NULL THEN
36      DBMS_OUTPUT.PUT(':NULL');
37     ELSIF V_RESULT THEN
38      DBMS_OUTPUT.PUT(':TRUE');
39     ELSE
40      DBMS_OUTPUT.PUT(':FALSE');
41     END IF;
42     DBMS_OUTPUT.NEW_LINE;
43    END;
44
45   BEGIN
46    V_BOOL1(1) := TRUE;
47    V_BOOL1(2) := FALSE;
48    V_BOOL1(3) := NULL;
49    V_BOOL2 := V_BOOL1;
50    FOR I IN 1..V_BOOL1.COUNT LOOP
51     FOR J IN 1..V_BOOL2.COUNT LOOP
52      P(V_BOOL1(I), V_BOOL2(J), 'AND');
53      P(V_BOOL1(I), V_BOOL2(J), 'OR');
```

```
 54      END LOOP;
 55    END LOOP;
 56  END;
 57  /
TRUE AND TRUE:TRUE
TRUE OR TRUE:TRUE
TRUE AND FALSE:FALSE
TRUE OR FALSE:TRUE
TRUE AND NULL:NULL
TRUE OR NULL:TRUE
FALSE AND TRUE:FALSE
FALSE OR TRUE:TRUE
FALSE AND FALSE:FALSE
FALSE OR FALSE:FALSE
FALSE AND NULL:FALSE
FALSE OR NULL:NULL
NULL AND TRUE:NULL
NULL OR TRUE:TRUE
NULL AND FALSE:FALSE
NULL OR FALSE:NULL
NULL AND NULL:NULL
NULL OR NULL:NULL
PL/SQL 过程已成功完成。
```

由于 NULL 是未知的，所以 NULL AND NULL、NULL OR NULL、NULL AND TRUE 和 NULL OR FALSE 的值都是未知的，结果仍然是 NULL。

为什么 NULL AND FALSE 和 NULL OR TRUE 得到了一个确定的结果呢？这需要从 NULL 的概念来考虑。NULL 的定义是未知的，但是目前 NULL 的类型是布尔类型，虽然 NULL 的值不确定，但是 NULL 所在的类型确定了值的范围，因此 NULL 只有可能是 TRUE 或者 FALSE 中的一个。

根据前面的表格，TRUE AND FALSE 和 FALSE AND FALSE 的结果都是 FALSE，也就是说，不管 NULL 的值是 TRUE 还是 FALSE，它与 FALSE 进行 AND 的结果一定是 FALSE。

同理，TRUE OR TRUE 和 FALSE OR TRUE 的结果都是 TRUE，所以不管 NULL 取何值，NULL 和 TRUE、OR 的结果都是 TRUE。

AND 操作图表如表 8-3 所示。

表 8-3

AND	TRUE	FALSE	NULL
TRUE	TRUE	FALSE	NULL
FALSE	FALSE	FALSE	FALSE
NULL	NULL	FALSE	NULL

OR 操作图表如表 8-4 所示。

表 8-4

OR	TRUE	FALSE	NULL
TRUE	TRUE	TRUE	TRUE
FALSE	TRUE	FALSE	NULL
NULL	TRUE	NULL	NULL

下面来看一个例子。

```
SQL> SELECT * FROM TAB;
TNAME                    TABTYPE  CLUSTERID
------------------------ -------- ----------
PLAN_TABLE               TABLE
T                        TABLE
T1                       TABLE
T2                       TABLE
T3                       TABLE
TEST                     TABLE
TEST1                    TABLE
TEST_CORRUPT             TABLE
T_TIME                   TABLE
已选择 9 行。
SQL> SELECT * FROM TAB WHERE TNAME IN ('T', 'T1', NULL);
TNAME                    TABTYPE  CLUSTERID
------------------------ -------- ----------
T                        TABLE
T1                       TABLE
SQL> SELECT * FROM TAB WHERE TNAME NOT IN ('T', 'T1', NULL);
未选定行
```

对于 IN、NOT IN 与 NULL 的关系前面并没有说明，不过可以对其进行简单的变形。

表达式 TNAME IN（'T'，'T1'，NULL）等价于 TNAME = 'T' OR TNAME = 'T1' OR TNAME = NULL，根据前面的布尔运算结果，当查询到 T 或 T1 这两条记录时，WHERE 条件相当于 TRUE OR FALSE OR NULL，其结果是 TRUE，因此返回了两条记录。

表达式 TNAME NOT IN（'T'，'T1'，NULL）等价于 TNAME != 'T' AND TNAME != 'T1' AND TNAME != NULL，这时 WHERE 条件相当于 TRUE AND TRUE AND NULL 或 TRUE AND FALSE AND NULL，其最终结果不是 NULL 就是 FALSE，所以查询不会返回记录。

下面讨论一下 NULL 的布尔值运算 NOT。对于 TRUE 和 FALSE 的 NOT 运算很简单，NOT TRUE=FALSE，NOT FALSE=TRUE，那么如果包含 NULL 的情况呢？

```
SQL> SET SERVEROUT ON SIZE 100000
SQL> DECLARE
  2    TYPE T_BOOLEAN IS TABLE OF BOOLEAN INDEX BY BINARY_INTEGER;
  3    V_BOOL T_BOOLEAN;
  4
  5    PROCEDURE P(P_IN BOOLEAN) AS
  6     V_RESULT BOOLEAN;
```

```
  7  BEGIN
  8    IF P_IN IS NULL THEN
  9      DBMS_OUTPUT.PUT('NOT NULL');
 10    ELSIF P_IN THEN
 11      DBMS_OUTPUT.PUT('NOT TRUE');
 12    ELSE
 13      DBMS_OUTPUT.PUT('NOT FALSE');
 14    END IF;
 15
 16    V_RESULT := NOT P_IN;
 17
 18    IF V_RESULT IS NULL THEN
 19      DBMS_OUTPUT.PUT(':NULL');
 20    ELSIF V_RESULT THEN
 21      DBMS_OUTPUT.PUT(':TRUE');
 22    ELSE
 23      DBMS_OUTPUT.PUT(':FALSE');
 24    END IF;
 25    DBMS_OUTPUT.NEW_LINE;
 26  END;
 27
 28  BEGIN
 29    V_BOOL(1) := TRUE;
 30    V_BOOL(2) := FALSE;
 31    V_BOOL(3) := NULL;
 32    FOR I IN 1..V_BOOL.COUNT LOOP
 33      P(V_BOOL(I));
 34    END LOOP;
 35  END;
 36  /
NOT TRUE:FALSE
NOT FALSE:TRUE
NOT NULL:NULL
PL/SQL 过程已成功完成。
```

现在看到了一个很有趣的结果，NOT NULL 的结果仍然是 NULL。可能很多人对此并不理解，下面从 NULL 的基本概念来解释。

NULL 表示未知，而增加一个 NOT 操作后，并不能使 NULL 变为一个确定的值，如果 NULL 的值是 TRUE，NOT TRUE 将变为 FALSE，如果值是 FALSE，NOT FALSE 将变为 TRUE。因此即使进行了 NOT 操作，NULL 本身的不确定性是仍然存在的，这就是最终结果仍然是 NULL 的原因。

这里需要注意：这个 NOT NULL 是一个布尔操作，要和 SQL 中的 NOT NULL 约束区分开。NOT NULL 约束是一个定性的描述，表示列中的数据不允许为 NULL。而这里的布尔操作是在求值，要得到对 NULL 取非后的结果，所以仍然得到 NULL，如表 8-5 所示。

表 8-5

NOT TRUE	NOT FALSE	NOT NULL
FALSE	TRUE	NULL

8.4.3 NULL 的默认数据类型

Oracle 的 NULL 的含义是不确定,那么不确定的东西也会有确定的数据类型吗？换个说法，NULL 在 Oracle 中的默认数据类型是什么？下面通过两个例子来说明这个问题。

NULL 的默认类型是字符类型，确切地说是 VARCHAR2 类型。我们知道一个字段不管是何种类型的，都可以插入 NULL 值，也就是说，NULL 可以随意地转换为任意的类型。

绝大部分的函数输入值为 NULL，返回的结果也为 NULL，这就使我们不能通过函数的返回结果判断 NULL 的类型。最常用来分析数据的 Dump 函数，这次也失效了。

```
SQL> SELECT Dump(NULL) FROM DUAL;
Dump
----
NULL
```

试图通过 CREATE TABLE AS 来判定 NULL 的类型也是不可能的。

```
SQL> CREATE TABLE T AS SELECT TNAME, NULL COL1 FROM TAB;
CREATE TABLE T AS SELECT TNAME, NULL COL1 FROM TAB
                                     *
ERROR 位于第 1 行:
ORA-01723: 不允许长度为 0 的列
```

发现 NULL 的数据类型的过程比较偶然，下面通过一个例子来简单说明。

```
SQL> create table t (id number);
表已创建。
SQL> insert into t values (1);
已创建 1 行。
SQL> insert into t values (8);
已创建 1 行。
SQL> insert into t values (0);
已创建 1 行。
SQL> insert into t values (15);
已创建 1 行。
SQL> commit;
提交完成。
```

返回结果需要按照 T 中的 ID 的升序显示数据，SQL 如下。

```
SQL> select * from t order by id;
        ID
----------
         0
         1
         8
        15
```

需求还有一点额外的要求，返回结果中 0 值比较特殊，其他结果正常排序，但是 0 排在所有非 0 值的后面。实现的方法有很多，比如使用 UNION ALL 将非 0 值和 0 值分开，或者将 0

值转换为一个很大的数值。这两种方法都有小缺点，前者需要扫描表两次，后者无法解决 ID 最大值不确定的情况。因此选择在排序的时候将 0 转化为 NULL 的方法,这样利用排序时 NULL 最大的原理，得到了希望的结果。

```
SQL> select * from t order by decode(id, 0, null, id);
        ID
----------
         1
        15
         8
         0
```

0 确实排在了最后，但是返回结果并不正确，15 居然排在了 8 的前面。这种结果感觉似乎是根据字符类型排序得到的。

检查排序的 DECODE 函数。

```
SQL> select decode(id, 0, null, id) from t;
DECODE(ID,0,NULL,ID)
----------------------------------------
1
8
15
```

看到 DECODE 函数的结果，就知道问题所在了。字符类型结果在 SQLPLUS 显示左对齐，而数值类型是右对齐，当前的结果说明 DECODE 函数的结果是字符类型。

现在处理的是数值类型，为什么会得到字符类型的输出呢？在 DECODE 函数中，输入的 4 个参数，两个 ID 和 0 都是 NUMBER 类型，只有 NULL 这一个输入参数类型是不确定的，导致问题的原因就是 NULL。

为了验证 NULL 是导致问题的原因，检查标准包中 DECODE 函数的定义。

下面的 DECODE 函数定义是从 STANDARD 中摘取出来的部分内容。

```
function DECODE (expr NUMBER, pat NUMBER, res NUMBER) return NUMBER;
function DECODE (expr NUMBER,
                 pat NUMBER,
                 res VARCHAR2 CHARACTER SET ANY_CS)
        return VARCHAR2 CHARACTER SET res%CHARSET;
function DECODE (expr NUMBER, pat NUMBER, res DATE) return DATE;
function DECODE (expr VARCHAR2 CHARACTER SET ANY_CS,
                 pat VARCHAR2 CHARACTER SET expr%CHARSET,
                 res NUMBER) return NUMBER;
function DECODE (expr VARCHAR2 CHARACTER SET ANY_CS,
                 pat VARCHAR2 CHARACTER SET expr%CHARSET,
                 res VARCHAR2 CHARACTER SET ANY_CS)
        return VARCHAR2 CHARACTER SET res%CHARSET;
function DECODE (expr VARCHAR2 CHARACTER SET ANY_CS,
                 pat VARCHAR2 CHARACTER SET expr%CHARSET,
                 res DATE) return DATE;
```

```
function DECODE (expr DATE, pat DATE, res NUMBER) return NUMBER;
function DECODE (expr DATE,
                 pat DATE,
                 res VARCHAR2 CHARACTER SET ANY_CS)
    return VARCHAR2 CHARACTER SET res%CHARSET;
function DECODE (expr DATE, pat DATE, res DATE) return DATE;
```

观察上面的定义不难发现，虽然 Oracle 对 DECODE 函数进行了大量的重载，且 DECODE 函数支持各种数据类型。但是 DECODE 函数具有一个规律，就是 DECODE 函数的返回值的类型和 DECODE 函数的输入参数中第一个用来返回的参数的数据类型一致。这句话可能不易理解，举个简单的例子。

```
SQL> select decode(id, 1, '1', 2) from t;
D
-
1
2
2
2
SQL> select decode(id, '1', 1, '2') from t;
DECODE(ID,'1',1,'2')
--------------------
         1
         2
         2
         2
```

从这两个简单的例子可以看出，DECODE 的返回值的数据类型和 DECODE 函数中第一个表示返回的参数的数据类型一致。

那么可以推断，NULL 的默认数量类型是字符类型，这导致 DECODE 的结果变成了字符串，而查询根据字符串规则进行排序比较，因此 15 小于 8。

解决排序被转为字符类型比较的方法很多，下面举例说明。

```
SQL> select * from t order by decode(id, 1, 1, 0, null, id);
        ID
----------
         1
         8
        15
         0
SQL> select * from t order by to_number(decode(id, 0, null, id));
        ID
----------
         1
         8
        15
         0
SQL> select * from t order by decode(id, 0, cast(null as number), id);
        ID
----------
```

```
            1
            8
           15
            0
SQL> select * from t order by decode(id, 0, to_number(null), id);
           ID
        ----------
            1
            8
           15
            0
```

回到前面讨论的问题，根据 DECODE 类型的返回结果，可以得到一个结论：当 Oracle 根据 NULL 来判断返回类型时，Oracle 给出的结果是字符串类型。

有人可能会认为这是 DECODE 函数的特性而已，下面还有一个比较特别的例子，同样可以说明这个问题。

```
SQL> CREATE VIEW V_NULL AS
  2  SELECT NULL N
  3  FROM DUAL;
视图已创建。
SQL> DESC V_NULL
 名称         是否为空? 类型
 -------- -------- --------
 N                  VARCHAR2
```

虽然建表时使用 NULL 会报错，但是创建视图并不会报错，而且观察视图的定义可以发现，Oracle 把 NULL 当作 VARCHAR2 类型来处理。

8.4.4　空字符串'' 与 NULL 的关系

很多人对空字符串'' 不是很清楚，这里做简单总结。

以前我总说空字符串'' 等价于 NULL，确切地说，空字符串'' 是 NULL 的字符类型的表现格式。

证明空字符串就是 NULL 是很容易的。

```
SQL> SELECT 1 FROM DUAL WHERE '' = '';
未选定行
SQL> SELECT 1 FROM DUAL WHERE '' IS NULL;
            1
        ----------
            1
SQL> SELECT Dump(''), Dump(NULL) FROM DUAL;
Dump Dump
---- ----
NULL NULL
```

上面 3 个 SQL 语句，任意一个都足以证明空字符串''就是 NULL。

8.4 Oracle 中的 NULL 剖析

有些人可能会说，既然''就是 NULL，为什么不能进行 IS ''的判断呢？

```
SQL> SELECT 1 FROM DUAL WHERE '' IS '';
SELECT 1 FROM DUAL WHERE '' IS ''
                                *
第 1 行出现错误：
ORA-00908: 缺失 NULL 关键字
```

从上面的错误信息就可以看到答案。原因是 IS NULL 是 Oracle 的语法，在 Oracle 运行的时刻''是 NULL，但是现在 Oracle 还没有运行这条 SQL，就由于语法不正确被 SQL 分析器挡住了。Oracle 的语法并不包含 IS ''的写法，所以，这一点并不能成为''不是 NULL 的理由。

为什么要说''是 NULL 的字符表示形式呢？因为''和 NULL 还确实不完全一样。对于 NULL 来说，它表示了各种数据类型的 NULL 值。而对于空字符串''来说，虽然它也具有 NULL 的可以任意转化为其他任何数据类型的特点，但是无论是从形式上，还是从本质上，它都表现出了字符类型的特点。

下面通过一个例子来证明''本质是字符类型的 NULL。

```
SQL> CREATE OR REPLACE PACKAGE P_TEST_NULL AS
  2    FUNCTION F_RETURN (P_IN IN NUMBER) RETURN VARCHAR2;
  3    FUNCTION F_RETURN (P_IN IN VARCHAR2) RETURN VARCHAR2;
  4  END;
  5  /
程序包已创建。
SQL> CREATE OR REPLACE PACKAGE BODY P_TEST_NULL AS
  2
  3    FUNCTION F_RETURN (P_IN IN NUMBER) RETURN VARCHAR2 AS
  4    BEGIN
  5      RETURN 'NUMBER';
  6    END;
  7
  8    FUNCTION F_RETURN (P_IN IN VARCHAR2) RETURN VARCHAR2 AS
  9    BEGIN
 10      RETURN 'VARCHAR2';
 11    END;
 12
 13  END;
 14  /
程序包体已创建。
SQL> SELECT P_TEST_NULL.F_RETURN(3) FROM DUAL;
P_TEST_NULL.F_RETURN(3)
-----------------------------------------------------------
NUMBER
SQL> SELECT P_TEST_NULL.F_RETURN('3') FROM DUAL;
P_TEST_NULL.F_RETURN('3')
-----------------------------------------------------------
VARCHAR2
SQL> SELECT P_TEST_NULL.F_RETURN('') FROM DUAL;
P_TEST_NULL.F_RETURN('')
-----------------------------------------------------------
```

```
VARCHAR2
SQL> SELECT P_TEST_NULL.F_RETURN(NULL) FROM DUAL;
SELECT P_TEST_NULL.F_RETURN(NULL) FROM DUAL
                *
第 1 行出现错误：
ORA-06553: PLS-307: 有太多的 'F_RETURN' 声明与此次调用相匹配
```

利用重载的原理，字符类型输出 VARCHAR2，数值类型输出 NUMBER。输入为空字符串时，输出为 VARCHAR2。从这一点上可以看出，''实际上已经具备了数据类型。所以我将''表述为空字符串是 NULL 的字符类型表现形式。

上面根据重载的特性证明了空字符就是 NULL 的字符表现形式。下面简单描述字符串合并操作||的特殊性。

根据 NULL 的定义，NULL 是不确定、未知的含义，那么为什么字符类型的 NULL 是一个空字符呢？对于 NULL 的加、减、乘、除等操作的结果都是 NULL，为什么字符串合并操作||，当输入字符串有一个为空时，不会得到结果 NULL？

```
SQL> SELECT NULL || 'A', 'B' || NULL, NULL || NULL FROM DUAL;
NU ' N
-- - -
A  B
```

上面两个问题需要从 NULL 的存储格式上解释。Oracle 在存储数据时，先是存储这一列的长度，然后存储列数据本身。对于 NULL，只包含一个 FF，没有数据部分。简单地说，Oracle 用长度 FF 来表示 NULL。

由于 Oracle 在处理数据存储时尽量避免 0 的出现，因此，认为 FF 表示的长度为 0 也是有一定道理的。或者从另一方面考虑，NULL 只有一个长度，没有数据部分。

对于字符串来说，不管是长度为 0 的字符串还是没有任何数据的字符串，代表的都是一个空字符串。空字符串就是 NULL 也有一定的道理。如果认为空字符串是字符形式的 NULL，那么||操作的结果就不难理解了。

最后需要说明的是，不要将 ORACLE 里面的空字符串''与 C 语言中的空字符串""混淆。C 语言中的空字符串并非不包含任何数据，里面包含了一个字符串结束符\0。C 语言中的空字符串""对应 Oracle 中 ASCII 表中的 0 值，即 CHR（0）。

但 CHR（0）是一个确定的值，显然不是 NULL。

```
SQL> SELECT * FROM DUAL WHERE CHR(0) = CHR(0);
D
-
X
SQL> SELECT * FROM DUAL WHERE CHR(0) IS NULL;
未选定行
```

8.4.5 NULL 和索引

NULL 类型比较容易出错，索引则让 NULL 又一次成为问题的焦点。

有一句很有名的话：索引不存储 NULL 值。这句话其实并不严谨。如果采用比较严谨的方式来说，应为：B 树索引不存储索引列全为空的记录。如果把这句话用在单列索引上，就是 B 树索引不存储 NULL。

索引分为 BTREE 和 BITMAP 两种，BTREE 索引不存储 NULL 值，而 BITMAP 索引是存储 NULL 值的。

从索引列的个数来划分，索引非为单列索引和复合索引。单列索引很简单，如果一条记录中索引字段为空，那么索引不会保存这条记录的信息。对于复合索引，由于存在着多个列，如果某一个索引列不为空，那么索引就会包含这条记录，即使索引中其他的所有列都是 NULL 值。

```
SQL> CREATE TABLE T AS SELECT * FROM DBA_OBJECTS;
表已创建。
SQL> DESC T
 名称                                      是否为空? 类型
 ----------------------------------------- -------- ----------------
 OWNER                                              VARCHAR2(30)
 OBJECT_NAME                                        VARCHAR2(128)
 SUBOBJECT_NAME                                     VARCHAR2(30)
 OBJECT_ID                                          NUMBER
 DATA_OBJECT_ID                                     NUMBER
 OBJECT_TYPE                                        VARCHAR2(19)
 CREATED                                            DATE
 LAST_DDL_TIME                                      DATE
 TIMESTAMP                                          VARCHAR2(19)
 STATUS                                             VARCHAR2(7)
 TEMPORARY                                          VARCHAR2(1)
 GENERATED                                          VARCHAR2(1)
 SECONDARY                                          VARCHAR2(1)
SQL> CREATE INDEX IND_T_OBJECT_ID ON T (OBJECT_ID);
索引已创建。
SQL> EXEC DBMS_STATS.GATHER_TABLE_STATS(USER, 'T', CASCADE => TRUE)
PL/SQL 过程已成功完成。
SQL> SET AUTOT ON EXP
SQL> SELECT COUNT(*) FROM T;
  COUNT(*)
----------
     50297

执行计划
----------------------------------------------------------
Plan hash value: 2966233522
----------------------------------------------------------
| Id  | Operation       | Name  | Rows  | Cost (%CPU)| Time     |
----------------------------------------------------------
```

```
|   0 | SELECT STATEMENT  |   |     1 |    41   (3)| 00:00:01 |
|   1 |  SORT AGGREGATE   |   |     1 |            |          |
|   2 |   TABLE ACCESS FULL| T | 50297 |    41   (3)| 00:00:01 |
--------------------------------------------------------------
SQL> SELECT /*+ INDEX(T IND_T_OBJECT_ID) */ COUNT(*) FROM T;
  COUNT(*)
----------
     50297

执行计划
----------------------------------------------------------
Plan hash value: 2966233522
--------------------------------------------------------------
| Id  | Operation          | Name | Rows | Cost (%CPU)| Time     |
--------------------------------------------------------------
|   0 | SELECT STATEMENT   |      |    1 |   41   (3) | 00:00:01 |
|   1 |  SORT AGGREGATE    |      |    1 |            |          |
|   2 |   TABLE ACCESS FULL| T    | 50297|   41   (3) | 00:00:01 |
--------------------------------------------------------------
```

Oracle 的优化器在确定是否使用索引时，第一标准是能否得到一个正确的结果。由于 OBJECT_ID 可以为空，而索引列不包含为空的记录。因此，通过索引扫描无法得到一个正确的结果，这就是 SELECT COUNT（*）FROM T 不会使用 OBJECT_ID 上的索引的原因。

对于 BITMAP 索引，则是另外的情况。

```
SQL> DROP INDEX IND_T_OBJECT_ID;
索引已删除。
SQL> CREATE BITMAP INDEX IND_B_T_DATA_ID ON T (DATA_OBJECT_ID);
索引已创建。
SQL> SELECT COUNT(*) FROM T;
  COUNT(*)
----------
     50297
执行计划
----------------------------------------------------------
Plan hash value: 3051411170
--------------------------------------------------------------------
| Id  | Operation                  | Name            | Rows | Cost (%CPU)|
--------------------------------------------------------------------
|   0 | SELECT STATEMENT           |                 |    1 |   2   (0)  |
|   1 |  SORT AGGREGATE            |                 |    1 |            |
|   2 |   BITMAP CONVERSION COUNT  |                 |50297 |   2   (0)  |
|   3 |    BITMAP INDEX FULL SCAN  | IND_B_T_DATA_ID |      |            |
--------------------------------------------------------------------
SQL> SELECT COUNT(*) FROM T WHERE DATA_OBJECT_ID IS NULL;
  COUNT(*)
----------
     46452
执行计划
----------------------------------------------------------
Plan hash value: 2587852253
--------------------------------------------------------------------
| Id  | Operation          | Name      | Rows  | Bytes| Cost (%CPU)|
```

8.4　Oracle 中的 NULL 剖析

```
--------------------------------------------------------------------------
| 0 | SELECT STATEMENT          |               |    1 |   2  | 2  (0)|
| 1 |  SORT AGGREGATE           |               |    1 |   2  |       |
| 2 |   BITMAP CONVERSION COUNT |               |46452 |92904 | 2  (0)|
|* 3|    BITMAP INDEX SINGLE VALUE| IND_B_T_DATA_ID|    |      |       |
--------------------------------------------------------------------------
Predicate Information (identified by operation id):
---------------------------------------------------
   3 - access("DATA_OBJECT_ID" IS NULL)
```

从上面的结果不难看出 BITMAP 索引中包含 NULL。

下面看复合索引的情况。

```
SQL> DROP INDEX IND_B_T_DATA_ID;
索引已删除。
SQL> CREATE INDEX IND_T_OBJECT_DATA ON T(OBJECT_ID, DATA_OBJECT_ID);
索引已创建。
SQL> EXEC DBMS_STATS.GATHER_TABLE_STATS(USER, 'T', METHOD_OPT => 'FOR ALL INDEXED COLUMNS')
PL/SQL 过程已成功完成。
SQL> SELECT OBJECT_ID, DATA_OBJECT_ID FROM T WHERE OBJECT_ID = 135;
 OBJECT_ID DATA_OBJECT_ID
---------- --------------
     135
执行计划
----------------------------------------------------------
Plan hash value: 1726226519
--------------------------------------------------------------------------
| Id | Operation         | Name              | Rows | Bytes | Cost (%CPU)|
--------------------------------------------------------------------------
|  0 | SELECT STATEMENT  |                   |   1  |   7   |   1  (0)   |
|* 1 |  INDEX RANGE SCAN | IND_T_OBJECT_DATA |   1  |   7   |   1  (0)   |
--------------------------------------------------------------------------
Predicate Information (identified by operation id):
---------------------------------------------------
   1 - access("OBJECT_ID"=135)
```

虽然结果中包含 NULL 值，但是 Oracle 并没有读取表，仅通过索引扫描返回了最终结果，这证实了复合索引中可以包含 NULL 值。

这里说明了索引和 NULL 值的关系。但没有对反键索引（reverse）、逆序索引（desc）、函数索引（FBI）和 CLUSTER 索引进行说明。

原因是这些索引并没有脱离 BTREE 索引和 BITMAP 索引的范畴。不必关心索引是否倒序或者反键，只要是 BTREE 索引，就不会存储全 NULL 记录；反之，只要是 BITMAP 索引就会存储 NULL 值。

需要注意的有：函数索引，与普通索引的区别是，函数索引的真正索引列是函数的计算结果，而不是行记录中的数据。

域索引，由于域索引的实现本身可能会很复杂，Oracle 可能在内部是用一套表和过程来实现的，因此域索引是否存储 NULL，要根据域索引的实现具体分析。

下面来看 NULL 与索引使用的关系。

很多人有一些错误的观点，认为指定 IS NULL 或者 IS NOT NULL 后无法使用索引，事实上很多和 NULL 相关的观点是 RBO 时代遗留下来的，已经不适用于 CBO 优化器了。

观点一：判断一个列 IS NOT NULL 不会使用索引。

这个观点解释不通，因为 B 树索引本身不存储键值全为 NULL 的记录，所以通过索引扫描得到的结果一定满足 IS NOT NULL 的要求。

```
SQL> CREATE TABLE T AS SELECT * FROM DBA_OBJECTS;
表已创建。
SQL> CREATE INDEX IND_T_DATAID ON T(DATA_OBJECT_ID);
索引已创建。
SQL> EXEC DBMS_STATS.GATHER_TABLE_STATS(USER, 'T')
PL/SQL 过程已成功完成。
SQL> SET AUTOT TRACE
SQL> SELECT COUNT(*) FROM T WHERE DATA_OBJECT_ID IS NOT NULL;
Execution Plan
----------------------------------------------------------
   0      SELECT STATEMENT Optimizer=CHOOSE (Cost=2 Card=1 Bytes=2)
   1    0   SORT (AGGREGATE)
   2    1     INDEX (FULL SCAN) OF 'IND_T_DATAID' (NON-UNIQUE) (Cost=26 Card=2946 Bytes=5892)
Statistics
----------------------------------------------------------
          0  recursive calls
          0  db block gets
          5  consistent gets
          4  physical reads
          0  redo size
        377  bytes sent via SQL*Net to client
        503  bytes received via SQL*Net from client
          2  SQL*Net roundtrips to/from client
          0  sorts (memory)
          0  sorts (disk)
          1  rows processed
```

索引的存储特性和 IS NOT NULL 访问本身没有冲突，因此很容易通过索引来得到相应的结果。

观点二：判断一个列 IS NULL 不会使用索引。

这里就不用 BITMAP 索引来举例了，即使是 B 树索引，这个观点也是不正确的。B 树索引不存储键值全为空的记录，所以 IS NULL 操作无法使用单列索引。但复合索引可能存储一部分 NULL 值，所以 IS NULL 操作也并非不可能使用索引。

```
SQL> ALTER TABLE T MODIFY OWNER NOT NULL;
表已更改。
SQL> UPDATE T SET OBJECT_ID = NULL WHERE ROWNUM = 1;
已更新 1 行。
SQL> CREATE INDEX IND_T_OBJECT_OWNER ON T (OBJECT_ID, OWNER);
```

```
索引已创建。
SQL> EXEC DBMS_STATS.GATHER_TABLE_STATS(USER,'T',METHOD_OPT => 'FOR ALL INDEXED COLUMNS SIZE 200')
PL/SQL 过程已成功完成。
SQL> SET AUTOT TRACE
SQL> SELECT * FROM T WHERE OBJECT_ID IS NULL;
Execution Plan
----------------------------------------------------------
   0      SELECT STATEMENT Optimizer=CHOOSE (Cost=3 Card=1 Bytes=93)
   1    0   TABLE ACCESS (BY INDEX ROWID) OF 'T' (Cost=3 Card=1 Bytes=93)
   2    1     INDEX (RANGE SCAN) OF 'IND_T_OBJECT_OWNER' (NON-UNIQUE) (Cost=2 Card=1)

Statistics
----------------------------------------------------------
          0  recursive calls
          0  db block gets
          3  consistent gets
          0  physical reads
          0  redo size
       1156  bytes sent via SQL*Net to client
        503  bytes received via SQL*Net from client
          2  SQL*Net roundtrips to/from client
          0  sorts (memory)
          0  sorts (disk)
          1  rows processed
```

从上面的两个例子可以看到，Oracle 的 CBO 不会因为 SQL 语句中指定了 IS NOT NULL 或 IS NULL 操作就不再使用索引。CBO 选择索引只需满足结果正确和代价最小这两个条件。

8.4.6 NULL 的其他方面特点

NULL 的一个显著特点就是两个 NULL 不相等。无法通过等号来判断两个 NULL 是否相等，利用唯一约束的特点也可以证实这一点，对于建立了唯一约束的列，Oracle 允许插入多个 NULL 值，这是因为 Oracle 不认为这些 NULL 是相等的。

```
SQL> CREATE TABLE T (ID NUMBER, CONSTRAINT UN_T UNIQUE(ID));
表已创建。
SQL> INSERT INTO T VALUES (1);
已创建 1 行。
SQL> INSERT INTO T VALUES (1);
INSERT INTO T VALUES (1)
*
ERROR 位于第 1 行:
ORA-00001: 违反唯一约束条件 (YANGTK.UN_T)

SQL> INSERT INTO T VALUES (NULL);
已创建 1 行。
SQL> INSERT INTO T VALUES (NULL);
已创建 1 行。
```

但有时，Oracle 会认为 NULL 是相同的，比如在 GROUP BY 和 DISTINCT 操作中，Oracle 会认为所有的 NULL 都是一类的。

另一种情况是在 DECODE 函数中,如果表达式为 DECODE(COL, NULL, 0, 1),当 COL 的值为 NULL 时,Oracle 会认为输入的 NULL 与第二个参数的 NULL 值相匹配,DECODE 的结果会返回 0。不过这里只是给人感觉 NULL 值是相等的,Oracle 在实现 DECODE 函数时,仍然通过 IS NULL 的方式进行判断。

对于大多数的常用函数来说,如果输入为 NULL,则输出也是 NULL。NVL、NVL2、DECODE 和||操作是例外。它们在输入参数为 NULL 时,结果可能不是 NULL。因为这些函数都有多个参数,当多个参数不全为 NULL 时,结果可能不是 NULL,如果输入参数均为 NULL,那么得到的输出结果也是 NULL。

NULL 的另一特点是一般聚集函数不会处理 NULL 值。不管是 MAX、MIN、AVG,还是 SUM,都不会处理 NULL。注意"不会处理 NULL",是指聚集函数会直接忽略 NULL 值记录的存在。当是聚集函数处理的列中包含的全部记录都是 NULL 时,聚集函数会返回 NULL 值。

```
SQL> DELETE T WHERE ID = 1;
已删除 1 行。
SQL> SELECT NVL(TO_CHAR(ID), 'NULL') FROM T;
NVL(TO_CHAR(ID),'NULL')
---------------------------------------
NULL
NULL
SQL> SELECT MAX(ID) FROM T;
   MAX(ID)
----------

SQL> SELECT AVG(ID) FROM T;
   AVG(ID)
----------

SQL> INSERT INTO T VALUES (1);
已创建 1 行。
```

聚集函数中比较特殊的是 COUNT,COUNT 不会返回 NULL 值,即使表中没有记录,或者 COUNT(COL)中,COL 列的记录全为 NULL,COUNT 也会返回 0 值而不是 NULL。此外,COUNT 可以计算包含 NULL 记录在内的记录总数。

```
SQL> SELECT COUNT(*), COUNT(1), COUNT('A'), COUNT(ID), COUNT(NULL) FROM T;
  COUNT(*)  COUNT(1) COUNT('A')  COUNT(ID) COUNT(NULL)
---------- --------- ---------- ---------- -----------
         3         3          3          1           0
```

最后简单说明 AVG,AVG(COL)等价于 SUM(COL)/COUNT(COL),不等价于 SUM(COL)/COUNT(*)。

```
SQL> SELECT AVG(ID), SUM(ID)/COUNT(ID), SUM(ID)/COUNT(*) FROM T;
 AVG(ID) SUM(ID)/COUNT(ID) SUM(ID)/COUNT(*)
-------- ----------------- ----------------
       1                 1        .333333333
```

第 9 章 诊断 Cache buffers chains 案例一则

——刘旭

> **题记**
>
> 这是某移动运营商在 SQL 线下审核项目中，协助开发商完善数据库性能的过程。以往开发商遇到此问题总是怀疑是数据库的 Bug，试图尝试重启 Tuxedo、Weblogic，严重时甚至重启实例来缓解问题。经过下面的详细分析，你会发现事实并非如此。

9.1 详细诊断过程

这是对于两个节点的 RAC 环境，数据库版本为 11.2.0.4 for HP-UX IA（64-bit）。在 2014 年 11 月 5 日 16 点至 18 点间，节点一的 CPU 使用率从平时的 40% 增长到 60% 左右，部分业务办理缓慢甚至超时。经过详细分析，发现是一个低效的、高并发的核心业务的 SQL 语句引起的。

通过询问业务人员得知，业务系统从 17:00 至 17:30 感觉慢得更为明显，因此我们导出了该时间段节点二的 AWR 报告。

图 9-1 显示了实例名为 crmdb21 的采样时间、数据库版本、CPU 个数和内存大小等概要信息，通过简单的换算 DB Time 和 Elapsed 可知（2807.24/29.7=94），这台 64 Cores 的小型机的确很忙。

如图 9-2 所示，Load Profile 中的 Logical read（blocks）973915.2/Per Second 表明平均每秒产生的逻辑读 blocks 数约为 97 万，每秒的逻辑读约有 7.6GB（973915*db_block_size=7.6GB/s），

一般来说，逻辑读高 CPU 的使用率也会随之升高，通常会在 Top 10 中出现诸如 latch: cache buffers chains、db file scattered read 等事件。

DB Name	DB Id	Instance	Inst num	Startup Time	Release	RAC
CRMDB2	3949975471	crmdb21	1	19-Sep-14 04:09	11.2.0.4.0	YES

Host Name	Platform	CPUs	Cores	Sockets	Memory (GB)
ncrmdb21	HP-UX IA (64-bit)	64	64	32	511.49

	Snap Id	Snap Time	Sessions	Cursors/Session	Instances
Begin Snap:	79707	05-Nov-14 17:00:29	6559	2.3	2
End Snap:	79708	05-Nov-14 17:30:12	6785	2.2	2
Elapsed:		29.70 (mins)			
DB Time:		2,807.24 (mins)			

图 9-1

Load Profile	Per Second	Per Transaction	Per Exec	Per Call
DB Time(s):	94.5	0.1	0.00	0.00
DB CPU(s):	22.1	0.0	0.00	0.00
Redo size (bytes):	9,134,467.7	7,612.5		
Logical read (blocks):	973,915.2	811.6		
Block changes:	42,477.2	35.4		
Physical read (blocks):	9,720.7	8.1		
Physical write (blocks):	1,111.7	0.9		
Read IO requests:	4,447.6	3.7		
Write IO requests:	442.3	0.4		
Read IO (MB):	75.9	0.1		
Write IO (MB):	8.7	0.0		
Global Cache blocks received:	1,050.0	0.9		
Global Cache blocks served:	461.2	0.4		
User calls:	37,994.5	31.7		
Parses (SQL):	17,110.1	14.3		
Hard parses (SQL):	9.6	0.0		
SQL Work Area (MB):	24.2	0.0		
Logons:	12.5	0.0		
Executes (SQL):	19,216.4	16.0		
Rollbacks:	52.9	0.0		
Transactions:	1,199.7			

图 9-2

如图 9-3 所示，在 Top 10 Foregrand Events by Total Wait Time 的部分，可以看到在该时间段的主要等待事件是 latch:cache buffers chains 和 db file sequential read。它们的%DB Time 分别占到 38.9%和 24.3%。后者代表单块读，是一种比较常见的物理 IO 等待事件，通常在数据块从磁盘读入到相连的内存空间中时发生，也可能是 SQL 语句使用了 selectivity 不高的索引，从而导致访问了过多不必要的索引块或者使用了错误的索引，这些等待说明 SQL 语句的执行计划可能不是最优的。前者是导致数据库逻辑读高的根本原因，由此推断某个或者某些 SQL 语句出现了性能衰变。

在接下来的 SQL Statistics 部分，分别截取了 SQL ordered by Elapsed Time 和 SQL ordered by Gets。如图 9-4 和 9-5 所示，可以发现 SQL Id=g5z291fcmwz08 的语句分别占了 42.50%的 DB Time 和 35.19%的逻辑读，而其他 SQL 所占的 DB Time 和逻辑读分别在 0.1% ~ 5%。由此可以确定，

就是该 SQL 语句影响系统性能，但还需要详细了解该 SQL 的执行计划、绑定变量和当时的逻辑读等信息。图 9-4 显示了该 SQL 的文本。

Top 10 Foreground Events by Total Wait Time

Event	Waits	Total Wait Time (sec)	Wait Avg(ms)	% DB time	Wait Class
latch: cache buffers chains	4,930,076	65.5K	13	38.9	Concurrency
db file sequential read	6,205,767	40.9K	7	24.3	User I/O
DB CPU		39.3K		23.4	
log file sync	2,079,804	8708.2	4	5.2	Commit
db file scattered read	616,321	4580.4	7	2.7	User I/O
gc buffer busy acquire	501,254	2600.3	5	1.5	Cluster
gc cr grant 2-way	2,741,705	2148.1	1	1.3	Cluster
db file parallel read	89,869	1056	12	.6	User I/O
gc cr multi block request	462,968	1030	2	.6	Cluster
read by other session	163,480	1005.1	6	.6	User I/O

图 9-3

SQL ordered by Elapsed Time

- Resources reported for PL/SQL code includes the resources used by all SQL statements called by the code.
- % Total DB Time is the Elapsed Time of the SQL statement divided into the Total Database Time multiplied by 100
- %Total - Elapsed Time as a percentage of Total DB time
- %CPU - CPU Time as a percentage of Elapsed Time
- %IO - User I/O Time as a percentage of Elapsed Time
- Captured SQL account for 69.7% of Total DB Time (s): 168,434
- Captured PL/SQL account for 1.2% of Total DB Time (s): 168,434

Elapsed Time (s)	Executions	Elapsed Time per Exec (s)	%Total	%CPU	%IO	SQL Id	SQL Module	SQL Text
71,581.38	2,281	31.38	42.50	7.46	0.01	g5z291fcmwz08	tpengine@wcrmtux1 (TNS V1-V3)	SELECT TO_CHAR(A.CUSTID) CUST...
4,122.36	1,065,717	0.00	2.45	21.53	72.79	a6bkjndw6rquh	JDBC Thin Client	SELECT USERSTATUS FROM (SELECT...
3,539.33	170,984	0.02	2.10	8.59	86.84	f70f38az80sfd	tpengine@winftux4 (TNS V1-V3)	Select to_char(REGION) REGION,...
3,273.87	42,759	0.08	1.94	7.07	84.29	2cy5xqmxkrmhd	tpengine@winftux4 (TNS V1-V3)	Select to_char(REGION) REGION,...
2,618.83	21,128	0.12	1.55	6.53	91.04	cs2hxvs39f00d	tpengine@wcrmtux2 (TNS V1-V3)	UPDATE subs_spservice SET stat...
2,568.68	22,267	0.12	1.53	4.81	93.41	62kbdd8s4wnnw	tpengine@winftux4 (TNS V1-V3)	UPDATE CS_CU_SUBS_SPSERVICE_LI...
1,953.98	159,450	0.01	1.16	6.65	87.45	6mf11z0d64m6x	tpengine@winftux4 (TNS V1-V3)	SELECT /*+ INDEX(CS_SUBS_PROD...
1,906.46	45,740	0.04	1.13	6.11	88.06	ddfpuf22c8gjy	JDBC Thin Client	UPDATE SUBS_SERVICE_AUDIT SET ...
1,902.00	166,724	0.01	1.13	8.41	87.33	4xtb57uh0fgdu	tpengine@winftux4 (TNS V1-V3)	SELECT TO_CHAR(REGION) REGION...
1,775.98	0		1.05	8.56	85.38	8s03ug4ytgh1c	sqlplus@ncrmdb21 (TNS V1-V3)	select count(*), CUSTTYPE from...

图 9-4

SQL ordered by Gets

- Resources reported for PL/SQL code includes the resources used by all SQL statements called by the code.
- %Total - Buffer Gets as a percentage of Total Buffer Gets
- %CPU - CPU Time as a percentage of Elapsed Time
- %IO - User I/O Time as a percentage of Elapsed Time
- Total Buffer Gets: 1,735,585,100
- Captured SQL account for 62.1% of Total

Buffer Gets	Executions	Gets per Exec	%Total	Elapsed Time (s)	%CPU	%IO	SQL Id	SQL Module	SQL Text
610,749,218	2,281	267,755.03	35.19	71,581.38	7.5	0	g5z291fcmwz08	tpengine@wcrmtux1 (TNS V1-V3)	SELECT TO_CHAR(A.CUSTID) CUST...
73,238,635	1,065,717	68.72	4.22	4,122.36	21.5	72.8	a6bkjndw6rquh	JDBC Thin Client	SELECT USERSTATUS FROM (SELECT...
62,670,991	3	20,890,330.33	3.61	383.22	84.5	6.4	btd2rn5dw8gt2	JDBC Thin Client	select decode(a.region, '11', ...
41,797,282	85,557	488.53	2.41	286.72	98	.1	4c6p2nxbz1j5b	tpengine@wcrmtux5 (TNS V1-V3)	SELECT DISTINCT(A.FUNCTIONID)...
28,642,433	2,196	13,043.00	1.65	1,167.49	98	0	dsu9h2y0knb36	tpengine@wbgapp1 (TNS V1-V3)	select a.SERVNUMBER, a.REGION ...
25,427,652	1,076,015	23.63	1.47	293.36	90.7	0	6sjfpns5sa3bz	JDBC Thin Client	SELECT OUTER_PARAM_ID ITEMCODE...
23,682,111	40,782	580.70	1.36	617.04	79.2	0	5h6nxmjpj73jx	UnifyFileDeal_@ncrminf1 (TNS V1-V3)	update int_uf_log set FileProc...
15,535,210	54,469	285.21	0.90	549.12	24.5	54.1	agpjs9nqm3hyb	tpengine@wcctux1 (TNS V1-V3)	SELECT * FROM DUAL WHERE (EXI...
14,372,084	2,195	6,547.65	0.83	357.99	93.9	0	32tavn5jxprr9		SELECT TELNUM, DATA_PRI FROM (...
14,370,321	2,194	6,549.83	0.83	353.28	94.6	0	843hfcz2n81yy		SELECT TELNUM, DATA_PRI FROM (...

图 9-5

```
SELECT TO_CHAR(A.CUSTID) CUSTID,
       TO_CHAR(A.REGION) REGION,
       A.CUSTNAME CUSTNAME,
       A.SHORTNAME SHORTNAME,
       A.CUSTTYPE CUSTTYPE,
```

```sql
       A.VIPTYPE VIPTYPE,
       TO_CHAR(A.FOREIGNER) FOREIGNER,
       A.CUSTCLASS1 CUSTCLASS1,
       A.CUSTCLASS2 CUSTCLASS2,
       A.NATIONALITY NATIONALITY,
       A.ADDRESS ADDRESS,
       A.CERTID CERTID,
       A.CERTTYPE CERTTYPE,
       A.CERTADDR CERTADDR,
       A.LINKMAN LINKMAN,
       A.LINKPHONE LINKPHONE,
       A.HOMETEL HOMETEL,
       A.OFFICETEL OFFICETEL,
       A.MOBILETEL MOBILETEL,
       A.POSTCODE POSTCODE,
       A.LINKADDR LINKADDR,
       A.EMAIL EMAIL,
       A.HOMEPAGE HOMEPAGE,
       TO_CHAR(A.ISMERGEBILL) ISMERGEBILL,
       A.ORGID ORGID,
       A.CREATEDATE CREATEDATE,
       A.NOTES NOTES,
       A.STATUS STATUS,
       A.STATUSDATE STATUSDATE,
       A.RESPONSECUSTMGR RESPONSECUSTMGR,
       A.CURRENTCUSTMGR CURRENTCUSTMGR,
       A.CREDITLEVEL CREDITLEVEL,
       TO_CHAR(A.INLEVEL) INLEVEL,
       A.REGSTATUS REGSTATUS,
       A.OWNERAREAID OWNERAREAID,
       A.CERTSTARTDATE CERTSTARTDATE,
       A.CERTENDDATE CERTENDDATE,
       A.COUNTRYID COUNTRYID
  FROM CUSTOMER A, PERSON_CUSTOMER B
 WHERE A.CUSTID = B.CUSTID
   AND A.CUSTTYPE = :CUSTTYPE
   AND A.CERTID = :CERTID
   AND A.CERTTYPE IN ('BusinessLicence')
 ORDER BY DECODE(A.STATUS, 'stcmNml', 0, 1) ASC
```

通过分析可以导出该 SQL Report。如图 9-6 所示，记录了故障时间段 SQL Id=g5z291fcmwz08 的 SQL 概要信息。

如图 9-7 所示，可以发现该 SQL 在 30 分钟内执行了 2281 次，单次的逻辑读在 267755.03，由此很可能认为这个 SQL 始终没有走到合适的索引或者是全表扫描。值得关注的是 267755.03 这个数字，它是一个平均值，很有可能被平均，换句话说，有时单次逻辑读很小、有时可能还会大于这个平均值。如图 9-8 所示的执行计划中，先通过索引 IDX_CUSTOMER_CERTID 过滤出 2 条数据，之后回 CUSTOMER 表得到 1 条记录，再通过结果集去驱动查询主键 PK_CM_CU_INDIVIDUAL，最后通过嵌套循环返回结果集。看上去每一步的 Rows 和 Cost 都非常理想，实际上存在 4 点值得关注的隐患。

WORKLOAD REPOSITORY SQL Report

Snapshot Period Summary

DB Name	DB Id	Instance	Inst num	Startup Time	Release	RAC
CRMDB2	3949975471	crmdb21	1	19-Sep-14 04:09	11.2.0.4.0	YES

	Snap Id	Snap Time	Sessions	Cursors/Session
Begin Snap:	79707	05-Nov-14 17:00:29	6559	2.3
End Snap:	79708	05-Nov-14 17:30:12	6785	2.2
Elapsed:		29.70 (mins)		
DB Time:		2,807.24 (mins)		

图 9-6

Plan Statistics

- % Total DB Time is the Elapsed Time of the SQL statement divided into the Total Database Time multiplied by 100

Stat Name	Statement Total	Per Execution	% Snap Total
Elapsed Time (ms)	71,581,380	31,381.58	42.50
CPU Time (ms)	5,339,060	2,340.67	13.57
Executions	2,281		
Buffer Gets	610,749,218	267,755.03	35.19
Disk Reads	2,469	1.08	0.01
Parse Calls	2,281	1.00	0.01
Rows	28,380	12.44	
User I/O Wait Time (ms)	9,683		
Cluster Wait Time (ms)	1,769,251		
Application Wait Time (ms)	0		
Concurrency Wait Time (ms)	64,402,308		
Invalidations	0		
Version Count	4		
Sharable Mem(KB)	62		

图 9-7

Execution Plan

Id	Operation	Name	Rows	Bytes	Cost (%CPU)	Time	Pstart	Pstop
0	SELECT STATEMENT				49 (100)			
1	SORT ORDER BY		1	206	49 (3)	00:00:01		
2	NESTED LOOPS		1	206	48 (0)	00:00:01		
3	PARTITION LIST SINGLE		1	196	6 (0)	00:00:01	KEY	KEY
4	TABLE ACCESS BY LOCAL INDEX ROWID	CUSTOMER	1	196	6 (0)	00:00:01	KEY	KEY
5	INDEX RANGE SCAN	IDX_CUSTOMER_CERTID	2		4 (0)	00:00:01	KEY	KEY
6	PARTITION RANGE ALL		1	10	42 (0)	00:00:01	1	14
7	INDEX RANGE SCAN	PK_CM_CU_INDIVIDUAL	1	10	42 (0)	00:00:01	1	14

图 9-8

（1）Rows 或 Cost 仅代表一个估算值。

（2）谓词条件中用到的索引列都是绑定变量，在遇到绑定变量+窥视关闭（该数据库窥视关闭）时，即使有直方图信息，优化器在估算时总是认为数据分布是平均的。

（3）在扫描索引 PK_CM_CU_INDIDUAL 时，扫描了 1 到 14 个分区。

（4）CUSTOMER 表返回的记录数稍有偏差，将会增加与 PK_CM_CU_INDIDUAL 索引循环的次数，也会影响整个查询的性能。

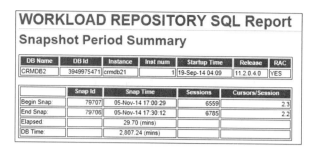

在继续往下分析之前，开发人员抛出一个疑问，为什么同一个 SQL 在其他两个库都正常，而在这个库却出现了问题？通过以下查询不难发现，这个 SQL 在每一个库的单次平均逻辑读都相似，见图 9-9。从执行次数上看，其他两个库基本维持在几十或上百次，而在故障库中峰值高达 2281 次，这同样是故障点的执行次数。这表明该 SQL 的性能是低效的，对数据库造成的性能冲击也是随着执行次数的增加而愈演愈烈的。

```
select *
  from (select BEGIN_INTERVAL_TIME,
               a.instance_number,
               plan_hash_value,
               EXECUTIONS_DELTA exec,
               round(BUFFER_GETS_DELTA / EXECUTIONS_DELTA) per_get,
               round(ROWS_PROCESSED_DELTA / EXECUTIONS_DELTA, 1) per_rows,
               round(ELAPSED_TIME_DELTA / EXECUTIONS_DELTA / 1000000, 2) time_s,
               round(DISK_READS_DELTA / EXECUTIONS_DELTA, 2) per_read
      from dba_hist_SQLstat a, DBA_HIST_SNAPSHOT b
     where a.snap_id = b.snap_id
       and EXECUTIONS_DELTA <> 0
       and a.instance_number = b.instance_number
       and a.SQL_id =  'g5z291fcmwz08'
     order by 1 desc)
 where rownum < 30;
```

crmdb11:

BEGIN_INTERVAL_TIME	I	EXEC	PER_GET	PER_ROWS	TIME_S
31-OCT-14 06.00.17.751 PM	2	419	149105	7	0.64
31-OCT-14 05.30.03.445 PM	2	339	240970	10.7	1.01
31-OCT-14 05.00.47.614 PM	2	113	403945	17.3	1.7
31-OCT-14 04.30.25.662 PM	2	50	576596	25.2	3.19
31-OCT-14 04.00.06.151 PM	2	68	1024521	43.3	4.21
31-OCT-14 03.30.06.859 PM	2	42	1887467	78.7	7.97
31-OCT-14 03.00.29.984 PM	2	38	2086138	86.9	8.77
31-OCT-14 02.30.09.929 PM	2	35	2264954	94.5	9.27
31-OCT-14 02.00.32.728 PM	2	26	1755498	73.8	8.22
31-OCT-14 11.00.13.699 AM	2	28	2402213	100	9.69
31-OCT-14 10.30.17.733 AM	2	40	1921781	80.2	7.86
31-OCT-14 10.00.03.259 AM	2	44	1801671	75.2	7.41
31-OCT-14 09.30.24.515 AM	2	30	2081925	86.8	8.61
31-OCT-14 09.00.12.952 AM	2	35	2264945	94.3	9.01
31-OCT-14 08.30.02.414 AM	2	12	2402214	100	12.34
30-OCT-14 06.00.29.979 PM	2	173	222248	11	0.87
30-OCT-14 05.30.16.203 PM	2	90	774101	33.4	3.1
30-OCT-14 05.00.35.627 PM	2	76	1011505	43.1	4.14
30-OCT-14 04.30.09.578 PM	2	66	1128344	47.7	4.65
30-OCT-14 04.00.34.487 PM	2	50	15374		
30-OCT-14 03.30.20.709 PM	2	35	22837		
30-OCT-14 03.00.07.952 PM	2	38	20950		
30-OCT-14 02.30.22.643 PM	2	39	20070		
30-OCT-14 02.00.12.276 PM	2	36	20668		
30-OCT-14 11.00.01.002 AM	2	38	20229		
30-OCT-14 10.30.47.360 AM	2	40	19217		
30-OCT-14 10.00.34.003 AM	2	28	10303		
30-OCT-14 09.30.09.877 AM	2	27	20463		
30-OCT-14 09.00.14.626 AM	2	31	23247		

crmdb21:

BEGIN_INTERVAL_TIME	I	EXEC	PER_GET	PER_ROWS	TIME_S
05-NOV-14 09.00.40.543 PM	1	20	1628504	68.1	6.15
05-NOV-14 08.30.26.854 PM	1	353	663432	29.9	3.23
05-NOV-14 08.00.12.747 PM	1	565	781964	34.3	6.37
05-NOV-14 07.30.24.857 PM	1	844	702189	32.5	8.22
05-NOV-14 07.00.12.189 PM	1	1323	426515	21.6	6.51
05-NOV-14 06.30.04.255 PM	1	1218	430483	22.1	5.4
05-NOV-14 06.00.21.009 PM	1	1598	292567	17.1	3.75
05-NOV-14 05.30.12.004 PM	1	2263	284468	15.4	38.78
05-NOV-14 05.00.29.934 PM	1	2281	267755	12.4	31.38
05-NOV-14 04.30.11.785 PM	1	606	177101	9.4	15.57
05-NOV-14 04.00.01.453 PM	1	1865	382885	18.1	18.04
05-NOV-14 03.30.29.082 PM	1	334	295827	11.4	28.49
05-NOV-14 03.00.17.771 PM	1	1122	505486	17.1	122.14
05-NOV-14 02.30.03.981 PM	1	1129	610392	23.6	58.23
05-NOV-14 02.00.24.779 PM	1	1682	428338	17.8	15.99
05-NOV-14 01.30.15.629 PM	1	1648	527988	23	13.19
05-NOV-14 01.00.10.186 PM	1	1145	0	22.9	7.89

crmdb31:

BEGIN_INTERVAL_TIME	I	EXEC	PER_GET	PER_ROWS	TIME_S
03-NOV-14 11.00.28.864 AM	2	120	540662	31.7	2.62
03-NOV-14 10.30.15.115 AM	2	79	841688	38.9	3.94
02-NOV-14 12.00.35.173 PM	2	28	1438402	60.8	5.97
02-NOV-14 08.30.11.474 AM	2	63	1439481	59	6.52
02-NOV-14 08.00.03.147 AM	2	83	1712738	69.9	8.5
01-NOV-14 06.30.26.791 PM	2	98	600292	25.6	2.79
01-NOV-14 05.30.02.619 PM	2	97	1238147	50.9	6.65
01-NOV-14 03.00.05.802 PM	2	107	641627	31.7	2.9
01-NOV-14 12.00.01.320 PM	2	77	891296	36.9	6.01
01-NOV-14 11.00.23.968 AM	2	177	387779	16.9	1.97
01-NOV-14 10.30.09.352 AM	2	242	506464	22	2.97
01-NOV-14 09.00.27.419 AM	2	80	742587	31.8	3.63
01-NOV-14 08.30.17.193 AM	2	69	1128279	45.2	5.14
01-NOV-14 08.00.03.474 AM	2	39	1256926	51.4	6.23

图 9-9

谓词条件中用到了绑定变量，需要找到当时的绑定变量值来看真实的执行计划。

```
select name,value_string,last_captured From dba_hist_SQLbind
Where SQL_id= ' g5z291fcmwz08'  Order by last_captured desc
NAME                VALUE_STRING                LAST_CAPTURED
-------------       ----------------------      -----------------
```

```
:CERTID            1xxx1159440(20141025)    2014/11/5 18:27
:CUSTTYPE          1                        2014/11/5 18:27
:CUSTTYPE          1                        2014/11/5 18:17
:CERTID            3xxxxxxxx0402040         2014/11/5 18:05
:CUSTTYPE          1                        2014/11/5 18:05
:CERTID            3xxxxxxxx402044          2014/11/5 17:47
:CUSTTYPE          1                        2014/11/5 17:47
:CERTID            3xxxxxxxx402045          2014/11/5 16:50
:CUSTTYPE          1                        2014/11/5 16:50
:CERTID            3xxxxxxxx402056          2014/11/5 16:00
:CUSTTYPE          1                        2014/11/5 16:00
:CERTID            3xxxxxxxx402045          2014/11/5 15:50
:CUSTTYPE          1                        2014/11/5 15:50
```

在上述分析中提到，"如果 CUSTOMER 表返回的记录数稍有偏差，将会增加与 PK_CM_CU_INDIDUAL 索引循环的次数，从而影响整个查询的性能"。在回表之前先通过索引字段 CERTID 过滤数据，该字段的数据分布非常重要。即该字段的值是否存在数据倾斜至关重要，下面对其进行统计。

```
select CERTID, count(*)
  from tbcs.CUSTOMER partition(p_l_1) a
 where CERTID in ('1xxxxxx00402045',
                  '1xxxxxx9440(20141025)',
                  '3xxxxxx00402040',
                  '3xxxxxx00402056',
                  '3xxxxxx00402045',
                  '3xxxxxx00402044',
                  '1xxxxxx4473(20141025)')
 group by CERTID;
CERTID                         COUNT(*)
----------------------         ----------
1xxxxxx9440(20141025)                 1
1xxxxxx4473(20141025)                 1
3xxxxxx00402040                      45
3xxxxxx00402044                      14
3xxxxxx00402045                   59084
3xxxxxx00402056                       1
3xxxxxx00402045                       4
```

结果很明显，该字段使用过的值出现了严重的数据倾斜。我们把 3xxxxxx00402045 代入 WHERE 条件，通过/*+gather_plan_statistics*/查看优化器的估计值与真实值之间的差异。在扫描 IDX_CUSTOMER_CERTID 索引时，估算值是 1，而实际值是 59106，回表之后的记录数依然是 59106，这表明至少要和下面的索引关联 59106 次，被驱动表恰巧共有 14 个分区，循环的次数就变成了 59106×14 次，这也是单次逻辑高达 2503731 的原因，如图 9-10 所示。

说明，/*+gather_plan_statistics*/这个提示会记录每一步操作中真实返回的行数（A-Rows）、逻辑读（Buffers）和耗费的时间（A-Time）。

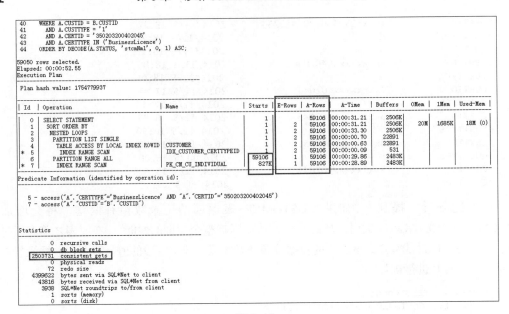

图 9-10

同时也测试了一个不倾斜的值进行对比（3xxxxxx00402044），发现逻辑读只有 760，这说明了该 SQL 被执行的次数越多，逻辑读就会被平均得更小，如图 9-11 所示。

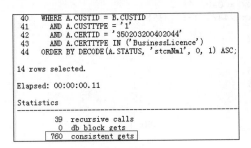

图 9-11

经过以上分析可知，数据倾斜、全分区扫描、执行次数三者相加之和的增加使得该 SQL 的性能影响了整个系统。接下来从两个方面对其进行优化。

（1）从 SQL 方面。

编写一个新 SQL 专门针对倾斜值，用 HASH JOIN 替代默认的 NEST LOOP。

（2）从业务方面。

1）检查倾斜数据的合理性。

2）在访问 PERSON_CUSTOMER 表时，可以增加分区字段加以限制，从而避免扫描多余

3）梳理业务场景。

通过和开发人员、业务人员反复沟通确认，该条 SQL 在业务场景和数据倾斜上都是可以修正和清理的。在完全满足业务场景的前提下，开发人员对 SQL 做了调整，通过改写 union all 和 rownum <2，有效合理地控制了返回的记录数，逻辑读也从 26 万降低到了 1303，性能提升了近 200 倍，如图 9-12 所示。这个案例的优化过程也是围绕技术和业务反复优化的过程。

```
select ….
  FROM CUSTOMER A, PERSON_CUSTOMER B
WHERE A.CUSTID = B.CUSTID
  AND A.CUSTTYPE = :CUSTTYPE
  AND A.CERTID = :CERTID
  AND A.CERTTYPE IN ('BusinessLicence')
  AND A.STATUS = 'stcmNml'
  AND ROWNUM < 2
union all
select …
  FROM CUSTOMER A, PERSON_CUSTOMER B
WHERE A.CUSTID = B.CUSTID
  AND A.CUSTTYPE = :CUSTTYPE
  AND A.CERTID = :CERTID
  AND A.CERTTYPE IN ('BusinessLicence')
  AND A.STATUS != 'stcmNml'
  AND ROWNUM < 2
```

```
Predicate Information (identified by operation id):
---------------------------------------------------

  2 - filter(ROWNUM<2)
  5 - filter("A"."CERTTYPE"='BusinessLicence' AND "A"."STATUS"='stcmNml')
  6 - access("A"."CERTID"='350203200402045')
  8 - access("A"."CUSTID"="B"."CUSTID")
  9 - filter(ROWNUM<2)
 12 - filter("A"."CERTTYPE"='BusinessLicence' AND "A"."STATUS"<>'stcmNml')
 13 - access("A"."CERTID"='350203200402045')
 15 - access("A"."CUSTID"="B"."CUSTID")


Statistics
----------------------------------------------------
          0  recursive calls
          0  db block gets
       1303  consistent gets
          0  physical reads
          0  redo size
       3794  bytes sent via SQL*Net to client
        520  bytes received via SQL*Net from client
          2  SQL*Net roundtrips to/from client
          0  sorts (memory)
          0  sorts (disk)
          2  rows processed
```

图 9-12

9.2 总结

在移动运营商 SQL 审核项目的交付过程中，和大家分享以下几点。

（1）在 AWR、SQL Report 中 Per 所对应的往往是平均值，需要结合执行计划和数据分布来分析该平均值是否真实可靠。

（2）当全分区扫描出现在执行计划中时，需要结合业务判断其合理性。

（3）绑定变量的历史值和其数据分布能够提高分析问题的精确性。

（4）提高 SQL 的性能存在着多种优化方法，随着系统本身和系统架构的推进，SQL 优化需要结合技术、数据特点、业务场景等要素综合分析，从而才能给客户提出一套完整、全面、可执行的优化方案。

第 10 章　戒骄戒躁、细致入微

——我的 SQL 生涯笔记（怀晓明）

> **题记**
>
> 怀晓明是 ITPUB 论坛版主，同样成长于论坛的网络时代，他的开发经历以及卓越的个人学习能力，成就了他精深的 SQL 造诣。作为 ITPUB 论坛几届 SQL 大赛的评委，他在组织方面的能力也得以显现。在本章的内容中，我们汇集了他在成长过程中的学习、思考、技术积累和总结分享等经验，希望他的思路和方法，可以为大家的成长提供一定借鉴。

10.1　我的职业生涯

我 2001 年大学毕业，起初在大型国有 IT 企业工作，主要做电子政务方面的项目。第一家公司做了 6 年，从售后工程师到程序员再做到项目经理，第二家公司和第三家公司都担任项目经理兼系统架构师。那时候我看到移动互联网的迅猛发展，希望接受更多的挑战。2011 年，我毅然投入到互联网的大环境里，加入了一家创业公司。在此期间，我主要负责全面的技术管理工作，兼做数据库开发方面的技术工作，目标是使用较低的技术成本实现高效率的系统。

随后我加入了云和恩墨公司，主要职责是利用自己丰富的 SQL 能力、架构能力帮助用户进行系统优化。现在更丰富的实践积累让我有能力帮助更多的用户实现更优化的架构。

从 2002 年注册 ITPUB 开始，我的 Oracle 方面的技术知识基本上来自官方文档和论坛（包括在积极为网友解决问题时学到的知识），ITPUB 见证了我的成长和梦想。我在论坛上的网名是"lastwinner",有人说是"最终赢家"的意思,其实不是。这个名字的含义是"只有坚持才能成为最终的胜利者!"——To last long to be the last winner! last 除了"最后的"含义外,还可以作为动词"坚持"去使用。我更喜欢这个含义。这句话也送给大家,希望大家都能在自己的岗位上通过坚持不懈的努力取得更好更优的成绩!

10.2 运维的现状及发展

目前运维面临一个重大的问题,很多公司并没有意识到提高开发技术可以有效地提高系统性能,即使有的公司意识到了也没有资源去做。我认为这个现象的本质是老板们没意识到提高开发技术其实是可以降低开发和后期运营维护成本的。有可靠的统计数据指出,80%的数据库性能问题出在攻城狮编写的 SQL 上。而在这现象背后更根本的原因是,没有可胜任数据库开发工作的攻城狮!一旦出现系统性能问题,大家第一反应就是去找调优高手来优化 SQL,久而久之,这就成了一个习惯。老外有句谚语———天一个苹果,你就不需要医生了,这说的就是预防为主。转换下思路,如果提高了攻城狮们的开发水平,甚至是配备了专职的数据库开发工程师,那写出较高质量的 SQL 就不是什么难事儿。这样就避免了多数性能方面的隐患,自然就降低了后期出现性能问题的概率,也免去了大量请人做调优的成本,而提高攻城狮开发水平的成本并不是特别高,何乐而不为?ISO-9000 告诉我们,质量是生产出来的,不是检测出来的。同样,高质量的 SQL 应该是开发写出来的,而不应总是通过 DBA 去调优出来。无论公司是否意识到、是否有资源去做,提高开发技术尤其是数据库端的开发技术都是大势所趋,不去迎面解决问题而装鸵鸟的做法是不可取的。

10.3 如何提高数据库的开发水平

提高数据库的开发水平是一个系统性的工作。简单说就是理论与实践充分结合,只懂理论或只会实践是不可取的。要学会用理论指导实践,通过实践验证理论,在实践过程中不断丰富理论知识,在理论指引下不断提高实践能力。我认为一个好的数据库开发者应该具备以下品质。

● 掌握 SQL 基础知识和数据库基本理论。这有助于理解 SQL 是如何运作的,什么样的 SQL 会跑得更快。可以通过学习相关白皮书或者技术文档掌握。

● 学会提问。提问是一门艺术,无论学什么都需要掌握这门艺术。能够准确地描述问题其实是解决问题的关键。

● SQL 中高级知识。这能让 SQL 成为你的有力工具。可以通过阅读官方文档,经常参与 Itpub 的数据库开发版块学习从而提高自己的水平。学习时不要想当然,就像 trim 并不等于 rtrim + ltrim,认真读文档的人都知道。

● 掌握至少一门相关的开发语言。Java、PHP 等都行,这有助于你从另一个视角来认识数据库开发。

- 一定的数学能力，最好具备高中以上的数学知识。良好的数学素养可以为你带来新的思路和方法，有助于提高开发能力，并能帮助你理解。
- 一定的科学素养。类似于"某月有 5 个周五、周六和周日，这种现象 823 年才出现一次"的论调，或"1582 年 10 月 5 日~14 日这 10 天是不存在的"等要一眼就要能看穿是假的（或者会通过程序去证伪）。

当然，学习任何技术都是没有捷径的，最重要的是勤奋和努力，要静心学习，切忌浮躁。脚踏实地才能学得真知。做到以上几点，假以时日，成为高手不是梦。

10.4 DBA 面临的挑战

如果技术进步，部分 DBA 失业，那只有一个原因——落伍。活到老就要学到老，而在技术的世界，学到老才能活到老。技术进步只会减少重复性的劳动，而且多数还是"体力"劳动。数据库方面的技术进步，将让 DBA 更加专注于业务方面的考量，发挥出 DBA 本身应有的更大的价值，但同时也对 DBA 提出了更高的要求。以前你擅长的工作现在可以轻易被新技术所取代，那你就必须学会新的本领。现实是很残酷的，尽管我们处于文明社会中，"弱肉强食"的丛林法则照样适用。但是，机遇总是和挑战并行的，敞开心扉，纳入新知，新时代的弄潮儿非你莫属！要知道，技术进步也会促进新职位的出现，因为技术进步会促进产业分工明晰，这时候 DBA 必然会分化到各个不同的细分领域，比如"产品 DBA""调优 DBA"和"应用开发 DBA"等。

综上所述，我认为，"应用开发 DBA"的群体会越来越壮大，因为这部分群体有强烈的市场需求。而作为常规意义上的 DBA，多掌握一些开发技术，会让你的数据库管理更加游刃有余，轻松许多！轻松应对将来的职业挑战。

当然这并不意味着运维 DBA 的价值会减少。相反，性能问题和安全问题一直存在，在大数据与云计算的时代里，这些问题越发突出。**以数据驱动决策的时代里，数据就是企业的生命，因此越来越多的企业开始关注与重视数据**。在这样的环境下，能管好数据的 DBA 自然会身价不菲。

时代的发展不会淘汰谁，自己不发展才会被淘汰。

10.5 数据库优化的思考

针对数据库的优化，不同层次的开发者考虑问题的思路和角度是不完全一样的。作为我个人而言，会先考虑业务上为什么要这样做，其次考虑业务是否可以简化，再次考虑是通过 SQL、PL/SQL 还是前端程序去实现才能达到最好的效果（需要综合效率、可维护性、实现复杂度等多方面去考虑），再后考虑具体的算法，最后才去考虑具体算法的实现。

而在具体的实现上，先依据自身的知识和经验选择一个合适的算法，并通过合适的 SQL 去实现要求，检查无误后再考虑 SQL 优化的事情。一般在经历过大量的数据库开发工作后，基本上一出手 SQL 的质量都不会差，通常也就没什么优化空间了。需要注意的是，有时候 SQL 优化仅仅是某几个 SQL 优化的事，而有时候 SQL 优化却需要改动技术方案以支撑。技术方案基于开发技术但又高于开发技术，所以掌握了扎实的开发技术，才能将技术方案和相关设计工作这样重要的事情做好。

我举几个工作中的案例。

- 数据迁移。需要从 A 服务器上的一张将近有一亿条记录的表中抽取数据，将其导入到 B 服务器中。当时用 java 来当搬运工，每次生成 2 000 条数据的 xml 文件。实测发现每次生成 xml 文件花的时间大约在 2 分钟左右。调整为每次生成 5 000 条数据的 xml 文件，花的时间大约在 4 分钟左右，速度没有太明显的改善（规范的上限是 5 000 条/xml 文件，不能再多了）。这个速度很要命，一天才 1 440 分钟，全部导出完成差不多要俩月时间，这还是在速度恒定的情况下。可事实是随着数据不停地导出，导出速度会越来越慢。经过不断地改进开发方法，包括调整技术开发方案，最终我们将导出的速度优化为每次生成 2 000 条数据的 xml 文件只需要 2 秒，这样一天多点我们就可以将一亿条记录的表中的数据都抽取出来了，这是程序的优化。

- 数据检索。某 SQL 需要提取出某博文的前 200 字作为摘要，将其显示在作者博客的首页上。结果开发人员 SQL 是这么写的。

SELECT blog_id, blog_title, blog_text from blog where …

然后在前端程序里截取 blog_text 的前 200 字展示出来。在访问量大的时候，这导致了严重的系统性能问题。一查原因，网络 I/O 负载过大。调优很简单，SQL 改为：

SELECT blog_id, blog_title, substr(blog_text, 1, 200) as blog_text from blog where …

性能问题马上消除了。为什么？因为改写后的 SQL 只从数据库中取出少量信息发送到前端，而不是像之前那样发送全部博文到前端，自然就降低了网络 I/O，性能自然而然也就提高了。当然，在用户量和访问量进一步提高的时候，就得采取其他的方法了，比如加缓存之类的，这是另外的话题，就不细说了。这是 SQL 的优化。

- 并行导入。当时我们和别的公司都需要往数据库里导入 300G 的数据，我采取了多进程的方式导入，结果最快的一次不到 7 小时就全部导入完成了。而另外那家公司 5 天了还没导完，双方的服务器配置都差不多，怎么时间差这么多？经过了解，他们采用单进程的方式。这并不是什么高深的技术，但知道和不知道的成本有时候就会相差很多。这是方法的优化。

所谓优化，其实就是在充分了解技术之后的灵活运用。

10.6 提问的智慧

前面我提到，提问也是一种技能、一种艺术。关于这个话题，还要从我刚开始学数据库的时候说起。

那时候，因为刚入门很多东西都不了解，所以**每次提问，我都尽量将问题描述清楚**。就算是一个多行 SQL 在 SQL*Plus 下执行的结果，我也会将每一行之前的行号去掉再贴出来，这样就方便别人复制粘贴去还原问题环境了。每次贴出的问题，也不会直接用系统中的数据和表结构来做演示，因为那时的论坛环境远不像现在这样，一是人没这么多，二是高手不经常上来。如果问题太复杂，别人想帮我也会觉得困难多多从而放弃。这我能理解，毕竟每个人的业余时间是有限的，所以我都会新建一些测试表和测试数据，用尽量简单的表结构和数据来重现问题。虽然这会推迟问题提出的时间，但事实上很多问题都能在较短的时间内得到解答。**我的经验告诉我，问题提出得快并不等于解决得也快，所谓"磨刀不误砍柴工"也**。

那时候也有不少问题的标题是"高手请进""这个 SQL 怎么写"等，这不能较为准确地展现所碰到问题的关键点。现在各种技术分享和提问的平台很多，但大部分人提问的方式还是将遇到的问题截图上传，问这个问题如何解决或者这是什么原因导致的，也不会将问题描述得清晰些，具体是哪里出错了或者自己卡在哪里不能理解。有些问题涉及的知识比较广泛或者比较深奥，别人也不可能一一去排除你的问题所在，更没有时间一一去确认，因此过于模糊的问题往往让热心人望而却步。

因此，在向别人提问之前，要做好自己该做的事情。比如出现系统宕机，若不能清晰描述问题所在，等高手指导你才去查告警日志，去生成 AWR 报告，很显然已经浪费了很多时间，而这种**常规的排错其实根本不需要别人的指导**。

浮躁心态一直都不鲜见，大家太急于立即成功，遇到问题不深入思考、不静下心来钻研。想着发贴自有高手解答何必麻烦。但忽略了一点，你若只是简单地将问题截图，很少描述或者描述得很模糊，恐怕连自己都不能搞清楚核心问题在哪里，那么别人如何帮你呢，最清楚问题的人就是你。若问题过于笼统没有重点，回答问题的热心人为了帮你解决问题，会尽量去猜测你未描述清楚的地方到底是什么样子。来回的沟通不但消耗了解决问题的时间，还消磨了热心人的耐性，对于你的问题的解决毫无益处。甚至当别人解决了你的问题，你得到了答案，但你可能仍然不懂，下次遇到类似的问题可能还是一头雾水，因为你没有思考过。**会不会提问，除了能够表现出你的思维逻辑方式之外，更能表现出你的技术水平、你的态度**。作为一个做技术的人，分析问题的能力很重要，甚至超过你解决问题的能力。当你遇到问题，要有很明确的思路，知道我要检查什么，这个问题可能跟什么有关，我是否遇到或者看到过类似的问题。当你在搜索完自己的知识体系后，做了相关的检查仍然没有找到答案的时候，其实你已经离答案很近了，只需要稍加指点即可。因此准确地描述问题，是解决问题的第一步也是至关重要的一步。

因为这不仅使别人更好地理解你的问题来帮助你，也让你自己对问题有清晰的思路，培养你的逻辑思维。

事实上，大多数人在提问时，如果能缓一缓，理清思路，准备好还原问题环境所需的脚本，让问题清晰明了，那你的问题也会更早得被解决。整个社会不光是做技术的在浮躁，其他各行各业各种职位也都很浮躁。但技术是很多工作的基石，而大家又都在从事技术工作，那就让我们从技术做起。**拒绝浮躁，第一步，就是学会提问。第二步，就是学会总结。**最简单的，问题解决后，将解决方法总结分享出来，以给后来人学习参考。要知道，只懂得索取而不付出的人，无论在工作还是生活中都是不受欢迎的，只有"我为人人"，才能"人人为我"。

10.7 细致入微方显价值——通过真实案例认识 SQL 审核

做任何事情，都需要细致认真，做得尽可能完美。作为数据运维的每一个人，这一点品质更是必不可少。

DevOps 理念逐渐为大众认可，SQL 审核作为协同开发和运营工作的一项服务，或者一种工作手段，也显得异常重要。通过 SQL 审核可以提前消除数据库隐患，为未来系统更稳定的运行打下良好的基础。

然而，如何将 SQL 审核做好，是一件很有挑战性的事情。我将通过两个具体案例，来展现 SQL 审核工作如何才能做到更好，更有价值。

10.7.1 案例一 不仅仅是 NULL 的问题

在某系统中，通过 SQL 审核的规则进行问题筛选，发现了一条违反规则的语句。这个 SQL 非常简单，语句如下。

```
delete from publckrec where RecKey = null
```

该语句违反了"索引全扫"规则，也就是执行计划选择了全索引扫描的执行方式。

首先我们看一下这条 SQL 的执行计划，如图 10-1 所示。

```
| Id | Operation          | Name        | Rows | Bytes | Cost (%CPU)| Time     |
-------------------------------------------------------------------------------
|  0 | DELETE STATEMENT   |             |    1 |     5 |    0   (0) |          |
|  1 |  DELETE            | PUBLCKREC   |      |       |            |          |
|  2 |   FILTER           |             |      |       |            |          |
|  3 |    INDEX FULL SCAN | SYS_C0035850| 200K |  976K |  421   (1) | 00:00:06 |
-------------------------------------------------------------------------------
Predicate Information (identified by operation id):
---------------------------------------------------
   2 - filter(NULL IS NOT NULL)
```

图 10-1

我们需要知道的是，Oracle 在执行原 SQL 的时候，并不会真正去做全索引扫描。为什么呢？因为执行计划中的第二步的 filter，其断言是 NULL IS NOT NULL，这是永远不会成立的条件，所以 Oracle 是不会去执行全索引扫描。尽管全索引扫描的成本是 421，但是其最终的 Cost 成本是 0，也就是执行计划数的子分支并不会被执行。

回到这个 SQL 的书写上，如果你对 Oracle 的 NULL 有一定的了解，那么你一定能知道，该语句的正确写法应该如下。

```
delete from publckrec where RecKey IS null
```

这是因为在 Oracle 中，对 NULL 的比较必须使用 IS NULL 或者 IS NOT NULL。

故事到这里还没有结束！

作为一名合格的 SQL 审核工程师，我们应当能发现执行计划第三步用到的索引的名字并不是用户自定义的名字，而是系统自动生成的约束的名字。

我们知道，Oracle 只会自动为主键和唯一键这两种约束添加同名的索引。如果这是唯一键约束，那么改写就是最终的方案，如果是主键约束，那该语句就是一个无用的语句。

通过查看如图 10-2 所示的元数据（图 10-2 来自云和恩墨的 SQL 审核产品——Z3），我们发现，这是一个主键约束。

图 10-2

因为该表的 RecKey 字段为主键，不会存在 NULL 值，所以从数据库的角度看，此 SQL 总是删除 0 条记录，实质上是一条无用语句。那么既然无用，为什么还会有语句被审核到？这连单元测试都通不过，开发不是应该在开发阶段就应该消灭这种错误的吗？

当问题反馈给开发后，得到开发的反馈是前台 JavaScript 有 Bug，导致传递的键值变成了 null，所以出现了如上问题。

所以最终的解决方案并不是不执行该 SQL，而是修正前台 JavaScript 的 Bug，并采用绑定变量的方式编码。

在数据运维的过程中，最忌讳经验论。遇到问题不应该想当然地去处理，而应该结合系统的具体情况进行深入分析。同样对于一些复杂、繁琐的工作，我们可以借助一些工具来减轻工作量，提高效率。

10.7.2 案例二 意想不到的优化方式

某电信客户通过 Top SQL 抓取，发现一消耗资源过高的问题语句，具体如下。

```
SELECT * FROM
  (SELECT * FROM
    (SELECT j.order_id, j.jms_xml
     FROM jms_cent j, om_order oo
     WHERE oo.service_id IN (355, 597)
     AND j.order_id        = oo.id
     AND j.state           = '-1'
     ORDER BY j.in_time ASC )
  WHERE rownum <= (10 - (SELECT COUNT(1)
    FROM
      (SELECT j.order_id, j.jms_xml
       FROM jms_cent j, om_order oo
       WHERE oo.service_id NOT IN (355, 597)
       AND j.order_id        = oo.id
       AND j.state           = '-1'
       AND rownum            <= 10
ORDER BY j.in_time ASC)
)))
UNION
SELECT * FROM
  (SELECT * FROM
    (SELECT j.order_id, j.jms_xml
     FROM jms_cent j, om_order oo
     WHERE oo.service_id NOT IN (355, 597)
     AND j.order_id        = oo.id
     AND j.state           = '-1'
     ORDER BY j.in_time ASC)
  WHERE rownum <=
    (SELECT COUNT(1)
     FROM
       (SELECT j.order_id, j.jms_xml
        FROM jms_cent j, om_order oo
        WHERE oo.service_id NOT IN (355, 597)
        AND j.order_id        = oo.id
        AND j.state           = '-1'
        AND rownum            <= 10
        ORDER BY j.in_time ASC
        )))
```

在生产环境中，该 SQL 平均每次消耗 10 万逻辑读，单次执行时间超过 2 秒，是非常影响整体性能的一条典型 SQL，如图 10-3 所示。

其执行计划如图 10-4 所示（测试环境，生产环境的执行计划与之类似）。

如果你有一定的开发经验，当看到该 SQL 及其执行计划时，你的第一反应应该是该 SQL 存在写法不当的问题（ORDER BY 与 rownum 用法不当）。该问题可以通过 WITH 来改写 SQL，

10.7 细致入微方显价值——通过真实案例认识 SQL 审核

改写后的 SQL 逻辑读，能降到原先的 1/4 左右。

Stat Name	Statement Total	Per Execution	% Snap Total
Elapsed Time (ms)	89,421,299	2,394.40	4.99
CPU Time (ms)	89,368,825	2,393.00	7.75
Executions	37,346		
Buffer Gets	3,853,773,909	103,191.08	10.67
Disk Reads	5,500	0.15	0.00
Parse Calls	23,973	0.64	0.01
Rows	103,437	2.77	
User I/O Wait Time (ms)	19,509		
Cluster Wait Time (ms)	0		
Application Wait Time (ms)	0		
Concurrency Wait Time (ms)	196		
Invalidations	0		
Version Count	224		
Sharable Mem(KB)	12,516		

图 10-3

```
| Id | Operation                          | Name        | Rows | Bytes | Cost (%CPU) | Time     |
|  0 | SELECT STATEMENT                   |             |   2  | 4030  | 36863  (56) | 00:07:23 |
|  1 |  SORT UNIQUE                       |             |   2  | 4030  | 36863  (56) | 00:07:23 |
|  2 |   UNION-ALL                        |             |      |       |             |          |
|  3 |    VIEW                            |             |   1  | 2015  | 18430  (11) | 00:03:42 |
|* 4 |     COUNT STOPKEY                  |             |      |       |             |          |
|  5 |      VIEW                          |             |   1  | 2015  |  9215  (11) | 00:01:51 |
|* 6 |       SORT ORDER BY STOPKEY        |             |   1  |  120  |  9215  (11) | 00:01:51 |
|  7 |        NESTED LOOPS                |             |   1  |  120  |  9214  (11) | 00:01:51 |
|* 8 |         TABLE ACCESS FULL          | JMS_CENT    |   1  |  110  |  9212  (11) | 00:01:51 |
|  9 |         TABLE ACCESS BY INDEX ROWID| OM_ORDER    |   1  |   10  |     2   (0) | 00:00:01 |
|*10 |          INDEX UNIQUE SCAN         | PK_OM_ORDER |   1  |       |     1   (0) | 00:00:01 |
| 11 |      SORT AGGREGATE                |             |   1  |       |             |          |
| 12 |       VIEW                         |             |   1  |       |  9215  (11) | 00:01:51 |
| 13 |        SORT ORDER BY               |             |   1  |  120  |  9215  (11) | 00:01:51 |
|*14 |         COUNT STOPKEY              |             |      |       |             |          |
| 15 |          NESTED LOOPS              |             |   1  |  120  |  9214  (11) | 00:01:51 |
|*16 |           TABLE ACCESS FULL        | JMS_CENT    |   1  |  110  |  9212  (11) | 00:01:51 |
| 17 |           TABLE ACCESS BY INDEX ROWID| OM_ORDER  |   1  |   10  |     2   (0) | 00:00:01 |
|*18 |            INDEX UNIQUE SCAN       | PK_OM_ORDER |   1  |       |     1   (0) | 00:00:01 |
| 19 |      SORT AGGREGATE                |             |   1  |       |             |          |
| 20 |       VIEW                         |             |   1  |       |  9215  (11) | 00:01:51 |
| 21 |        SORT ORDER BY               |             |   1  |  120  |  9215  (11) | 00:01:51 |
|*22 |         COUNT STOPKEY              |             |      |       |             |          |
| 23 |          NESTED LOOPS              |             |   1  |  120  |  9214  (11) | 00:01:51 |
|*24 |           TABLE ACCESS FULL        | JMS_CENT    |   1  |  110  |  9212  (11) | 00:01:51 |
| 25 |           TABLE ACCESS BY INDEX ROWID| OM_ORDER  |   1  |   10  |     2   (0) | 00:00:01 |
|*26 |            INDEX UNIQUE SCAN       | PK_OM_ORDER |   1  |       |     1   (0) | 00:00:01 |
| 27 |    VIEW                            |             |   1  | 2015  | 18430  (11) | 00:03:42 |
|*28 |     COUNT STOPKEY                  |             |      |       |             |          |
| 29 |      VIEW                          |             |   1  | 2015  |  9215  (11) | 00:01:51 |
|*30 |       SORT ORDER BY STOPKEY        |             |   1  |  120  |  9215  (11) | 00:01:51 |
| 31 |        NESTED LOOPS                |             |   1  |  120  |  9214  (11) | 00:01:51 |
|*32 |         TABLE ACCESS FULL          | JMS_CENT    |   1  |  110  |  9212  (11) | 00:01:51 |
| 33 |         TABLE ACCESS BY INDEX ROWID| OM_ORDER    |   1  |   10  |     2   (0) | 00:00:01 |
|*34 |          INDEX UNIQUE SCAN         | PK_OM_ORDER |   1  |       |     1   (0) | 00:00:01 |
| 35 |      SORT AGGREGATE                |             |   1  |       |             |          |
| 36 |       VIEW                         |             |   1  |       |  9215  (11) | 00:01:51 |
| 37 |        SORT ORDER BY               |             |   1  |  120  |  9215  (11) | 00:01:51 |
|*38 |         COUNT STOPKEY              |             |      |       |             |          |
| 39 |          NESTED LOOPS              |             |   1  |  120  |  9214  (11) | 00:01:51 |
|*40 |           TABLE ACCESS FULL        | JMS_CENT    |   1  |  110  |  9212  (11) | 00:01:51 |
| 41 |           TABLE ACCESS BY INDEX ROWID| OM_ORDER  |   1  |   10  |     2   (0) | 00:00:01 |
|*42 |            INDEX UNIQUE SCAN       | PK_OM_ORDER |   1  |       |     1   (0) | 00:00:01 |
| 43 |      SORT AGGREGATE                |             |   1  |       |             |          |
| 44 |       VIEW                         |             |   1  |       |  9215  (11) | 00:01:51 |
| 45 |        SORT ORDER BY               |             |   1  |  120  |  9215  (11) | 00:01:51 |
|*46 |         COUNT STOPKEY              |             |      |       |             |          |
| 47 |          NESTED LOOPS              |             |   1  |  120  |  9214  (11) | 00:01:51 |
|*48 |           TABLE ACCESS FULL        | JMS_CENT    |   1  |  110  |  9212  (11) | 00:01:51 |
| 49 |           TABLE ACCESS BY INDEX ROWID| OM_ORDER  |   1  |   10  |     2   (0) | 00:00:01 |
|*50 |            INDEX UNIQUE SCAN       | PK_OM_ORDER |   1  |       |     1   (0) | 00:00:01 |
```

图 10-4

但是云和恩墨为这个客户做了一段时间的 SQL 审核与优化，已经熟悉了开发商的开发风格，该系统的绑定变量编码其实一直做得不错，为什么出现了 oo.service_id NOT IN（355，597）和 j.state = '-1' 这样没有用绑定变量方式编码的代码呢？

实际上，jms_cent.state 上的数据分布非常不均匀。

```
ST      COUNT(*)
-----   --------
-1             3
 0         15623
 1         35782
```

经过与开发沟通，确认 state 为 "-1" 的数据的确是很少数。并且根据业务，该语句总是要查 state 为 "-1" 的值。然而该列上并没有索引，于是我们在测试环境上我们新建一个索引。

```
create index idx_jms_state_new on jms_cent(state);
```

创建索引后，再看此时 SQL 的执行计划，如图 10-5 所示。

图 10-5

可见，建立索引后 Cost 大幅下降。如果我们没有深入分析，而是直接就上手改写 SQL，原来 SQL 的成本 36 863 和现在加索引后的 22 相比，天壤之别。所以，在 SQL 审核与优化工作中，找对改进或优化的方向是更为重要的事。

而我更想说的是，事情到此其实才刚开了一个头，精彩在后面……继续分析，我们会注意到，该表上建立了好几个索引，其中一个索引为。

```
INDEX_NAME      COLLIST
-----------     ----------
IND_JMS_CENT    SYS_NC00007$
```

索引列只有一列，但命名是 SYS_NC00007$，这意味着什么呢？答案是："SYS_NC" 的前缀暗示着这很可能是一个函数索引。

使用 DBMS_METADATA.GET_DDL 获得该索引的创建语句核心部分为。

```
CREATE INDEX IND_JMS_CENT ON JSM_CENT('STATE');
```

此刻，真相大白了。原来并不是该列上没有索引，而是当初建立索引时由于一个疏忽，将一个普通索引建立成了一个函数索引，一个单引号将 STATE 转换成了一个字符串！

通过对库内所有索引进行检查，并未发现其他索引有类似问题存在，所以最终在正式库上执行的优化改进脚本如下。

```
DROP INDEX IND_JMS_CENT;
CREATE INDEX IND_JMS_CENT ON JSM_CENT(STATE);
```

案例至此圆满结束。

结合以上两个案例，接下来我和大家分享一下 SQL 审核工作要如何才能做得更好。

要做好这项工作，细致认真是必不可少的，然而工具也是极其重要的，所谓"工欲善其事，必先利其器"。SQL 审核工作如果需要做到完美，那并不是一项简单的事。我们需要"由点及面"，找出问题的真正原因，才能真正将这项工作做好。

做好 SQL 审核，可以让技术和运营团队形成更紧密的协作关系，有助于提高应用系统的稳定和效率，保障业务顺畅进行，而这也是 DevOps 在数据库领域落地的最佳方式之一。

10.8　号段选取应用的 SQL 技巧

实际生活中，我们会碰到很多与号段选取有关的应用。比如连续销售收入超过 100 万元的月份是从几月到几月？一组世博会门票的号码是 0001～9999，现在已经随机卖出去了一些，问剩下的票号区间是哪些？公司拿顺序编了号的乒乓球做抽奖道具，但路上不慎遗失了几个，到底是丢了哪几个呢？……

诸如以上这些问题，我们都可以对其进行建模。说简单点就是对类似问题中具有的共同特征抽取并进行描述，即：已知一组连续的数，去掉中间一些数，如何求出剩下的数的区间（即号段）？知道号段的起止，如何求出该号段内所有的数？知道一个大的号段范围和已经取过的号段，如何求出可用的号段？利用 Oracle 提供的强大的层次查询功能以及分析函数，我们可以很轻松地解决上述问题。

10.8.1　问题的提出

在实际工作中，我们常常会碰到号段选取的问题，常见问题如下。

- 一组连续的数，去掉中间一些数，要求出剩下的数的区间（即号段）。

例如，一串数字为 1，2，3，4，7，9，10，则号段为 1～4，7～7，9～10。

- 知道号段的起止，要求出该号段内所有的数。

例如，号段为 1～3，15～15，则号段内所有的数为 1，2，3，15。

- 一组数，中间可能有断点，要求出缺失的数。

例如，一串数字为 1，2，3，4，7，9，10，则缺失的数为 5，6，8。

- 已知大号段范围及已用号段范围，求剩余可用号段范围。

例如，大号段范围 0～999，已用号段范围 0～200 和 399～599，则剩余可用号段范围为 201～398 和 600～999。

10.8.2 相关基础知识

要解决这类问题，最常用到的知识就是层次查询和分析函数中的 lead / lag 函数了。下面我们先做下热身运动，回顾一下如何运用它们。

1. 伪列 rownum 和 level

伪列是并非在表中真正存在的列，但是它们又可以像表中的列那样被操作，只不过操作只能是 SELECT，不能是 INSERT、UPDATE 和 DELETE。伪列像一个不带参数的函数。针对结果集中的每一条记录，通常一个不带参数的函数总会返回固定值，而伪列则返回不一样的值。所谓结果集，即一个 SELECT 语句的查询结果。

rownum 标识结果集中每一条记录的顺序，对于结果集中的第一条记录，其 rownum 值为 1，第二条记录，rownum 值为 2，依此类推。

level 则标识了一个层次查询结果集中各个节点的等级，根节点的 level 为 1，其下一级节点的 level 为 2，再下一级节点的 level 为 3，依此类推。

这里强调一点，**伪列 rownum 和 level** 是针对结果集而言的。

2. 利用层次查询构造连续的数

- 产生 5～8 这 4 个连续的数。

```
SELECT * FROM (SELECT rownum+4 FROM dual CONNECT BY rownum<5);
SELECT * FROM (SELECT level+4 FROM dual CONNECT BY level<5);
```

在这里我们可以看到，对于 dual 表来说，使用 rownum 和 level 得到的结果是一样的。以 level 为例的结果如下：

```
      LEVEL+4
      ----------
           5
           6
           7
           8
```

- 某大学本科均为四年制，当前在校学生的入学年份是以 8 月为界来计算的。例如对于 2005 年 8 月 1 日来说，在此日期前的在校学生入学年份为 2001 年~2004 年，在此日期和在此日期之后的在校学生入学年份为 2002 年~2005 年。求当前日期下的在校学生入学年份。

```
SELECT * FROM (
    SELECT to_number(to_char(add_months(sysdate, 4), 'yyyy')) - rownum EntryYear
        FROM dual CONNECT BY rownum<5
);
```

通过在当前时间上加上 4 个月（add_months（sysdate, 4）），巧妙地避免了跨年度需要判断带来的麻烦，以 2016 年 9 月 15 日为例，结果如下。

```
ENTRYYEAR
----------
     2009
     2008
     2007
     2006
```

注意，上述 SQL 语句，均在最外层套了一个"SELECT * FROM ……"，这是为了保证该语句在 Oracle 9i 下能正常显示所有的结果（内层 SQL 在 Oracle 9i 下通常只会显示第一条记录）。在 12C 环境测试的时候可以不用这么麻烦。

3. 用 Lead 和 Lag 获得相邻行的字段值

Lead 和 Lag 是 Oracle 的分析函数，这里列出这两个函数的语法，方便对照。

Lead 函数的语法如图 10-6 所示。

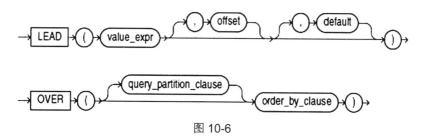

图 10-6

Lag 函数的语法如图 10-7 所示。

Syntax

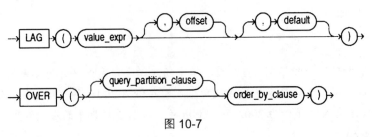

图 10-7

我们看下面的例子。

```
SELECT rown, lag(rown)OVER(ORDER BY rown) previous,
       lead(rown)OVER(ORDER BY rown) next
  FROM (SELECT rownum+4 rown FROM dual CONNECT BY rownum<5);
```

运行结果如下。

```
ROWN  PREVIOUS  NEXT
----- --------- -------
  5              6
  6      5       7
  7      6       8
  8      7
```

简单地说，Lag 是获得前一行的内容，而 Lead 是获得后一行的内容。当然，它们也可以获得前（后）n 行的内容，请看下例。

```
SELECT rown, lag(rown,2,-1)OVER(ORDER BY rown) previous,
       lead(rown,2,-1) OVER(ORDER BY rown) next
  FROM (SELECT rownum+4 rown FROM dual CONNECT BY rownum<5);
```

运行结果如下。

```
ROWN  PREVIOUS  NEXT
----- --------- -------
  5      -1      7
  6      -1      8
  7       5     -1
  8       6     -1
```

这里，通过指定第二个参数（即 offset 参数）来获得两行前的内容和两行后的内容，第三个参数（即 default 参数）表示超出范围后的默认值。如果 offset 超出范围并且未设定默认值-1，那么系统会自动将其值设为 NULL。

10.8.3 解决问题

做完热身运动，下面开始解决问题。每一个问题都包含题例、解答和扩展三部分内容。题例，即问题样例，为此类问题中的代表性问题；解答是针对该类问题给出解答思路、解答，并对解答做必要的讲解；扩展则是对与此问题类似的或在其基础上形成的更复杂的问题进行概括并提供解决方案，以 Q&A 的形式展现。

1. 已知号码求号段

题例

有一个表，其结构及数据如下。

```
fphm,kshm
2014,00000001
2014,00000002
2014,00000003
2014,00000004
2014,00000005
2014,00000007
2014,00000008
2014,00000009
2013,00000120
2013,00000121
2013,00000122
2013,00000124
2013,00000125
```

注意，第二个字段内可能是连续的数据，也可能存在断点。

怎样能查询出来这样的结果——查询出连续的记录来，具体如下。

```
2014,00000001,00000005
2014,00000007,00000009
2013,00000120,00000122
2013,00000124,00000125
```

解答

解答思路：首先要根据 fphm 进行分组，利用 lag 取得同组中前一行的 kshm，然后和本行的 kshm 相比，如果差值为 1，说明这一行和上一行是连续的。由于首尾的特殊性，故而需要先用 max 和 min 来获得首尾点。

SQL 如下。

```
SELECT fphm,
       nvl(lag(e)OVER(PARTITION BY fphm ORDER BY s),minn) st,
       nvl(s, maxn) en
  FROM (SELECT fphm,
               lag(kshm,1) OVER(PARTITION BY fphm ORDER BY kshm) s, kshm e,
               min(kshm)OVER(PARTITION BY fphm) minn,
               max(kshm) OVER(PARTITION BY fphm) maxn
          FROM t
       )
 WHERE nvl(e-s-1,1)<>0;
```

运行结果如下。

```
FPHM     ST          EN
-------  ----------  ----------
2013     00000120    00000122
2013     00000124    00000125
2014     00000001    00000005
2014     00000007    00000009
```

通过在分析函数 lag 的 OVER 部分中使用 PARTITION BY 关键字实现分组。请注意，此例中的 min 和 max 函数因带有 OVER 部分，也均为分析函数，而不是我们通常所用作的聚合函数。我们看一下内层 SQL 的执行结果。

```
SELECT fphm,
       lag(kshm,1) OVER(PARTITION BY fphm ORDER BY kshm) s, kshm e,
       min(kshm)OVER(PARTITION BY fphm) minn,
       max(kshm) OVER(PARTITION BY fphm) maxn
  FROM t;
```

运行结果如下。

```
FPHM     S           E             MINN            MAXN
-------  ---------   -----------   -------------   -----------
2013                 00000120      00000120        00000125
2013     00000120    00000121      00000120        00000125
2013     00000121    00000122      00000120        00000125
2013     00000122    00000124      00000120        00000125
2013     00000124    00000125      00000120        00000125
2014                 00000001      00000001        00000009
2014     00000001    00000002      00000001        00000009
2014     00000002    00000003      00000001        00000009
2014     00000003    00000004      00000001        00000009
2014     00000004    00000005      00000001        00000009
2014     00000005    00000007      00000001        00000009
2014     00000007    00000008      00000001        00000009
2014     00000008    00000009      00000001        00000009
```

选择了 13 行。

相邻两行的 kshm 的差值是否为 1 通过 nvl（e-s-1，1）来判断：如果差值为 1 则该表达式值为 0，说明这一行和上一行是连续的；如果差值不为 1，则该表达式的值不为 0，说明这一行和上一行是不连续的。

```
SELECT fphm, s, e, minn, maxn
  FROM (SELECT fphm,
               lag(kshm,1) OVER(PARTITION BY fphm ORDER BY kshm) s, kshm e,
               min(kshm)OVER(PARTITION BY fphm) minn,
               max(kshm) OVER(PARTITION BY fphm) maxn
          FROM t3;
       )
 WHERE nvl(e-s-1,1)<>0;
```

10.8 号段选取应用的 SQL 技巧

运行结果如下。

```
   FPHM  S           E           MINN        MAXN
-------  ----------  ----------  ----------  ----------
   2013              00000120    00000120    00000125
   2013  00000122    00000124    00000120    00000125
   2014              00000001    00000001    00000009
   2014  00000005    00000007    00000001    00000009
```

已选择 13 行。

于是,再通过 lag 函数,我们就可以轻而易举地得到最终结果。

扩展

Q:本例中有一个字段作为分组字段,如果没有分组字段或者有多个分组字段该怎么办?

A:如果没有分组字段,则直接将 lag(kshm,1) OVER(PARTITION BY fphm ORDER BY kshm)改为 lag(kshm,1) OVER(ORDER BY kshm)即可;如果有多个分组字段,假设还有 fphm1 和 fphm2 两个分组字段,则 lag(kshm,1) OVER(PARTITION BY fphm ORDER BY kshm)写为 lag(kshm,1) OVER(PARTITION BY fphm, fphm1, fphm2 ORDER BY kshm)。

Q:本例中,kshm 被隐式转换为数字了,如果 kshm 不是数字,是字母或者是日期,该如何处理?

A:OK,你说得很对,本例中确实对 kshm 做了隐式转换。对于非数字类型的字段,首先要定义什么是连续,然后根据定义给该字段上的值设定一个顺序值并进行排序,最后就可以套用上面的方法啦。比如你定义 26 个大写字母按字母顺序是连续的,请看下例。

```
SCOTT@lw.lw> SELECT kshm FROM t;

K
-
A
B
C
F
I
J
K
U
Y
Z
G
```

已选择 11 行。

```
with v as (SELECT kshm, ascii(kshm) kv FROM t)
SELECT nvl(lag(e)OVER(ORDER BY s),minn) st, nvl(S, maxn) en
```

```
    FROM (SELECT  lag(kshm,1) OVER(ORDER BY kv) s,
                  kshm e, lag(kv,1) OVER(ORDER BY kv) sv,
                  kv ev, min(kshm)OVER() minn,
                  max(kshm) OVER() maxn
           FROM v
)
WHERE nvl(EV-SV-1,1)<>0;
```

运行结果如下。

```
S E
- -
A C
F G
I K
U U
Y Z
```

对于日期型的，也是同样的方法。当然，首先你得定义差距 1 秒算连续还是差距 1 小时算连续或是其他的。具体就不细说了，留给各位去思考吧。

Q：这题不用分析函数能解吗？

A：当然可以！下面是两个解决该问题的 SQL，巧妙地利用 rownum 解决了问题。

```
/*SQL 1*/
SELECT b.fphm, min(b.kshm),max(b.kshm)
FROM (
    SELECT a.*,to_number(a.kshm - rownum) cc
       FROM (
              SELECT * FROM t ORDER BY fphm, kshm
           ) a
) b
GROUP BY b.fphm, b.cc
/

/*SQL 2*/
SELECT max(fphm),min(kshm),max(kshm)
FROM t
GROUP BY fphm||(kshm - rownum )
/
```

2．根据号段求出包含的数

题例

有表及测试数据如下。

```
CREATE TABLE T20
(
ID NUMBER(2),
S NUMBER(5),
E NUMBER(5)
);
```

10.8 号段选取应用的 SQL 技巧

```
INSERT INTO T20 ( ID, S, E ) VALUES ( 1, 10, 11);
INSERT INTO T20 ( ID, S, E ) VALUES ( 2, 1, 5);
INSERT INTO T20 ( ID, S, E ) VALUES ( 3, 88, 92);
COMMIT;
```

S 为号段起点，E 为号段终点，求出起点和终点之间的数（包括起点和终点）。

解答

解答思路：这与上一道题类似，只不过是将结果给还原回去。这该怎么做呢？实际上解决起来很简单，并没有刚才那么复杂。仔细回顾一下利用层次查询构造连续的数，你就会找到答案。

很明显，这需要构造序列来解决问题。

```
SELECT a.id, a.s, a.e, b.dis, a.S+b.dis-1 h
  FROM t20 a,
       (SELECT rownum dis
          FROM (SELECT max(e-s)+1 gap FROM t20)
         CONNECT BY rownum<=gap) b
 WHERE a.e>=a.s+b.dis-1
 ORDER BY a.id, 4
```

运行结果如下。

```
        ID          S          E        DIS          H
---------- ---------- ---------- ---------- ----------
         1         10         11          1         10
         1         10         11          2         11
         2          1          5          1          1
         2          1          5          2          2
         2          1          5          3          3
         2          1          5          4          4
         2          1          5          5          5
         3         88         92          1         88
         3         88         92          2         89
         3         88         92          3         90
         3         88         92          4         91
         3         88         92          5         92
```

首先，我们用 SELECT max（e-s）+1 gap FROM t20 找出首尾差最大是多少。其次通过 CONNECT BY 构造出一组序列，再通过简单的加减法，就可以将要求的数字算出来了。

这里，我们需要注意一下 rownum 的用法，请看下面改写的一种写法。

```
SELECT a.id, a.s, a.e,rownum, a.S+rownum-1 h
  FROM t20 a ,
       (SELECT id, e-s+1 gap FROM t20 WHERE id=2) b
 WHERE a.id=b.id
CONNECT BY rownum<=gap
```

运行结果如下。

```
ID          S           E         ROWNUM        H
----------  ----------  --------  ----------  ----------
2           1           5         1           1
2           1           5         2           2
2           1           5         3           3
2           1           5         4           4
2           1           5         5           5
```

这样得到的结果也是正确的,但这仅仅是针对一个 id 得到的结果。若我们把粗斜体字部分去掉后,看看结果是什么样。

```
ID          S           E         ROWNUM        H
----------  ----------  --------  ----------  ----------
1           10          11        1           10
1           10          11        2           11
2           1           5         3           3
2           1           5         4           4
2           1           5         5           5
2           1           5         6           6
3           88          92        7           94
```

这样的结果,显然不是我们想要的。更何况,这是错误的。由此我们应能更深入地理解,**伪列 rownum** 是只针对结果集的(**level** 也同样是这样的)。

扩展

Q:这里的 ID 是分组的标识吧?如果要增加分组标识字段或者没有分组,该怎么做?

A:对,这里的 ID 是分组标识。如果没有分组标识字段,那还是按照原方法去做就可以了。不同的是,如果源表中有且仅有一条记录,那 SQL 就可以简化一下。

```
SELECT a.s, a.e, a.S+rownum-1 h
  FROM t20 a
CONNECT BY rownum<=a.e-a.s;
```

如果增加分组字段,那没有关系,还是按照原方法去做就可以。你只需要在 SELECT 和 ORDER BY 子句中将分组标识字段逐个列出来即可。

Q:非数字型的该如何处理呢?

A:和已知号码求号段一样,你需要定义什么叫连续,并且给每个字符型或者日期型数据都赋予明确的顺序值。当然就像已知号码求号段中使用 ASCII 函数将字母翻译为数字一样,这里你需要一个函数将数字翻译为字符型或者日期型数据。ASCII 可以将大写字母"A"翻译为数字 65,如果你想将 65 翻译为大写字母"A",那么请用 CHR 函数。

3. 求缺失的号

题例

我想查询一下 serial_no 这个字段的不连续的值。

10.8 号段选取应用的 SQL 技巧

例如。

```
serial_no
1
2
3
4
6
8
10
```

我想一个 SQL 语句查出来缺失的号码，显示结果为。

```
5
7
9
```

解答

解答思路：对于这样的问题，首先是要进行排序。然后对于每一个数 B，找出它前面的数 A 并进行比较，如果 B-A=1（数按从小到大进行排序），则说明这两个数是连续的，否则说明它们之间是有空档的，存在缺失的数。找出所有存在空档的一对数，考虑到缺失的数字可能是多个，故而最后得用层次查询来构造出中间缺失的一个或者多个数字。解答 SQL 如下。

```sql
SELECT DISTINCT s+level-1 rlt
  FROM (SELECT lag(serial_no,1) OVER(ORDER BY serial_no)+1 s,
               serial_no-1 e FROM t2
       )
START WITH e-s>=0    -- WHERE e-s>=0 也能达到同样的效果，
                     -- 但用 START WITH 可以过滤掉很多不必要的冗余数据
CONNECT BY level<=e-s+1
ORDER BY 1;
```

运行结果如下。

```
    RLT
----------
         5
         7
         9
```

分析函数 lag 再次派上用场，注意，此地用了 level 而没有用 rownum，简单地说这是因为这里可能出现多个空档，如果用 rownum 则无法达到同样的效果，除非采用类似于根据号段求出包含的数中的方式。采用 level 虽然简化了写法，但会带来重复数据，所以在最终结果出来前，需要用 DISTINCT 关键字进行过滤。

分别看一下用 WHERE 和 START WITH 产生的原始数据的数量。为了使效果变得明显一些，我们加入一条新的数据——50：

```
select * from t2 order by 1;
 SERIAL_NO
```

```
----------
         1
         2
         3
         4
         6
         8
        10
        50
```
```
/*使用 START WITH 过滤*/
SELECT s+level-1 rlt
  FROM (SELECT lag(serial_no,1) OVER(ORDER BY serial_no)+1 s,
               serial_no-1 e FROM t2
       )
START WITH e-s>=0
CONNECT BY level<=e-s+1
/
```

运行结果如下（限于篇幅，中间部分数据已省略）。

```
       RLT
----------
         5
        12
        13
        14
       ......
        47
        48
        49
         7
        12
        13
        14
       ......
        47
        48
        49
         9
        12
        13
        14
       ......
        47
        48
        49
```
```
/*使用 WHERE 过滤*/
SELECT s+level-1 rlt
  FROM (SELECT lag(serial_no,1) OVER(ORDER BY serial_no)+1 s,
               serial_no-1 e FROM t2
       )
WHERE e-s>=0
CONNECT BY level<=e-s+1
/
```

10.8 号段选取应用的 SQL 技巧

运行结果如下（限于篇幅，中间部分数据已省略）。

```
     RLT
----------
       5
      12
      13
      14
     ......
      47
      48
      49
       7
      12
      13
      14
     ......
      47
      48
      49
       9
      12
      13
      14
     ......
      47
      48
      49
```

已选择 308 行。

尽管使用 START WITH 代替 WHERE 已经为我们减少了很多不必要的数据。但很明显，上面用 DISTINCT 过滤重复数据的方法，会导致大量不必要的开销。当我尝试再往表里加了两个三位数后，我在一分钟之内都没看到结果，性能问题成了此种解法的瓶颈，必须进行改进。所以我们这里根据号段求出包含的数中的方式来写解决该问题的 SQL。

```
with tt as (SELECT s, e
        FROM (SELECT serial_no, serial_no-1 e,
                   lag(serial_no,1) OVER(ORDER BY serial_no)+1 s
              FROM t2
             )
        WHERE e-s>=0
)
SELECT a.s+b.dis-1 h
FROM tt a,
    (SELECT rownum dis FROM
      (SELECT max(e-s+1) gap FROM tt)
    CONNECT BY rownum<=gap) b
WHERE a.s+b.dis-1<=a.e
ORDER BY 1;
```

新写的这个 SQL，一秒钟之内就能得到答案了，记住这个高效的写法。

扩展

Q：分组标识的问题以及非数字型数据的问题是不是参考前两部分的做法就可以？

A：是的，你说的完全没错。而且从这题的解决之道你完全可以看出，是由前两部分综合而成的。

Q：如果第一个数就缺失了怎么办？

A：你问到点子上了，本题的解法是在数据中第一个和最后一个数都存在的情况下成立的。如果第一个数就不存在，假设是-1，那你需要手动将-2 这个数插入表中，或者构造出一个等价的视图，比如这样。

```
SELECT * FROM t
UNION
SELECT -2 FROM dual
```

同样，如果最后一个数不在表中，依葫芦画瓢即可。

4．求尚未使用的号段

题例

表 A 结构如下。

Q varchar2（1），　-- bill_type_id，类型 ID。

A number，　-- bill_start，开始号码。

B number，-- bill_end，结束号码。

C number --office_level，等级。

数据如下。

```
A 0 999 1
A 100 199 2
A 300 499 2
A 555 666 2
```

SQL 目的是取出包含在 office_level 为级别 1 里的，还没有录入 office_level 为级别 2 的号段。

创建表及测试数据的脚本如下。

```
CREATE TABLE T8
(
A NUMBER(4),
B NUMBER(4),
C NUMBER(4),
Q VARCHAR2(1 BYTE)
);
```

10.8 号段选取应用的 SQL 技巧

```
Insert into T8(A, B, C, Q)Values(0, 999, 1, 'A');
Insert into T8(A, B, C, Q)Values(100, 199, 2, 'A');
Insert into T8(A, B, C, Q)Values(300, 499, 2, 'A');
Insert into T8(A, B, C, Q)Values(555, 666, 2, 'A');
COMMIT;
```

解答

解答思路：本题看起来很像是第一个和第三个问题的逆问题，将大号段的边界与小号段的边界相比，从大号段中将小号段"挖"掉，这样剩下的就是可用号段了。

解答 SQL 如下。

```
SELECT s, e FROM
(
SELECT nvl2(lag(a)OVER(PARTITION BY q ORDER BY a), b+1,
        min(a)OVER(PARTITION BY q)) s,
     nvl(lead(a)OVER(PARTITION BY q ORDER BY a)-1,
        max(b)OVER(PARTITION BY q)) e
 FROM t8 START WITH c=1
CONNECT BY c-1 = PRIOR c AND q= PRIOR q
)
WHERE s<=e;
```

运行结果如下。

```
         S          E
---------- ----------
         0         99
       200        299
       500        554
       667        999
```

由于既有头也有尾，所以分析函数 lag 和 lead 都上场了。取得每段数据的开始（s）和结束（e）之后，就可以直接得到可用的号段了。至于外层的过滤条件 s<=e，则是一个容错处理手段，可将因出现重叠号段而出现的不正确的结果过滤掉。

```
Insert into T8(A, B, C, Q) Values(280, 340, 2, 'A');
COMMIT;

SELECT nvl2(lag(a)OVER(PARTITION BY q ORDER BY a), b+1,
        min(a)OVER(PARTITION BY q)) s,
     nvl(lead(a)OVER(PARTITION BY q ORDER BY a)-1,
        max(b)OVER(PARTITION BY q)) e
 FROM t8 START WITH c=1
CONNECT BY c-1 = PRIOR c AND q= PRIOR q
/
```

运行结果如下。

```
         S          E
---------- ----------
         0         99
```

```
        200        279
        341        299
        500        554
        667        999
```

```sql
SELECT s, e FROM
(
SELECT nvl2(lag(a)OVER(PARTITION BY q ORDER BY a), b+1,
        min(a)OVER(PARTITION BY q)) s,
      nvl(lead(a)OVER(PARTITION BY q ORDER BY a)-1,
        max(b)OVER(PARTITION BY q)) e
  FROM t8 START WITH c=1
CONNECT BY c-1 = PRIOR c AND q= PRIOR q
)
WHERE s<=e;
```

运行结果如下。

```
         S          E
---------- ----------
         0         99
       200        279
       500        554
       667        999
```

可见，加上 s<=e 这个过滤条件后，内层 SQL 产生的非粗体字部分的结果，均被过滤掉了。

当然，为了避免超出大号段范围的数据的出现，我们还可以直接在 SQL 中加以限制。

```sql
SELECT q, s, e FROM
(
SELECT nvl2(lag(a)OVER(PARTITION BY q ORDER BY a),
        b+1, min(a)OVER(PARTITION BY q)) s,
      nvl(lead(a)OVER(PARTITION BY q ORDER BY a)-1,
        max(b)OVER(PARTITION BY q)) e,
    q , level
  FROM t8 START WITH c=1
CONNECT BY c-1 = PRIOR c AND q= PRIOR q AND a>= PRIOR a AND b<= PRIOR b
)
WHERE s<=e;
```

扩展

Q：如果我另外有一组数怎么办？

A：解答中的列 Q 就是用来分组的，你可以指定另外一个标识来计算新的分组中的可用号段。

Q：如果我插入一条数据：990, 1099, 2, 'A'，会怎样？

A：超出的部分会被自动过滤（通过条件 WHERE s<=e），最终结果显示如下。

```
         S          E
---------- ----------
         0         99
```

```
        200       299
        500       554
        667       989
```

Q: office_level 为 2 的号段数据出现交叉会怎样？

A: 出现交叉也没关系，同样会被自动过滤（通过条件 WHERE s<=e）。

Q: office_level 为 1 的号段数据多出来一条会怎样？

A: 大号段数据只允许有一个存在，多了会导致结果不正确的。

Q: 既然用到 CONNECT BY，而且有一个条件是 c-1 = PRIOR c，那我增加一条 office_level 为 3 的数据，会怎样？

A: 尽管代码还可以执行，但这条数据的意义并未事先定义，因而得到的结果未必是你想要的。这里用层次查询的目的仅仅是将同组的数据串在一起，并可由此实现对号段范围合法性的校验。

Q: 好像不用层次查询也能解决这个问题哦？

A: 没错，层次查询的作用在上面的 Q&A 已经说过了。实际上由于数据中本身就有 Q 列来标明 level，因此我们可以根据这个字段来做排序，达到将同组数据从外层到内层串起来的目的。

```
SELECT q, s, e FROM
(
SELECT  nvl2(lag(a)OVER(PARTITION BY q ORDER BY c, a), b+1
           , min(a)OVER(PARTITION BY q)
           ) s,
        nvl(lead(a)OVER(PARTITION BY q ORDER BY c, a)-1,
            max(b)OVER(PARTITION BY q)
           )e,
        q
  FROM t8
)
WHERE s<=e;
```

10.8.4　小结

在本节中，我们通过使用层次查询和分析函数，解决了常见的号段选取相关应用的问题。当然，有些问题还可以用其他的方法实现。所谓条条大路通罗马，能快速有效解决问题的办法，都是好办法。

10.9　connect by 的作用与技巧

或许你已经见过"connect by"这哥们了，也或许你还对它没啥了解，不过这并不妨碍我们去深入了解它的强大功能。说深入，其实也就是个人这些年的较为深刻的体会，冀望能借此文章与大家分享，同时更期望能起到抛砖引玉的效果，让我们看到 connect by 更精彩的表现。

下面就让我们一步一步地体验一下 connect by 在 Oracle 开发中能为你做什么，以便在今后的开发中能熟练运用它。

以下例子在 10g 和 12c 中的环境验证结果一致。

10.9.1　connect by 是什么

在了解 connect by 可以做什么之前，我们先来看看它是什么。

"connect by"是**层次查询（Hierarchical Queries）**子句中的关键字，层次查询子句的语法如图 10-8 所示。

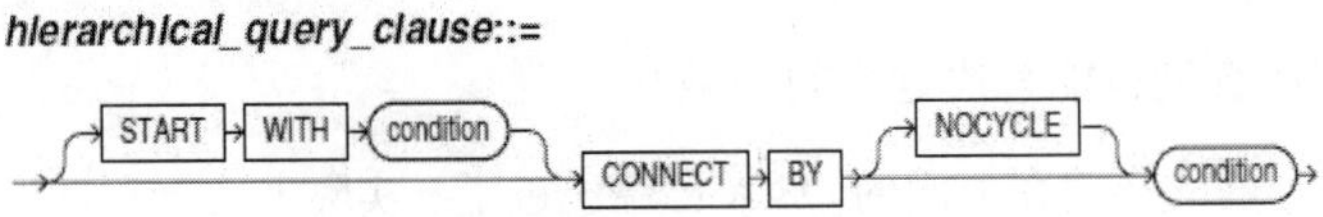

图 10-8

从语法路线图上来看，层次查询可以包含 START WITH、CONNECT BY 和 NOCYCLE 关键字。其中只有 CONNECT BY 是必选的，其余二者均为可选项，所以 CONNECT BY 是层次查询的核心所在。它的含义在官方文档上的解释为："CONNECT BY specifies the relationship between parent rows and child rows of the hierarchy"，即 CONNECT BY 指明了父记录和子记录之间的层次关系。

10.9.2　connect by 可以做什么

好了，现在步入正题，connect by 到底可以做什么？

从传统的描述来看，将有上下级关联的数据以树状形式展现是其基本功能，这一点没任何问题。自上而下展现数据没问题，那自下而上展现数据有没有问题？这问题等价于将一棵正树变成 N 棵倒树是否可能，我可以提前告诉你，也没问题。connect by 还可以干嘛呢？构造数列？求排列组合？对，您没听错，且跟我一步一步慢慢来看。

我们先从样例数据开始。

1．样例数据

本节所采用的样例数据为 SCOTT 用户下的 EMP 表，不是 HR 用户下的 EMPLOYEES 表。

该表表结构如下。

```
CREATE TABLE SCOTT.EMP
(
  EMPNO     NUMBER(4),
  ENAME     VARCHAR2(10 BYTE),
  JOB       VARCHAR2(9 BYTE),
  MGR       NUMBER(4),
  HIREDATE  DATE,
  SAL       NUMBER(7,2),
  COMM      NUMBER(7,2),
  DEPTNO    NUMBER(2)
);
```

数据如表 10-1 所示。

表 10-1

EMPNO	ENAME	JOB	MGR	HIREDATE	SAL	COMM	DEPTNO
7369	SMITH	CLERK	7902	1980-12-17	800		20
7499	ALLEN	SALESMAN	7698	1981-2-20	1600	300	30
7521	WARD	SALESMAN	7698	1981-2-22	1250	500	30
7566	JONES	MANAGER	7839	1981-4-2	2975		20
7654	MARTIN	SALESMAN	7698	1981-9-28	1250	1400	30
7698	BLAKE	MANAGER	7839	1981-5-1	2850		30
7782	CLARK	MANAGER	7839	1981-6-9	2450		10
7788	SCOTT	ANALYST	7566	1987-4-19	3000		20
7839	KING	PRESIDENT		1981-11-17	5000		10
7844	TURNER	SALESMAN	7698	1981-9-8	1500	0	30
7876	ADAMS	CLERK	7788	1987-5-23	1100		20
7900	JAMES	CLERK	7698	1981-12-3	950		30
7902	FORD	ANALYST	7566	1981-12-3	3000		20
7934	MILLER	CLERK	7782	1982-1-23	1300		10

2．列出上下级关系

从上面的数据中，我们怎么按职级高低列出上下级的人员关系呢？

从数据中我们可以看到，MGR 为 null 的记录，也就是 empno 为 7839 的记录，其 JOB 为 PRESIDENT（总裁），很明显就是最高层的领导。从他开始入手，可以顺藤摸瓜将其直接下属以及直接下属的下属等等，一个一个地列出来。这个动作看似复杂，但对 Oracle 数据库来说，其实只需一个简单的 SQL 就可以做到。

```
select empno, ename, job, mgr, deptno, level
  from emp
 start with mgr is null
connect by prior empno = mgr;
```

其中 level 为伪列，表示顺藤摸瓜的"藤"的层级。

查询结果如图 10-9 所示。

EMPNO	ENAME	JOB	MGR	DEPTNO	LEVEL
7839	KING	PRESIDENT		10	1
7566	JONES	MANAGER	7839	20	2
7788	SCOTT	ANALYST	7566	20	3
7876	ADAMS	CLERK	7788	20	4
7902	FORD	ANALYST	7566	20	3
7369	SMITH	CLERK	7902	20	4
7698	BLAKE	MANAGER	7839	30	2
7499	ALLEN	SALESMAN	7698	30	3
7521	WARD	SALESMAN	7698	30	3
7654	MARTIN	SALESMAN	7698	30	3
7844	TURNER	SALESMAN	7698	30	3
7900	JAMES	CLERK	7698	30	3
7782	CLARK	MANAGER	7839	10	2
7934	MILLER	CLERK	7782	10	3

图 10-9

这样的结果展现，不仔细看，还真看不出是一棵树！怎么办？要想使其和我们平常见的"树"长得差不多，只要做个简单的改造就行。

```
select empno,
       lpad(' ', level * 2 - 1, ' ') || ename ename,
       job,
       mgr,
       deptno,
       level
  from emp
 start with mgr is null
connect by prior empno = mgr;
```

这样查询结果就很清晰了，如图 10-10 所示。

EMPNO	ENAME	JOB	MGR	DEPTNO	LEVEL
7839	KING	PRESIDENT		10	1
7566	JONES	MANAGER	7839	20	2
7788	SCOTT	ANALYST	7566	20	3
7876	ADAMS	CLERK	7788	20	4
7902	FORD	ANALYST	7566	20	3
7369	SMITH	CLERK	7902	20	4
7698	BLAKE	MANAGER	7839	30	2
7499	ALLEN	SALESMAN	7698	30	3
7521	WARD	SALESMAN	7698	30	3
7654	MARTIN	SALESMAN	7698	30	3
7844	TURNER	SALESMAN	7698	30	3
7900	JAMES	CLERK	7698	30	3
7782	CLARK	MANAGER	7839	10	2
7934	MILLER	CLERK	7782	10	3

图 10-10

10.9 connect by 的作用与技巧

也许你会想，怎么 BLAKE 这哥们手下的销售人员的名字没按字母顺序排序？能不能给排下序？注意，我不要那种不顾层次关系而只按名字排序的结果。你这个想法很合理，要求不过分。

```
select empno,
       lpad(' ', level * 2 - 1, ' ') || ename ename,
       job,
       mgr,
       deptno,
       level
  from emp
 start with mgr is null
connect by prior empno = mgr
 order siblings by emp.ename;
```

注意其中的关键字 SIBLINGS，后面的 S 千万别漏掉了。同时注意一个细节，order siblings by 后面用名字进行排序，但是不能直接写成 ename，而是要加上表名作限定。这是因为 order 子句总是在最后才执行的，而 select 语句中为 lpad ('', level*2-1, '') ||ename 取了个别名为 ename，此时只写 ename 就表明用的是 lpad('', level*2-1, '')||ename 去排序。而 order siblings by 后面是不允许出现 LEVEL、PRIOR 和 ROWNUM 的。表达式中出现了 LEVEL，自然就被 Oracle 认为是非法的。加上表名限定后，Oracle 就会认为这是根据 ename 排序，而不是根据 lpad ('', level*2-1, '') ||ename 去排序。

为什么 order siblings by 后面不允许出现 LEVEL、PRIOR 和 ROWNUM 呢？这是因为 order siblings by 本身是在层次查询结果内对具有同一上级元素的同级元素进行排序。

另外，常规的 order by 后面可以跟数字，表示按第几列排序，但 order siblings by 后面不能跟数字。

查询结果如图 10-11 所示。

EMPNO	ENAME	JOB	MGR	DEPTNO	LEVEL
7839	KING	PRESIDENT		10	1
7698	BLAKE	MANAGER	7839	30	2
7499	ALLEN	SALESMAN	7698	30	3
7900	JAMES	CLERK	7698	30	3
7654	MARTIN	SALESMAN	7698	30	3
7844	TURNER	SALESMAN	7698	30	3
7521	WARD	SALESMAN	7698	30	3
7782	CLARK	MANAGER	7839	10	2
7934	MILLER	CLERK	7782	10	3
7566	JONES	MANAGER	7839	20	2
7902	FORD	ANALYST	7566	20	3
7369	SMITH	CLERK	7902	20	4
7788	SCOTT	ANALYST	7566	20	3
7876	ADAMS	CLERK	7788	20	4

图 10-11

也许你已经注意到了，不光是 BLAKE 的手下按名字字母顺序做了排序，连和 BLAKE 一样同属于 KING 管理的 CLARK 和 JONES 也按名字字母顺序排序了，而 JONES 的手下与 BLAKE 的手下由于不属于同一个上级，故而他们相互之间没有排序。由此可以深入体会一下 SIBLINGS 的含义。

3. 构造序列

以"树"的方式展示上下级关系只是 connect by 的基本功能，它还能构造序列。

```
select rownum row from dual connect by rownum<5;
```

运行结果如下。

```
SCOTT@lw.lw> select rownum row from dual connect by rownum<5;
     ROWN
----------
        1
        2
        3
        4
```

不用 rownum，换作 level 也是一样的结果。

```
select level lv from dual connect by level<5;
```

（1）序列的构造。

上面展示的是自然数截止到 4 的序列，也是最基本的序列构造方法。下面的例子展示扩展的序列构造方法。

```
起始不是从 1 开始：
select rownum+15 row from dual connect by rownum<5;--序列从 16 开始
select rownum-5 row from dual connect by rownum<5;--序列从-4 开始
序列是递减的：
select 10-rownum row from dual connect by rownum<5;--序列从 9 开始，序列为 9、8、7、6
序列间隔不为 1 的：
select 3*rownum-9 row from dual connect by rownum<5;--序列从-6 开始，序列为-6、-3、0、3
```

以上均为等差数列的构造。回想我们中学学过的函数的表达式：y=f（x）（表示 y 是关于自变量 x 的函数），我们可以将最后一个 SQL 语句用函数表达式的方式写为

$$y=f(x)=3x-9 \quad (x \text{ 为自然数，且 } x<5)$$

很显然，SQL 语句的运行结果和函数的求值结果是完全匹配的。

常见的等比数列也可这样来构造，比如 y=f（x）=2x（x 为自然数，且 x<5），用 SQL 语句来构造就可以这么写。

```
select power(2,rownum) row from dual connect by rownum<5;
```

运行结果如下。

10.9 connect by 的作用与技巧

```
SCOTT@lw.lw> select power(2,rownum) rown from dual connect by rownum<5;
      ROWN
----------
         2
         4
         8
        16
```

除了基本的加减乘除关系运算外，Oracle 还提供了丰富的函数，如三角函数 sin、cos、tan 等，对数函数 log、ln 等。你可以根据实际需要选用合适的函数来形成你自己的 f(x)，然后转化为 SQL 语句，让 Oracle 帮你求值。

这其中可能会有一个让你感到有些困惑的地方：rownum 不是从 1 开始，间隔为 1 这样递增的吗？如果我要求值的函数，其定义域（即 x 的取值范围）为[0.1, 0.8]，间隔为 0.01，那该怎么办？别忘记我们还有除法可以用，以 y=f(x)=sin(x) 为例。

```
select sin(rownum/100+0.09) rown from dual connect by rownum<=(0.8-0.1)/0.01;
```

需要注意的是，通过这种方式去计算 y 值，得到的 y 并不是连续的。这与数列本身就是有间隔的，是保持一致的。

（2）序列用于求解数学题。

构造序列可在许多实际应用中用上，比如求解多元一次的数学题（元其实就是未知数的意思，多元表示是多个未知数。一元的求解可以用上面 y=f(x) 的方式直接求出），下面我们就来看一个实际的例子。

问题如下。

有 11 种不同的方法可以将 20 表示成 8 个正奇数之和。

( ) + ( ) + ( ) + ( ) + ( ) + ( ) + ( ) + ( ) =20

列出所有的可能性。

解答：既然是"不同的方法"，那依据加法交换律调换两个数而形成的新答案就不算不同的方法了，所以我们在条件里要加以限制。依据加法交换律，我们总可以做到前一个括号中的数不大于其后任意一个括号中的数，这样就可以保证得到的解都是不同的了，于是我们可以写出下面的 SQL。

```
with t as
  (select rownum * 2 - 1 r from dual connect by rownum <= 7)
--正奇数最小是 1, 7 个 1 相加得 7,
                 --所以第 8 个数就是 13, 也就是最大的数只可能是 13
select *
  from t t1, t t2, t t3, t t4, t t5, t t6, t t7, t t8
 where t1.r + t2.r + t3.r + t4.r + t5.r + t6.r + t7.r + t8.r = 20 and t1.r <= t2.r and t2.r
<= t3.r and t3.r <= t4.r and t4.r <= t5.r and t5.r <= t6.r and t6.r <= t7.r and t7.r <= t8.r;
```

运行结果如下。

```
    R     R     R     R     R     R     R     R
----- ----- ----- ----- ----- ----- ----- -----
    1     1     3     3     3     3     3     3
    1     1     1     3     3     3     3     5
    1     1     1     1     3     3     5     5
    1     1     1     1     1     5     5     5
    1     1     1     1     3     3     3     7
    1     1     1     1     1     3     5     7
    1     1     1     1     1     1     7     7
    1     1     1     1     1     3     3     9
    1     1     1     1     1     1     5     9
    1     1     1     1     1     1     3    11
    1     1     1     1     1     1     1    13
```

（3）构造的序列，其记录之间存在父子关系吗？

至此，我提一个问题，根据官方文档中对 connect by 的定义，connect by 仅仅是指定父子记录之间的关系吗？从上面的例子中可以看出，connect by 只指定了记录之间的"连接条件"，但却并未明确指出记录之间的父子关系。而事实上，尽管没有使用 PRIOR 关键字，但 Oracle 对这样的连接结果，也为其中的每条记录都赋与了指明上下级关系的 LEVEL 值。

```
SCOTT@lw.lw>select rownum rown, level from dual connect by rownum<5;
```

运行结果如下。

```
      ROWN      LEVEL
---------- ----------
         1          1
         2          2
         3          3
         4          4
```

如果画出树状图，你会发现，这是一棵每个非叶子节点都有且仅有一个下级节点的树，是一棵特殊的树，用生活中的语言描述就是"代代单传"。

4．求排列组合

排列组合，看似陌生又很熟悉，在生活中可是会常常碰到的。彩票就是很典型的代表，千千万万种组合里，大奖就只有它一个。有些景点，票有好几种，不同组合的娱乐项目，票价不一样。这些都是排列组合在生活中的应用。

（1）排列组合的定义。

排列组合是组合学最基本的概念。所谓排列（permutation，现在多用 arrangement 表示），就是指从给定个数的元素中取出指定个数的元素进行排序。组合（combination）则是指从给定个数的元素中仅仅取出指定个数的元素，不考虑排序。排列组合的中心问题是研究给定要求的排列和组合可能出现的情况总数。排列组合与古典概率论关系密切。

排列组合的计算公式如下。

$$P_n^r = n(n-1)...(n-r+1) = \frac{n!}{(n-r)!}$$

$$C_n^r = \frac{P_n^r}{r!} = \frac{n!}{r!(n-r)!}$$

$$C_n^r = C_n^{n-r}$$

排列：公式 P 是排列公式，从 N 个元素取 R 个进行排列（即排序）。

组合：公式 C 是组合公式，从 N 个元素取 R 个不进行排列（即不排序）。

（2）求组合。

回到 connect by 上来，我们怎么用 SQL 获得排列组合的结果呢？比如，我们想取得 3 个元素的排列组合结果。我们先从组合开始入手。

仔细思考一下刚才说过的组合的定义，组合是从 N 个元素取 R 个，不进行排列（即不排序）。既然不排序，也就是说组合的结果是与元素的顺序无关的，举个例子就是说，{1，2，3} 和{2，3，1}对于组合来说是等价的。既然这样，为便于得到结果，那我们就给它指定一个顺序，比如，组合中每个元素都比其前一个元素要小（即从大到小的顺序）。

```
SCOTT@lw.lw>column xmlpath format a15
SCOTT@lw.lw>select rown, sys_connect_by_path(rown, ',') xmlpath from (select rownum rown from dual connect by level<4) connect by rown<prior rown;
```

运行结果如下。

```
 ROWN   XMLPATH
 -----  ----------
     3   ,3
     2   ,3,2
     1   ,3,2,1
     1   ,3,1
     2   ,2
     1   ,2,1
     1   ,1
```

已选择 7 行。

上例中，XMLPATH 就是我们所要的结果。1、2、3 这三个数中任取一个数，有{3}、{2}、{1}三种方式。任取两个数，有{3，2}、{3，1}、{2，1}三种方式。任取三个数，有{3，2，1}一种方式。当然，需要注意的是，根据组合的定义，还有从三个数中一个数也不取的做法，这时就只有{}一种方式（{}内什么都没有，表示没有元素，即无任何数字），但在 SQL 结果中不会展现。ROWN 是当前记录的 ROWN，其与 XMLPATH 的最后一个字符是一致的。

这里要提一下 sys_connect_by_path 这个从 Oracle 9i 开始就提供的函数。该函数的作用是将从根节点到当前节点的数据以指定的连接字符串连接起来,该函数只在包含 connect by 子句的 select 语句中使用。

为了能更清晰得理解查询结果,我们增加两个字段。

```
SCOTT@lw.lw>select connect_by_root(rown) rrown, rown, level, sys_connect_by_path(rown,
',') xmlpath from (select rownum rown from dual connect by level<4) connect by rown<prior rown;
```

运行结果如下。

```
RROWN  ROWN   LEVEL  XMLPATH
------ ------ ------ ---------
     3      3      1  ,3
     3      2      2  ,3,2
     3      1      3  ,3,2,1
     3      1      2  ,3,1
     2      2      1  ,2
     2      1      2  ,2,1
     1      1      1  ,1
```

已选择 7 行。

其中 RROWN 是用 connect_by_root 这个函数求出的。该函数求的是根节点的值,对应 XMLPATH,就是连接字符串","后的第一个数字。LEVEL 则不用赘述,表示当前记录是层次查询的第几层。从中你也可以看出,level 为 1 的就表示任取 1 个元素的取法,level 为 2 的就表示任取 2 个元素的取法,依此类推。

如果你不想取出全部的组合,只想取其中的一部分,比如找出任取 2 个元素,获得其取法的 SQL 如下。

```
SCOTT@lw.lw>select connect_by_root(rown) rrown, rown, level, sys_connect_by_path(rown,
',') xmlpath from (select rownum rown from dual connect by level<4) where level=2 connect by
rown<prior rown;
```

运行结果如下。

```
RROWN  ROWN  LEVEL  XMLPATH
-----  ----  -----  ---------
    3     2      2  ,3,2
    3     1      2  ,3,1
    2     1      2  ,2,1
```

由此我们可以推出,从 n 个不同的元素里取 r 个元素,找出其可能的组合的 SQL 如下。

```
select connect_by_root(rown) rrown,
       rown,
       level,
       sys_connect_by_path(rown, ',') xmlpath
  from (select rownum rown from dual connect by level < (&n + 1))
 where level = &r
connect by rown < prior rown;
```

注意,当 n 很大而 r 很小时,该查询会产生大量无用的备选结果,请想一想这是为什么?

我们将其做如下改造。

```
with para as (select &n n, &r r from dual)
select connect_by_root(rown) rrown, level lvl, rownum rown, sys_connect_by_path(rown, ',') xmlpath from
        (select rownum rown from para connect by level<(n+1)), para
        where level=r connect by rown<prior rown and level<=r
```

最后的 level<=r，这是为了将多余的备选结果筛除。所谓备选结果，我们从前面的查询中可以看出，为了得到从 3 个元素中任取 2 个元素的结果，我们需要将 3 个元素中任取 1 个、2 个和 3 个元素的方法都找出——这些都是备选结果，然后才进行筛选。显然，这种做法会产生大量无用的结果，尤其当 n 很大而 r 又很小的时候。level<=r 可以将 level 为 r+1, r+2, …, n 的备选结果都剔除掉，这能大大提高效率。至于为什么不用 level=r 来直接限制掉 level 为 1, 2, …, r–1 的备选结果，那是因为尽管 Oracle 很强大，但空中楼阁在 Oracle 上也是无法搭建的。

（3）求排列。

组合是不讲究顺序的，而排列是讲究顺序的。那讲究顺序如何表示呢？{1, 2, 3}、{2, 3, 1}、{3, 1, 2}等虽然都是相同的组合，但都被视为不同的排列。我们用不等号"<>"替换组合中的小于号"<"，就可以实现以不同的顺序排列元素，从而实现排列。

也许你会有疑问，用"<"可以确保元素不重复，用"<>"却不能保证，可能会出现"1<>2<>3<>1……"，这不就死循环了吗？对，你说的没错，所以我们要避免死循环的出现。为了避免进入无限连接且可以同时满足排列中的元素不重复的要求，我们隆重推出 NOCYCLE 关键字。

```
SCOTT@lw.lw>column xmlpath format a15
SCOTT@lw.lw>select level, sys_connect_by_path(rown, ',') xmlpath from (select rownum rown from dual connect by level<4) connect by NOCYCLE rown<>prior rown order by level, 2;
```

运行结果如下。

```
   LEVEL  XMLPATH
   -----  -------
       1  ,1
       1  ,2
       1  ,3
       2  ,1,2
       2  ,1,3
       2  ,2,1
       2  ,2,3
       2  ,3,1
       2  ,3,2
       3  ,1,2,3
       3  ,1,3,2
       3  ,2,1,3
       3  ,2,3,1
       3  ,3,1,2
       3  ,3,2,1
```

5．逆转求出下上级的关系路径

从"CONNECT BY specifies the relationship between parent rows and child rows of the hierarchy."这句话来看，也许你已经稍微有点微妙的感觉了。两条记录，谁是父记录，谁是子记录，实际上是由你定义的两条记录之间的层次关系来决定的。

（1）列出上下级关系路径。

还是来看一下 emp 表，我们列出总裁、各级管理者和员工之间的上下级关系路径。

```
select empno, ename, job, mgr, deptno, level lv, sys_connect_by_path(empno,'/') c1,
sys_connect_by_path(ename,'/') c2, connect_by_isleaf isleaf from emp start with mgr is
null connect by prior empno= mgr;
```

查询结果如图 10-12 所示。

EMPNO	ENAME	JOB	MGR	DEPTNO	LV	C1	C2	ISLEAF
7839	KING	PRESIDENT		10	1	/7839	/KING	0
7566	JONES	MANAGER	7839	20	2	/7839/7566	/KING/JONES	0
7788	SCOTT	ANALYST	7566	20	3	/7839/7566/7788	/KING/JONES/SCOTT	0
7876	ADAMS	CLERK	7788	20	4	/7839/7566/7788/7876	/KING/JONES/SCOTT/ADAMS	1
7902	FORD	ANALYST	7566	20	3	/7839/7566/7902	/KING/JONES/FORD	0
7369	SMITH	CLERK	7902	20	4	/7839/7566/7902/7369	/KING/JONES/FORD/SMITH	1
7698	BLAKE	MANAGER	7839	30	2	/7839/7698	/KING/BLAKE	0
7499	ALLEN	SALESMAN	7698	30	3	/7839/7698/7499	/KING/BLAKE/ALLEN	1
7521	WARD	SALESMAN	7698	30	3	/7839/7698/7521	/KING/BLAKE/WARD	1
7654	MARTIN	SALESMAN	7698	30	3	/7839/7698/7654	/KING/BLAKE/MARTIN	1
7844	TURNER	SALESMAN	7698	30	3	/7839/7698/7844	/KING/BLAKE/TURNER	1
7900	JAMES	CLERK	7698	30	3	/7839/7698/7900	/KING/BLAKE/JAMES	1
7782	CLARK	MANAGER	7839	10	2	/7839/7782	/KING/CLARK	0
7934	MILLER	CLERK	7782	10	3	/7839/7782/7934	/KING/CLARK/MILLER	1

图 10-12

C1 表示按 EMPNO 来组织关系路径，C2 表示按 ENAME 来组织关系路径。ISLEAF 是 connect_by_isleaf 这个伪列的别名，如果当前节点是叶子节点，则 connect_by_isleaf 返回值为 1，否则返回值为 0。

（2）列出下上级关系路径。

有了 ISLEAF，我们很容易可以想到，如果我们将现有的这棵树倒过来，从叶子节点开始逐级向上遍历，不就可以生成一棵从下到上的树了吗？于是我们可以很自然的想到如下 SQL。

```
select empno, ename, job, mgr, deptno, level lv, c1,c2, sys_connect_by_path(empno,'/') a1,
    sys_connect_by_path(ename,'/') a2, connect_by_isleaf isleaf2 from
(select empno, ename, job, mgr, deptno, level lv, sys_connect_by_path(empno,'/') c1,
```

10.9 connect by 的作用与技巧

```
sys_connect_by_path(ename,'/') c2, connect_by_isleaf isleaf from emp start with mgr is
null connect by prior empno= mgr
)
start with isleaf=1 connect by lv=prior lv-1 and
empno= prior mgr
/
```

该 SQL 我们无论怎么看都没有问题。但可惜的是，它查不出来任何结果（我在 Windows XP（32 位）、Windows 2003 Server（32 位）和 AIX5.3（64 位）下的 Oracle Database 10.2.0.4 上运行该 SQL，都是"未选定行"，但在 Windows 上的 Oracle 11g，可以运行出正确的结果）。

如果我们换一种思路，使用表代替嵌套子查询，则没有任何问题。

```
create table ctest as (select empno, ename, job, mgr, deptno, level lv,
sys_connect_by_path(empno,'/') c1, sys_connect_by_path(ename,'/') c2, connect_by_isleaf
isleaf from emp start with mgr is null connect by prior empno= mgr)
/
select empno, ename, job, mgr, deptno, level lv, c1,c2, sys_connect_by_path(empno,'/') a1,
    sys_connect_by_path(ename,'/') a2, connect_by_isleaf isleaf2 from ctest
start with isleaf=1 connect by lv=prior lv-1 and empno= prior mgr
/
```

换成表肯定是没问题的，这都不用细想。否则的话，列出上下级关系中的 SQL 也不应该能查出结果了。但换成表是不灵活的方法，于是我们换个思路尝试。

```
with t as (select empno, ename, job, mgr, deptno, level lv, sys_connect_by_path(empno,'/')
c1, sys_connect_by_path(ename,'/') c2, connect_by_isleaf isleaf from emp start with mgr is null
connect by prior empno= mgr)
    select empno, ename, job, mgr, deptno, level lv, c1,c2, sys_connect_by_path(empno,'/')
a1, sys_connect_by_path(ename,'/') a2, connect_by_isleaf isleaf2 from t start with isleaf=1
connect by lv=prior lv-1 and empno= prior mgr
/
```

运行结果如图 10-13 所示。

EMPNO	ENAME	JOB	MGR	DEPTNO	LV	C1	C2	A1	A2	ISLEAF2
7369	SMITH	CLERK	7902	20	1	/7839/7566/7902/7369	/KING/JONES/FORD/...	/7369	/SMITH	0
7902	FORD	ANALYST	7566	20	2	/7839/7566/7902	/KING/JONES/FORD	/7369/7902	/SMITH/FORD	0
7566	JONES	MANAGER	7839	20	3	/7839/7566	/KING/JONES	/7369/7902/7566	/SMITH/FORD/JONES	0
7839	KING	PRESIDENT		10	4	/7839	/KING	/7369/7902/7566/7839	/SMITH/FORD/JONES/KING	1
7499	ALLEN	SALESMAN	7698	30	1	/7839/7698/7499	/KING/BLAKE/ALLEN	/7499	/ALLEN	0
7698	BLAKE	MANAGER	7839	30	2	/7839/7698	/KING/BLAKE	/7499/7698	/ALLEN/BLAKE	0
7839	KING	PRESIDENT		10	3	/7839	/KING	/7499/7698/7839	/ALLEN/BLAKE/KING	1
7521	WARD	SALESMAN	7698	30	1	/7839/7698/7521	/KING/BLAKE/WARD	/7521	/WARD	0
7698	BLAKE	MANAGER	7839	30	2	/7839/7698	/KING/BLAKE	/7521/7698	/WARD/BLAKE	0
7839	KING	PRESIDENT		10	3	/7839	/KING	/7521/7698/7839	/WARD/BLAKE/KING	1
7654	MAR...	SALESMAN	7698	30	1	/7839/7698/7654	/KING/BLAKE/MARTIN	/7654	/MARTIN	0
7698	BLAKE	MANAGER	7839	30	2	/7839/7698	/KING/BLAKE	/7654/7698	/MARTIN/BLAKE	0
7839	KING	PRESIDENT		10	3	/7839	/KING	/7654/7698/7839	/MARTIN/BLAKE/KING	1
7844	TUR...	SALESMAN	7698	30	1	/7839/7698/7844	/KING/BLAKE/TURNER	/7844	/TURNER	0
7698	BLAKE	MANAGER	7839	30	2	/7839/7698	/KING/BLAKE	/7844/7698	/TURNER/BLAKE	0
7839	KING	PRESIDENT		10	3	/7839	/KING	/7844/7698/7839	/TURNER/BLAKE/KING	1

图 10-13

查询结果中，同一管理者将可能多次出现，尤其是总裁，出现的次数是最多的。这是因为树的结构决定了一个父节点可以拥有多个子节点，但一个子节点只能有一个父节点。当我们将一棵树倒着遍历的时候，如果原树有多个分支，倒着遍历就会形成多棵树。

实质上，表中的 job 字段表明了哪些员工处于最底层的字段，所以我们完全可以用 start with job in ('CLERK', 'SALESMAN') 来实现同样的目的。但对于没有明确字段来标识叶子节点，所以只能采用本小节所述的方法来达到倒着遍历树的目的。

实战

上面说的都是基础知识，下面就来看看 connect by 的一些实际应用情况吧！

辅助生成 SQL 脚本

不知你有没有想过，在前面求解 8 个正奇数之和为 20 的 11 种不同写法的问题中，解答的 SQL 中的一部分我也是用 connect by 来生成的呢？在此我们仅关注最后的一串小于等于的式子。

```
and t1.r<=t2.r and t2.r<=t3.r and t3.r<=t4.r and t4.r<=t5.r
and t5.r<=t6.r and t6.r<=t7.r and t7.r<=t8.r;
```

这段式子我们可以用 connect by 来生成，这样能避免因修改时复制粘贴可能导致的错误。

```
SCOTT@lw.lw> select sys_connect_by_path(nee,' and  ') e from (select rownum rown,
't'||rownum||'.r<=t'||(rownum+1)||'.r' nee from dual connect by rownum<8) where level=7 connect
by rown>prior rown;
```

运行结果如下。

```
E
------------------------------------------------------------
 and t1.r<=t2.r and t2.r<=t3.r and t3.r<=t4.r and t4.r<=t5.r and t5.r<=t6.r
and t6.r<=t7.r and t7.r<=t8.r
```

对于"t t1，t t2，t t3，t t4，t t5，t t6，t t7，t t8"和"t1.r+ t2.r+ t3.r+ t4.r+ t5.r+ t6.r+ t7.r+ t8.r=20"，还是复制粘贴再修改来得快。杀鸡焉用牛刀，在合适的时候选用合适的工具，才能使我们提高解决问题的效率。你说的没错，这两处其实跟上面的条件生成很类似，何况基础都打好了，你要是有兴趣，稍微改动下结果也就出来了——牛刀顺带可杀两只鸡，也挺划算的不是？

组合应用

组合常见的应用就是抽签，这跟古典概率论有密切联系。

假如现在一个有 20 名员工的小部门搞表彰会，会上要抽取 3 名幸运员工，颁发幸运奖。用 Oracle 就可以这么做：从 1 开始，给每个员工都编上号，然后运行下面的 SQL，输入 n 的值 20，r 的值 3，就可以得到幸运儿的编号，再对应找到编号对应的员工，就可以了。

```
with para as (select &n n, &r r from dual),
     t as (select level lvl, rownum rown, sys_connect_by_path(rown, ',') xmlpath from
          (select rownum rown from para connect by level<(n+1)), para
```

```
           where level=r connect by rown<prior rown and level<=r),
    s as (select count(*) cnt from t),
    rnd as (select ceil(dbms_random.value(0, (select cnt from s))) rnd_val from dual)
SELECT xmlpath FROM t WHERE rown = (select rnd_val from rnd);
```

其实抽签还有更简单而且高效的做法，具体如下。

```
with t as (select rownum rown from dual connect by rownum<=&n)
select rown from (select rown, rank()over(order by dbms_random.value() desc) rk from t)
where rk<=&r
```

拆分字符串

```
SCOTT@lw.lw>with t as (select 'lastwinner' str from dual)
select substr(str, rownum, 1) splitchar from t connect by rownum<=length(str)
/
```

运行结果如下。

```
SPLITCHAR
-----------------
l
a
s
t
w
i
n
n
e
r
```

辅助 exp 数据导出

DBA 备份的手段有很多，exp 是其中之一。虽然 exp 是冷备份中历史悠久的方式，但是我用 exp 依然达到了一个理想的速度。这其中，思路很重要，实现方法也不可忽视，分享我的一个案例及其思路。

通过实际观察发现，对于一个暂时无业务的、数据量在上亿级别、存储大约为 1.5TB 的数据库系统，单开一个 exp 进程的时候，服务器的资源消耗是不高的，网络 IO 相当空闲。当同时开到 5 个进程时，每个进程的执行速度几乎没有出现减慢的现象，系统资源的消耗有所增加，最明显的是网络 IO 成倍增长，但也未达到 80%的使用率。最后导出的数据，当然也是原来单进程的 5 倍了，而时间则几乎差不多。由此我得出一个结论，同时开多个 exp 进程导出数据，可以大大加快导出的速度。

总体思路确定了，那就开始做具体的分析。exp 导出的文件里，通常最占空间的就是表和快照的数据，而快照也会对应一张实体表，所以事实上最终表现出来的都是表。一个数据库中，一般不会所有表都很大，通常会有以下几种情况。

- 某几张表很大，单表导出为数个 1GB 的文件，其余表都很小，很多张表加起来导出的文件不到 1GB。

- 部分表都较大，单表导出为 1GB 以内的文件，其余表很小，很多张表加起来导出的文件不到 1GB。
- 上面两种情况的混合，某几张表很大，部分表都较大，其余表很小。

具体情况具体分析。现在以第二种情况为例，假设所有表的大小都与其记录数的多少成正比，每张表的记录数都不超过 1000 万。现在的要求是若几张表的记录数之和超过 2000 万，则这几张表就作为一组表来导出。

```
select   max(sys_connect_by_path(table_name,','))   keep(dense_rank   first   order   by
length(sys_connect_by_path(table_name,','))  desc) tablelist from
        (select table_name, ceil(sum(num_rows)over(order by table_name asc)/20000000) s,
              rank()over(order by table_name asc) rk from user_tables
        )
connect by rk=prior rk+1 and s= prior s
group by s
/
```

可能这个 SQL 中的 keep 让大家有点晕晕的感觉。有需要详细了解的请去查阅 Oracle 的官方文档，在此我就不做详解了，在这里我们可以换一种等价的写法。

```
select max(sys_connect_by_path(table_name,',')) tablelist from
(select table_name, ceil(sum(num_rows)over(order by table_name asc)/20000000) s,
        rank()over(order by table_name asc) rk from user_tables)
connect by rk=prior rk-1 and s= prior s
group by s;
```

该 SQL 得到的结果形如"，tablea，tableb，tablec，…"。将其稍做处理，就可将之写到 exp 的参数文件中去，然后就可以轻松地对其进行调用了。

注意，实际导出时还需要考虑表之间的关联性等问题，通常会将有主外键关联的表一起导出。但若表较大，则还是要分别导出，但导入时一定要注意顺序。当然，实际情况也有更复杂的例子，不过只要我们掌握了基本的思路，会很快解决那些问题的。

Connect by 功能强大，归纳起来，大致就是以上几种用途。它可以应用在很多场合中。实战小节中只是列举了其中的一些情况，它还可以结合分析函数可方便地解决号段选取中的各种问题等等。Connect by 是我们的一大实战利器，用好了可以事半功倍。

第四篇

诊 断 分 析

- 第 11 章　抽丝剥茧——一次特殊的 ORA-04030 故障处理
- 第 12 章　不积跬步，无以至千里
- 第 13 章　反思与总结：轻松从菜鸟到专家
- 第 14 章　勤奋与汗水

第 11 章 抽丝剥茧——一次特殊的 ORA-04030 故障处理

——黄宸宁

这是一则来自客户现场的真实案例，客户现场人员在通过 PL/SQL Developer 客户端工具创建索引时出现 ORA-04030 错误，导致索引创建失败。但是通过 SQL*Plus 重新执行创建索引语句却能够成功。为什么会出现这种错误和执行差异呢？请跟随我们的思路来一步步地了解这个案例的解析过程。

11.1 聚集数据的信息采集和分析

首先，了解 ORA-04030 错误的原因。ORA-04030 错误引起的原因大概有以下几种情况。

（1）是否有足够的可用内存？

查看操作系统内存使用情况。

```
+-topas_nmon--r=Resources--------Host=gisdata2-------Refresh=2 secs---10:03.50-----------+
| Memory                                                                                 |
|          Physical   PageSpace |    pages/sec  In    Out | FileSystemCache               |
|% Used      87.7%       2.6%   | to Paging Space 0.5  0.0 | (numperm)   4.6%             |
|% Free      12.3%      97.4%   | to File System  0.0  0.0 | Process    66.1%             |
|MB Used   55450.2MB   523.7MB  | Page Scans      0.0      | System     17.0%             |
|MB Free    7781.6MB 19956.3MB  | Page Cycles     0.0      | Free       12.3%             |
|Total(MB) 63231.9MB 20480.0MB  | Page Steals     0.0      |  ------                      |
|                               | Page Faults  2110.3      | Total     100.0%             |
|-------------------------------------------------------------| numclient  4.6%           |
|Min/Maxperm    3065MB(  5%)  9195MB( 15%) <--% of RAM        | maxclient 15.0%           |
|Min/Maxfree    3000    4000     Total Virtual    81.7GB      | User      67.7%           |
```

```
|Min/Maxpgahead        2        8      Accessed Virtual  50.8GB 62.2%| Pinned    19.8%         |
|                                                                    | lruable pages  15692064.0|
|--------------------------------------------------------------------------------------------|
```

查看当时出故障的服务器资源，可以看到剩余内存还有 7.7GB 左右，说明在操作系统层面还有足够多的可用内存。

（2）是否设置了 Oracle 的限制？

查看数据库中与 PGA 相关的设置。

```
SQL> show parameter PGA
NAME                                 TYPE             VALUE
------------------------------------ ---------------- ------------------------------
pga_aggregate_target                 big integer      8400M
SQL> select * from v$pgastat;
NAME                                                      VALUE UNIT
--------------------------------------------------- ----------- ------------
aggregate PGA target parameter                       8808038400 bytes
aggregate PGA auto target                            7288095744 bytes
global memory bound                                   880803840 bytes
total PGA inuse                                       720201728 bytes
total PGA allocated                                   985562112 bytes
maximum PGA allocated                                2692780032 bytes    ---实际分配
```
最大内存 2568.03516MB
```
total freeable PGA memory                             163905536 bytes
process count                                               414
max processes count                                         529
PGA memory freed back to OS                          2.4854E+11 bytes
total PGA used for auto workareas                      10177536 bytes
maximum PGA used for auto workareas                   642154496 bytes
total PGA used for manual workareas                           0 bytes
maximum PGA used for manual workareas                    537600 bytes
over allocation count                                         0
bytes processed                                      5.7321E+11 bytes
extra bytes read/written                             2125827072 bytes
cache hit percentage                                      99.63 percent
recompute count (total)                                   61401
SQL> SELECT x.ksppinm NAME, y.ksppstvl VALUE, x.ksppdesc describ
  2  FROM SYS.x$ksppi x, SYS.x$ksppcv y
  3  WHERE x.indx = y.indx
  4  AND x.ksppinm LIKE '%_pga_max_size%';
NAME                   VALUE          DESCRIB
---------------------- -------------- --------------------------------------------------
_pga_max_size          1761607680     Maximum size of the PGA memory for one process
SQL> SELECT x.ksppinm NAME, y.ksppstvl VALUE, x.ksppdesc describ
  2  FROM SYS.x$ksppi x, SYS.x$ksppcv y
  3  WHERE x.indx = y.indx
  4  AND x.ksppinm LIKE '%_smm_max_size%';
NAME                   VALUE          DESCRIB
---------------------- -------------- --------------------------------------------------
_smm_max_size          860160         maximum work area size in auto mode (serial)
```

从上面的内容可以看到 PGA 设置的大小为 8400M，根据单个会话使用 PGA 的期望尺寸（也

第 11 章 抽丝剥茧——一次特殊的 ORA-04030 故障处理

可以认为是实际分配的最大尺寸）计算公式是：min（5%*pga_aggregate_target，50%*_pga_max_size，_smm_max_size），可以简单计算下 min（5%*8400M，50%1680M，840M）=420M（其中_pga_max_size 的单位为 bytes，_smm_max_size 的单位为 kb），即单个会话能使用 PGA 的期望尺寸为 420M，那报错的会话是否超过了该限制呢？

查看 ORA-04030 报错的 trace 文件如下所示。

```
TOP 10 MEMORY USES FOR THIS PROCESS
---------------------------------------
98% 109 MB, 1797 chunks: "permanent memory      " SQL
        sort subheap    ds=1108b67f0 dsprt=11048a508
 1% 1493 KB, 1779 chunks: "free memory          "
        session heap    ds=11047a818 dsprt=11019eae0
 0%  193 KB,   26 chunks: "permanent memory     "
        pga heap        ds=110072eb0 dsprt=0
 0%   82 KB,    6 chunks: "frame segment        " SQL
        kxs-heap-f      ds=11048a8c8 dsprt=11047a818
 0%   45 KB,    4 chunks: "permanent memory     "
        session heap    ds=11047a818 dsprt=11019eae0
 0%   35 KB,   11 chunks: "kzctxhugi2           "
        session heap    ds=11047a818 dsprt=11019eae0
 0%   33 KB,    1 chunk : "free memory          "
        top call heap   ds=11019e8c0 dsprt=0
 0%   24 KB,   11 chunks: "koh-kghu session heap"
        session heap    ds=11047a818 dsprt=11019eae0
 0%   22 KB,    1 chunk : "Fixed Uga            "
        pga heap        ds=110072eb0 dsprt=0
 0%   20 KB,    5 chunks: "kxsFrame4kPage       "
        session heap    ds=11047a818 dsprt=11019eae0


=======================================
PRIVATE MEMORY SUMMARY FOR THIS PROCESS
---------------------------------------
******************************************************
PRIVATE HEAP SUMMARY Dump
111 MB total:
111 MB commented, 194 KB permanent
 45 KB free (0 KB in empty extents),
111 MB,   1 heap:   "session heap       "
------------------------------------------------------
Summary of subheaps at depth 1
111 MB total:
109 MB commented, 55 KB permanent
1513 KB free (1 KB in empty extents),
   2 KB uncommented freeable with mark,     110 MB, 1 heap:  "kxs-heap-w      "  1490 KB free held
------------------------------------------------------
Summary of subheaps at depth 2
109 MB total:
109 MB commented, 10 KB permanent
 14 KB free (4 KB in empty extents),
   0 KB uncommented freeable with mark,     109 MB, 1 heap:   "sort subheap    "
------------------------------------------------------
```

11.1 聚集数据的信息采集和分析

```
Summary of subheaps at depth 3
109 MB total:
0 KB commented, 109 MB permanent
2 KB free (0 KB in empty extents),
========================================
INSTANCE-WIDE PRIVATE MEMORY USAGE SUMMARY
------------------------
Top 10 processes:
------------------------
(percentage is of 693 MB total allocated memory)
16% pid 168: 111 MB used of 111 MB allocated  <= CURRENT PROC
 6% pid 3: 43 MB used of 43 MB allocated
 2% pid 24: 402 KB used of 12 MB allocated (12 MB freeable)
 2% pid 33: 532 KB used of 12 MB allocated (11 MB freeable)
 2% pid 7: 11 MB used of 11 MB allocated
 2% pid 8: 11 MB used of 11 MB allocated
 2% pid 9: 11 MB used of 11 MB allocated
 2% pid 10: 11 MB used of 11 MB allocated
 2% pid 11: 11 MB used of 11 MB allocated
 2% pid 12: 11 MB used of 11 MB allocated
```

从以上的 trace 文件中可以看到，报错的进程实际分配的进程只有 111MB，远远未达到 420MB，说明并非是由于 Oracle 自身的限制引起的 ORA-04030 报错。

（3）哪个进程需要的内存过多？

上一个是否是由于 Oracle 自身限制引起的解释中，已经可以从 trace 文件中看到，消耗最多内存的进程就是报 ORA-04030 的进程，消耗的内存为 110MB，并未发现其他更消耗内存的进程。

（4）是否设置了操作系统限制？

查看操作系统限制，Oracle 用户的限制。

```
$ ulimit -a
time(seconds)         unlimited
file(blocks)          unlimited
data(kbytes)          unlimited
stack(kbytes)         4194304
memory(kbytes)        unlimited
coredump(blocks)      unlimited
nofiles(descriptors)  unlimited
threads(per process)  unlimited
processes(per user)   unlimited
```

root 用户的限制。

```
gisdata2@root[/]ulimit -a
time(seconds)         unlimited
file(blocks)          unlimited
data(kbytes)          131072
stack(kbytes)         32768
memory(kbytes)        unlimited
```

```
coredump(blocks)          unlimited
nofiles(descriptors)      unlimited
threads(per process)      unlimited
processes(per user)       unlimited
```

从上面 root 和 oracle 的 limit 限制来看，root 用户的 data（kbytes）的限定值得关注，该属性的意义是 soft data segment size in blocks（进程数据段大小限制）。

11.2 聚焦疑点的跟踪测试与验证

现在做一个反向测试：利用 SQL*Plus 工具创建索引，创建成功。

问题来了，为何 SQL*Plus 会成功，PL/SQL Developer 却会失败？

通过 PL/SQL Developer 工具创建索引时报 ORA-04030 错误，但是通过 SQL*Plus 创建却能成功，两者除了使用的工具不同，还有就是连接的方式不同（PL/SQL Developer 是通过监听程序建立的进程连接，而 SQL*Plus 是在数据库服务器上直接创建创建的连接，绕过了监听程序建立的进程）。

从连接工具和方式的不同得到了不一样的结果，如何来验证到底是连接工具的问题还是连接方式的引起的报错？

由于 PL/SQL Developer 只能通过监听的方式进行连接，但是 SQL*Plus 可以通过监听或则直接连接两种方式进行，所以先对连接方式进行测试。

通过 SQL*Plus 以 tnsnames.ora 标签名的方式通过监听进行连接，并执行创建索引报错的语句，发现错误依然存在，但是如果不通过监听而直接连接是不会报错的，说明跟是否通过监听进行连接有很大的关系。

为何会受监听的影响？

在 Oracle RAC 环境中，由于 crs 的启停是通过 root 用户进行的。

```
gisdata2@root[/]ps -ef|grep init
    root     1         0    0   Aug 31      - 18:24 /etc/init
    root     5046686   1    0   Jul 18      - 0:00 /bin/sh /etc/init.crsd run
    root     5702082   27066830  0  Jul 18  - 0:00 /bin/sh /etc/init.cssd oclsomon
    root     6554098   27066830  0  Jul 18  - 0:00 /bin/sh /etc/init.cssd daemon
    root     24969600  1    0   Jul 18      - 0:00 /bin/sh /etc/init.evmd run
    root     27066830  1    1   Jul 18      - 8:07 /bin/sh /etc/init.cssd fatal
    root     40108312  18940178  0  10:57:45 pts/1 0:00 grep init
```

所以在 crs 会继承 root 用户的 limit 属性，当通过 crs 或则 srvctl 命令启动监听时，也会继承 root 用户相应的 limit 属性，即 data（kbytes）为 131072。如何验证该推断？

现在通过监听的形式进行连接。

```
$ SQLplus sys/password@gissc2 as sysdba
SQL*Plus: Release 10.2.0.5.0 - Production on Mon Jul 20 17:20:02 2015
Copyright (c) 1982, 2010, Oracle. All Rights Reserved.

SQL> select * from v$mystat where rownum<2;
     SID STATISTIC#     VALUE
---------- ---------- ----------
    1427         0         1
SQL> select spid from v$process where addr =(select paddr from v$session where sid=1427);
SPID
------------
10158446                -----该连接的操作系统进程号
```

通过 dbx 工具查看该进程的 limit 信息：

```
$ dbx -a 10158446
Waiting to attach to process 10158446 ...
Successfully attached to oracle.
warning: Directory containing oracle could not be determined.
Apply 'use' command to initialize source path.

Type 'help' for help.
reading symbolic information ...
stopped in read at 0x90000000002e294 ($t1)
0x90000000002e294 (read+0x274) e8410028         ld   r2,0x28(r1)
(dbx) proc rlimit
rlimit name:       rlimit_cur         rlimit_max         (units)
  RLIMIT_CPU:      (unlimited)        (unlimited)        sec
  RLIMIT_FSIZE:    (unlimited)        (unlimited)        bytes
  RLIMIT_DATA:      134217728         (unlimited)        bytes
  RLIMIT_STACK:      33554432          4294967296        bytes
  RLIMIT_CORE:     (unlimited)        (unlimited)        bytes
  RLIMIT_RSS:      (unlimited)        (unlimited)        bytes
  RLIMIT_AS:       (unlimited)        (unlimited)        bytes
  RLIMIT_NOFILE:      65534              65536           descriptors
  RLIMIT_THREADS:  (unlimited)        (unlimited)        per process
  RLIMIT_NPROC:    (unlimited)        (unlimited)        per user
(dbx) quit
```

从上面的内容可以看到 data 属性的 limit 值为 134217728 bytes 即 131072 kbytes 与 root 的 data（kbytes） 131072 值完全吻合（stack 的 33554432 bytes 即 32768 也与 root 的 stack（kbytes） 32768 一致），说明是通过监听建立连接进程的 limit 继承于 root 用户。

不通过监听进行连接：

```
$ SQLplus / as sysdba
SQL*Plus: Release 10.2.0.5.0 - Production on Mon Jul 20 17:16:09 2015
Copyright (c) 1982, 2010, Oracle. All Rights Reserved.
SQL> select * from v$mystat where rownum<2;
     SID STATISTIC#     VALUE
---------- ---------- ----------
    1375         0         1
SQL> select spid from v$process where addr =(select paddr from v$session where sid=1375);
```

```
SPID
------------
47710432
```

跟踪这个服务器进程。

```
$ dbx -a 47710432
Waiting to attach to process 47710432 ...
Successfully attached to oracle.
warning: Directory containing oracle could not be determined.
Apply 'use' command to initialize source path.

Type 'help' for help.
reading symbolic information ...
stopped in read at 0x90000000002e294 ($t1)
0x90000000002e294 (read+0x274) e8410028         ld   r2,0x28(r1)
(dbx) proc rlimit
rlimit name:         rlimit_cur         rlimit_max        (units)
  RLIMIT_CPU:        (unlimited)        (unlimited)        sec
  RLIMIT_FSIZE:      (unlimited)        (unlimited)        bytes
  RLIMIT_DATA:       (unlimited)        (unlimited)        bytes
  RLIMIT_STACK:      33554432           4294967296         bytes
  RLIMIT_CORE:       (unlimited)        (unlimited)        bytes
  RLIMIT_RSS:        (unlimited)        (unlimited)        bytes
  RLIMIT_AS:         (unlimited)        (unlimited)        bytes
  RLIMIT_NOFILE:     65534              (unlimited)        descriptors
  RLIMIT_THREADS:    (unlimited)        (unlimited)        per process
  RLIMIT_NPROC:      (unlimited)        (unlimited)        per user
(dbx) quit
$
```

从上面的内容可以看到,如果不通过监听连接数据库创建的进程,它的 data 限制为 unlimited 的即无限制。

最后查看 crs 中监听的启动日志:(/u01/oracle/product/10.2.0/db_1/log/gisdata2/racg 中的日志文件 ora.gisdata2.LISTENER_GISDATA2.lsnr.log)。

```
2015-07-18 14:16:54.676:  [    RACG][1] [28508196][1][ora.gisdata2.LISTENER_GISDATA2.lsnr]:
The command completed successfully

2015-07-18 14:47:09.078:  [    RACG][1] [29950132][1][ora.gisdata2.LISTENER_GISDATA2.lsnr]:
LSNRCTL for IBM/AIX RISC System/6000: Version 10.2.0.5.0 - Production on 18-JUL-2015 14:47:08

Copyright (c) 1991, 2010, Oracle.  All rights reserved.
Starting /u01/oracle/product/10.2.0/db_1/bin/tnslsnr: please wait...
2015-07-18 14:47:09.079:  [    RACG][1] [29950132][1][ora.gisdata2.LISTENER_GISDATA2.lsnr]:
TNSLSNR for IBM/AIX RISC System/6000: Version 10.2.0.5.0 - Production
  System parameter file is /u01/oracle/product/10.2.0/db_1/network/admin/listener.ora
  Log messages written to /u01/oracle/product/10.2.0/db_1/network/log/listener_gisdata2.log

2015-07-18 14:47:09.079:  [    RACG][1] [29950132][1][ora.gisdata2.LISTENER_GISDATA2.lsnr]:
Listening on: (DESCRIPTION=(ADDRESS=(PROTOCOL=tcp)(HOST=10.176.30.81)(PORT=1521)))
    Listening on: (DESCRIPTION=(ADDRESS=(PROTOCOL=tcp)(HOST=10.176.30.79)(PORT=1521)))
    Listening on: (DESCRIPTION=(ADDRESS=(PROTOCOL=ipc)(KEY=EXTPROC)))
```

从以上的内容可以发现监听是由 crs（或则是 srvctl 命令）启动的。

以及监听的运行时间：

```
$ lsnrctl status

LSNRCTL for IBM/AIX RISC System/6000: Version 10.2.0.5.0 - Production on 20-JUL-2015 16:31:03
Copyright (c) 1991, 2010, Oracle.  All rights reserved.
Connecting to (ADDRESS=(PROTOCOL=tcp)(HOST=)(PORT=1521))
STATUS of the LISTENER
------------------------
Alias                     LISTENER_GISDATA2
Version                   TNSLSNR for IBM/AIX RISC System/6000: Version 10.2.0.5.0 - Production
Start Date                18-JUL-2015 14:56:39
Uptime                    2 days 1 hr. 34 min. 24 sec
Trace Level               off
Security                  ON: Local OS Authentication
SNMP                      ON
Listener Parameter File   /u01/oracle/product/10.2.0/db_1/network/admin/listener.ora
Listener Log File         /u01/oracle/product/10.2.0/db_1/network/log/listener_gisdata2.log
```

监听启动的时间也与日志中的时间对应。

11.3 解析原理的问题总结与建议

那为什么 CRS 会使用 ROOT 用户的 limit 限制呢？

首先，得明白在 Linux 中，每触发任何一个事件时，系统都会为它定义一个进程，并且给予一个 ID，即 PID，同时会根据触发这个进程的用户与相关属性关系，给这个 PID 设置有效的权限，如图 11-1 所示。

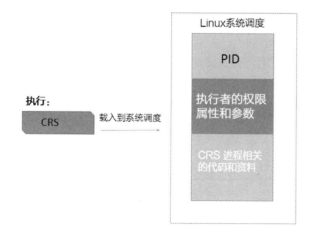

图 11-1

从上面可以知道，系统启动以后，CRS 会自动启动，启动主要由 /etc/init.d 的几个脚本完成。而这些脚本的执行用户和用户组是 root，也就是说当 CRS 启动时，Linux 系统会根据 UID/GID 来判断资源属性和环境变量，那么这个进程所衍生的进程，也会沿用这些属性。

由此可以得到结论，由于监听是通过 crs 进行的启动，继承了 root 用户的 limit 限制，每个会话所能持有的内存大小最大不能超过 128M，当通过监听进行数据库连接时，由监听创建的用户进程也将继承该 limit 限制，所以当通过 P/LSQL Developer 连接数据库（包括 SQL*Plus 等工具需要通过监听建立用户进程的情况），在创建索引过程中，当所请求的内存达到或非常接近该限制时，就会由于无法进一步申请更多的内存资源，抛出 ORA-04030 错误。

提示，如果是在 Linux 系统中，可以通过 cat /proc/PID/limits 命令查看单个进程的 limit 属性值，其中 PID 为要查看进程的进程号。

以 ORA-04030 案例进行延伸扩展，为避免该错误可参考以下建议。

（1）在安装 Oracle 软件创建数据库之前应该对主机层面的内核参数、limit 限制等进行规范的修改，以避免类似问题的发生。

（2）配置合理的内存，例如物理内存和交换空间。

（3）使用自动 PGA 内存管理可降低 ORA-04030 错误的概率。

第 12 章　不积跬步，无以至千里

——我的 DBA 手记（熊军）

> **题记**
>
> 有人说 DBA 行业的从业人员越老越值钱，其实值钱的不是他们的经验，而是他们的思维模式，只有经过长久的历练，分析问题才能一针见血。熊军十几年都在从事 Oracle 数据库专业技术工作，在此精选了他的成长历程、学习经验，以及分析问题的一些案例分享给大家，通过具体的案例我们可以看到一个运维的"老司机"如何精准有效地解读问题。

12.1　技术生涯有感

我最早是从事 IT 系统运维的，数据库、主机、网络都要维护。随着业务量和数据量的增加，系统也越来越慢，于是就尝试去优化系统。

优化系统最主要就是优化数据库，而优化数据库就需要全面深入地理解数据库。在优化的过程中，我不断地去学习数据库的知识，然后发现 Oracle 数据库提供了很多有趣的特性，利用这些特性来解决问题，给我带来了很多的满足感和成就感。这样我逐渐对 Oracle 数据库产生了很强的兴趣，就有一种内在的动力去更深入地学习 Oracle 数据库。

在学习 Oracle 数据库的过程中，逐步提高自己的技术水平主要把握以下几点。

（1）对 Oracle 数据库有浓厚的兴趣，这个是很重要的一方面。

（2）学习 Oracle 的官方文档，最基本的有《Oracle Database Concepts》《Oracle Database

Performance Tuning Guide》《Oracle Database Administrator's Guide》，这 3 份官方文档，对于各个版本我都反复去读。

（3）学习其他技术专家的书，包括 Tom Kyte、Eygle 的书等。

（4）在学习和实验的过程中，将一些心得体会、学习方法记录在个人技术博客上，在博客上写技术文章，是对文章中涉及的知识进行总结、梳理的过程，这对提高技术水平有极大的帮助。

（5）在工作中不断思考，例如解决一个问题，需要从原理上去解释为什么，一个问题是否有多个解决方案，每种解决方案各有什么优缺点。在问题解决后，再把这些案例涉及的知识进行梳理并写成文档，在这种方式下，解决一个问题就能获得技术进步。

（6）从 Oracle 数据库论坛（例如 ITPUB）中获得进步，在 ITPUB 上有一些很有价值的技术文章，以及有很多值得研究学习的案例，通过这些案例和文章学习其他人的思路和方法。

通过以上的一些方法，就逐渐地一步一步地提高了技术水平，并且通过在论坛上进行技术讨论、参与书籍的编写、通过个人博客进行分享，这些方式在一定程度上提高了自己的影响力，最后获得 ACE 称号也是水到渠成的事情。

ACE 是 Oracle 公司为那些在 Oracle 技术领域，除了数据库还包括其他产品（如中间件、JAVA 等）颁发的一个荣誉称号。以表彰那些在 Oracle 技术领域内具有很强技术水平又乐于向技术社区分享技术的人。所以要成为 ACE，需要具备两个条件，一是有有比较高的技术水平，二是乐于分享，让自己在技术社会里面有一定的影响力。

获得 ACE 以及 ACED 的称号对我的职业生涯的影响很大，这种影响首先在于获得一种认同感和肯定，同时也能够有更多的机会认识更多同行业中的朋友，让自己有机会进入到真正重视技术、以技术为竞争力的公司。

12.2　自我定位及规划

我对自己的定位还是一个技术工程师，一方面是因为这样能最大化地发挥我的价值和在技术方面的能力及作用，另一方面是出于我个人的兴趣和爱好仍然是技术。虽然现在承担了公司西区整个团队的领导工作，但是我在技术的道路上也从没有止步。

而且随着个人承担职责的变化，团队的成长和个人的成长就更加紧密地结合在一起，在现在的技术团队中，需要随着团队的进步而进步，也要引领团队的进步。

引领团队自然需要一直紧跟技术趋势，并且根据 IT 行业发展趋势，为客户提供更好的技术服务。例如随着企业信息化的方展，数据越来越重要。我们不光是围绕着数据库，还要围绕着数据来进行工作，挖掘数据本身的价值，这需要更深入的技术积累。

12.3 对数据库运维工作的认识

我个人认为数据库从业者工作中**最应该做的事情是**，把所有的操作和相关的数据都记录下来，这是一笔很宝贵的财富。最应该注意的事情自然是安全和测试。每一个操作，都尽可能地测试，因为一些操作看起来很简单，却可能产生不可预料的结果。对于安全，主要是指数据安全，要做好备份，这是很重要的一道防线，只有在有备份的情况下，操作错误时才有回退的可能。

最应该杜绝的想法是，对于这一点，我觉得是不要有工作有捷径的想法。工作的捷径来自于技术的积累和经验的积累。很多从业者喜欢从百度、Google 搜索到具体的操作方法并且不加思考地拿来就用，这样虽然能够解决问题，但是这会导致自己缺乏思考，缺乏对知识的系统掌握。而工作上的捷径应该是来自于自己对知识的全面掌握之后还有经验的积累，通过自动化或者是通过脚本、自行开发的工具来帮助提升工作效率。

12.4 学习理念分享

我个人在学习 Oracle 的一些方法这里我简单总结一下，供大家参考借鉴。我打一个比方，Oracle 的学习，就好比武侠小说中学武功，要从 3 个方面入手。

（1）**内功**：针以学习 Oracle 来说，内功就是对基本概念的掌握，Oracle 架构的深入理解，原理的掌握。如果有兴趣和时间，可以研究一下 Internal 的东西，这好比修习易筋经，需要极大的毅力和长期的坚持。

（2）**招式**：如果光有内功，没有招式，则会陷入空有高深内力，却无从发招的尴尬。学习 Oracle 也一样，还是需要掌握功能的使用，具体到 SQL 的使用，各个性能视图的使用，数据字典的使用。如果没有这些，在进行操作时，会有找不到无从下手的感觉。

（3）**实战经验**：武侠世界中的高手，都是从无数次战斗中取得经验，再武功大进。学习 Oracle 也一样，如果没有充分的实验，实际生产环境的实战，仍然只能说是只能入了 Oracle 的门，算不上登入大堂。

另外，学习过程中，多做笔记、多思考。做任何事，都需要多思考，学习 Oracle 也不例外。对 Oracle 的众多的功能和知识点，我们要经常思考，这个功能有什么好处，适用于什么地方，不适用于什么地方，每个知识点之间的联系等。甚至是要站在超越 Oracle 的高度，去思考 Oracle 为什么会这样设计。另外，好记性不如烂笔头，除非是天才，否则很多东西，久了就忘记了。

12.5 RAC 数据库频繁 hang 问题诊断案例

数据库的问题大体上可以分为两类，一类是数据库真的出了问题，不能正常运行，甚至影响到业务的。另一类是潜在的问题，也就是性能问题。对于这两类问题的分析，有相同也有不同之处。我们通过具体的案例来说明。

12.5.1 案例现象及概要

某客户的核心系统数据库是 2 节点的 Oracle RAC 数据库，版本为 10.2.0.4，数据库主机平台为 IBM AIX，数据量超过 10TB。这套数据库在最近一段时间经常出现系统挂起的严重故障，严重影响了生产的运行。在近一个月内，几乎每天可能都会出现故障，甚至是在业务不繁忙的时候。本文主要分析了系统的故障原因，并提出了对应的解决方案，希望能给读者在故障诊断时提供一些思路。

下面是 2014 年 1 月 7 日早上 8 点至 9 点时间段的故障日志（为保护案例数据库的信息，将实际的数据库名字替换为 orcl）。实例 1 日志信息。

```
Tue Jan  7 07:51:08 2014
ALTER SYSTEM ARCHIVE LOG
Tue Jan  7 07:51:08 2014
Thread 1 cannot allocate new log, sequence 75616
Checkpoint not complete
  Current log# 1 seq# 75615 mem# 0: /dev/rlv9
Tue Jan  7 08:15:52 2014
PMON failed to delete process, see PMON trace file
Tue Jan  7 08:23:34 2014
WARNING: inbound connection timed out (ORA-3136)
Tue Jan  7 08:24:53 2014
Errors in file /u01/app/oracle10g/admin/orcldb/udump/orcldb1_ora_10503176.trc:
ORA-00604: error occurred at recursive SQL level 1
ORA-01013: user requested cancel of current operation
ORA-06512: at line 2
......
Tue Jan  7 08:52:22 2014
Shutting down instance (abort)
License high water mark = 3510
Instance terminated by USER, pid = 5731764
Tue Jan  7 09:01:45 2014
Starting ORACLE instance (normal)
sskgpgetexecname failed to get name
```

早上 7:51 左右出现了"检查点未完成"而不能切换日志的信息，8:23 左右出现会话不能登录的信息，而到 8:52 强制关闭了实例然后重启。

在数据库关闭之前，对数据库做了 system state 转储，在跟踪文件 orcldb1_ora_5727520.trc 中有如下重要信息。

```
*** 2014-01-07 08:37:45.920
SYSTEM STATE
------------PROCESS 207:--------------
SO: 7000028079afb58, type: 2, owner: 0, flag: INIT/-/-/0x00
```

```
    (process) Oracle pid=207, calls cur/top: 70000275222f680/70000275222f680, flag: (0) -
    int error: 0, call error: 0, sess error: 0, txn error 0 (post info) last post received: 0 0 128
    last post received-location: kclrget last process to post me: 70000280f9952b0 1 6
    last post sent: 0 0 123 last post sent-location: kcrfw_redo_gen: wake LGWR after redo copy
    last process posted by me: 7000028079a3e58 1 6  (latch info) wait_event=0 bits=2
    holding     (efd=10) 7000027ce79ac58 Child cache buffers chains level=1 child#=40228
    Location from where latch is held: kcbgtcr: kslbegin excl:
    Context saved from call: 2160934423
    state=busy(exclusive) (val=0x20000000000000cf) holder orapid = 207
    waiters [orapid (seconds since: put on list, posted, alive check)]:
    40 (3851, 1389055081, 1)  waiter count=1
    Process Group: DEFAULT, pseudo proc: 70000280fb3a820
    O/S info: user: oracle, term: pts/2, ospid: 5297410
    OSD pid info: Unix process pid: 5297410, image: oracle@kjorcldb1 (TNS V1-V3)
    ......
    SO: 700002802e78798, type: 4, owner: 7000028079afb58, flag: INIT/-/-/0x00
      (session) sid: 4339 trans: 700002775cc8cc8, creator: 7000028079afb58, flag: (41) USR/-
BSY/-/-/-/-
    DID: 0001-00CF-00005C72, short-term DID: 0001-00CF-00005C73 txn branch: 0
    oct: 6, prv: 0, SQL: 7000026e943b618, pSQL: 70000268a5dd9d0, user: 49/ORCLADM
    service name: SYS$USERS
    O/S info: user: oracle, term: pts/2, ospid: 3114266, machine: kjorcldb1
    program: SQLplus@kjorcldb1 (TNS V1-V3)  application name: SQL*Plus, hash value=3669949024
    last wait for 'db file sequential read' blocking sess=0x0 seq=91 wait_time=26947 seconds
since wait started=4456
    file#=d5, block#=5b789, blocks=1
    Dumping Session Wait History
    for 'db file sequential read' count=1 wait_time=26947 file#=d5, block#=5b789, blocks=1
    for 'db file sequential read' count=1 wait_time=426
    ......
    LIBRARY OBJECT LOCK: lock=700002752e1b380 handle=7000026e943b618 mode=N
    call pin=0 session pin=0 hpc=0000 hlc=0000
    htl=700002752e1b400[700002687a44a90,70000272d437c30]        htb=700002687a44a90
ssga=700002687a44538
    user=700002802e78798 session=700002802e78798 count=1 flags=[0000] savepoint=0x52cad289
    LIBRARY OBJECT HANDLE: handle=7000026e943b618 mtx=7000026e943b748(1) cdp=1
    name=
    UPDATE   jl_jlzzda_13 a SET  zcbh =
     (SELECT   zcbh FROM   pc_dnb_jg_13 b
    WHERE    a.zzbh = b.zzbh AND a.zcbh <> b.zcbh
    AND ( TRIM (b.zcbh) = :"SYS_B_0" OR TRIM (b.zcbh) = :"SYS_B_1" OR TRIM (b.zcbh) = :"SYS_B_2"))
    WHERE   a.zcbh <> :"SYS_B_3" AND a.zcbh <> :"SYS_B_4" AND EXISTS
     (SELECT  :"SYS_B_5" FROM   pc_dnb_jg_13 b WHERE   a.zzbh = b.zzbh AND a.zcbh <> b.zcbh
    AND ( TRIM (b.zcbh) = :"SYS_B_6" OR TRIM (b.zcbh) = :"SYS_B_7" OR TRIM (b.zcbh) = :"SYS_B_8"))
```

12.5.2 故障详细分析

从 system dump 的信息，反映出以下几点。

（1）oracle 进程（orapid=207，ospid=5297410，session id=4339）持有一个 cache buffers chains child latch，子 latch 号为 40228，子 latch 的地址为：7000027ce79ac58。

(2)这个进程阻塞了一个进程(orapid=40),阻塞时间长为 3851 秒。

(3)这个进程上一个等待事件是磁盘单块读,到目前为止已经过了 4456 秒。system state dump 是从 2014 年 1 月 7 日 08:37:45 开始的,那么 4456 秒之前即 latch 开始持有的时间是在 2014 年 1 月 7 日 07:23:29 至 2014 年 1 月 7 日 07:28:29 之间(由于 system state dump 本身需要消耗一定的时间,因此会有一定的时间误差,这个时间误差在 5 分钟之内)。

(4)这个进程的是由数据库主机本机上的 SQLplus 连接上来的,正在执行的 SQL 语句如下。

```
UPDATE    jl_jlzzda_13 a
SET   zcbh =
(SELECT   zcbh
 FROM   pc_dnb_jg_13 b
 WHERE   a.zzbh = b.zzbh AND a.zcbh <> b.zcbh
 AND (    TRIM (b.zcbh) = :"SYS_B_0"
     OR TRIM (b.zcbh) = :"SYS_B_1"
     OR TRIM (b.zcbh) = :"SYS_B_2"))
WHERE   a.zcbh <> :"SYS_B_3" AND a.zcbh <> :"SYS_B_4"
AND EXISTS
(SELECT   :"SYS_B_5"
 FROM   pc_dnb_jg_13 b
 WHERE   a.zzbh = b.zzbh AND a.zcbh <> b.zcbh
 AND (    TRIM (b.zcbh) = :"SYS_B_6"
     OR TRIM (b.zcbh) = :"SYS_B_7"
     OR TRIM (b.zcbh) = :"SYS_B_8"))
```

这里很明显有一个异常,进程持有 latch 的时间过长。正常情况下 latch 持有时间在微秒级,而这个进程持有 latch 长达几千秒。这通常是由于进程异常或挂起所导致。

接下来,看看这个进程所阻塞的进程(orapid=40)的信息。

```
-------------PROCESS 40: ----------------------------------------
SO: 7000028069d2bf8, type: 2, owner: 0, flag: INIT/-/-/0x00
(process) Oracle pid=40, calls cur/top: 70000280012d810/70000280012d810, flag: (6) SYSTEM
int error: 0, call error: 0, sess error: 0, txn error 0  (post info) last post received:
0 0 114
    last post received-location: kcbbza last process to post me: 7000028039b5570 1 6
    last post sent: 504403329233300568 122 2 last post sent-location: kslges
    last process posted by me: 7000028039b35f0 1 14
    (latch info) wait_event=0 bits=0 Location from where call was made: kcbkzs:
    waiting for 7000027ce79ac58 Child cache buffers chains level=1 child#=40228
    Location from where latch is held: kcbgtcr: kslbegin excl:
    Context saved from call: 2160934423
    state=busy(exclusive) (val=0x20000000000000cf) holder orapid = 207
    waiters [orapid (seconds since: put on list, posted, alive check)]:
     40 (3837, 1389055067, 2) waiter count=1
    gotten 7812722 times wait, failed first 65133 sleeps 380 gotten 36115 times nowait, failed:
14223
    possible holder pid = 207 ospid=5297410
    on wait list for 7000027ce79ac58
```

12.5 RAC 数据库频繁 hang 问题诊断案例

```
    Process Group: DEFAULT, pseudo proc: 70000280fb3a820
    O/S info: user: oracle, term: UNKNOWN, ospid: 2508016
    OSD pid info: Unix process pid: 2508016, image: oracle@kjorcldb1 (CKPT)
    SO: 70000280112aa90, type: 4, owner: 7000028069d2bf8, flag: INIT/-/-/0x00
     (session) sid: 6567 trans: 0, creator: 7000028069d2bf8, flag: (51) USR/- BSY/-/-/-/-/-
    DID: 0001-0028-00000004, short-term DID: 0001-0028-00000005 txn branch: 0
    oct: 0, prv: 0, SQL: 0, pSQL: 0, user: 0/SYS
    service name: SYS$BACKGROUND
    waiting for 'latch: cache buffers chains' blocking sess=0x0 seq=26223 wait_time=0 seconds
since wait started=3835
    address=7000027ce79ac58, number=7a, tries=31f0
    Dumping Session Wait History
    for 'latch: cache buffers chains' count=1 wait_time=292977
```

从上面的信息可以确认以下几点：

（1）被阻塞的进程是 CKPT 进程，这是系统关键的后台进程——检查点进程。

（2）CKPT 进程由于等待 cache buffers chains latch 而被阻塞，被阻塞的时间长为 3837 秒，也就是在大约 2014 年 1 月 7 日 07:33:48 时被阻塞（由于 system state dump 需要消耗时间，所以可能有 5 分钟之内的时间误差）。

很显然，由于 CKPT 进程被长时间阻塞，将不能完成检查点工作。

```
    Tue Jan  7 07:18:53 2014
    Thread 1 advanced to log sequence 75613 (LGWR switch)
      Current log# 3 seq# 75613 mem# 0: /dev/rlv11
    Tue Jan  7 07:31:01 2014
    Thread 1 advanced to log sequence 75614 (LGWR switch)
      Current log# 2 seq# 75614 mem# 0: /dev/rlv10
    Tue Jan  7 07:43:04 2014
    Thread 1 advanced to log sequence 75615 (LGWR switch)
      Current log# 1 seq# 75615 mem# 0: /dev/rlv9
    Tue Jan  7 07:51:08 2014
    ALTER SYSTEM ARCHIVE LOG
    Tue Jan  7 07:51:08 2014
    Thread 1 cannot allocate new log, sequence 75616
    Checkpoint not complete
      Current log# 1 seq# 75615 mem# 0: /dev/rlv9
```

由于一个实例的在线日志文件只有 3 组，从 7:31 开始，经过 3 次切换，就不能再切换了。不能切换日志就意味着不能进行任何数据更改。由于数据库设置有登录触发器，对登录信息进行记录，会涉及数据插入。在这样的情况下，用户登录也不能成功，会一直挂起（或者直至超时）。

由于 CKPT 挂起，这导致了 SMON 这个关键后台进程也一直挂起。

```
    ---------------------------------------PROCESS 41:---------------------------------------
    SO: 7000028059c6be8, type: 2, owner: 0, flag: INIT/-/-/0x00
    (process) Oracle pid=41, calls cur/top: 70000280012dae8/70000280012dae8, flag: (16) SYSTEM
    int error: 0, call error: 0, sess error: 0, txn error 0  (post info) last post received:
0 0 154
    last post received-location: ktmpsm  last process to post me: 7000028039b35f0 1 14
```

```
last post sent: 0 0 90 last post sent-location: KJCS Post snd proxy to flush msg
last process posted by me: 7000028059c4488 1 6
(latch info) wait_event=0 bits=0
Process Group: DEFAULT, pseudo proc: 70000280fb3a820
O/S info: user: oracle, term: UNKNOWN, ospid: 103858
OSD pid info: Unix process pid: 103858, image: oracle@kjorcldb1 (SMON)
(session) sid: 6566 trans: 7000027734e5c60, creator: 7000028059c6be8, flag: (100051) USR/-
BSY/-/-/-/-/-
DID: 0001-0029-00000002, short-term DID: 0001-0029-00000003 txn branch: 0 oct: 0, prv:
0, SQL: 0, pSQL: 0, user: 0/SYS
service name: SYS$BACKGROUND
waiting for 'DFS lock handle' blocking sess=0x0 seq=59072 wait_time=0 seconds since wait
started=3835 type|mode=43490005, id1=1, id2=2
Dumping Session Wait History
for 'DFS lock handle' count=1 wait_time=488290 type|mode=43490005, id1=1, id2=2
for 'DFS lock handle' count=1 wait_time=488291 type|mode=43490005, id1=1, id2=2
……
----------------------------------------
SO: 7000027fd890990, type: 18, owner: 7000028003a2c80, flag: INIT/-/-/0x00
----------enqueue 0x7000027fd890990----------------------
lock version     : 3392757       Owner node      : 0
grant_level      : KJUSERNL      req_level       : KJUSEREX
bast_level       : KJUSERNL      notify_func     : 0
resp             : 7000025f95f7e50  procp         : 7000028003a2c80
    pid              : 103858    proc version    : 0
    oprocp           : 0         opid            : 0
    group lock owner : 0         xid             : 0000-0000-00000000
    dd_time          : 0.0       secs dd_count   : 0
    timeout          : 0.0       secs On_timer_q? : N
    On_dd_q?         : N         lock_state      : OPENING CONVERTING
    Open Options     : KJUSERPROCESS_OWNED
    Convert options  : KJUSERNODEADLOCKWAIT KJUSERNODEADLOCKBLOCK
    History          : 0x49a5149a    Msg_Seq     : 0x0
    res_seq          : 4090      valblk          : 0x00000000000000000000000000000000 .
    ----------resource 0x7000025f95f7e50---------------------
    resname     : [0x1][0x2],[CI]
    Local node  : 0    dir_node      : 1   master_node : 1 hv idx   : 98
    hv last r.inc : 4  current inc   : 4   hv status   : 0 hv master : 1
    open options : Held mode        : KJUSEREX
    Cvt mode    : KJUSERNL    Next Cvt mode : KJUSERNL
    msg_seq     : f0010       res_seq       : 4090
    grant_bits  : KJUSERNL KJUSERPR
    grant mode  : KJUSERNL KJUSERCR KJUSERCW KJUSERPR KJUSERPW KJUSEREX
    count       : 1        0        0        1        0        0
    val_state   : KJUSERVS_NOVALUE
    valblk      : 0x00000000000000000000000000000000 .
    access_node : 0    vbreq_state   : 0
    state       : x8   resp          : 7000025f95f7e50
    On Scan_q?  : N    Total accesses: 8014573    Imm. accesses: 7891534
    Granted_locks : 1  Cvting_locks  : 1  value_block: 00 00 00 00 00 00 00 00 00 00
00 00 00 00 00 00
        GRANTED_Q :
      lp 7000027fa0ce178 gl KJUSERPR rp 7000025f95f7e50 [0x1][0x2],[CI]
        master 1 pid 2508016 bast 0 rseq 4090 mseq 0 history 0x95614956
```

```
         open opt KJUSERPROCESS_OWNED
         CONVERT_Q:
         lp 7000027fd890990 gl KJUSERNL rl KJUSEREX rp 7000025f95f7e50 [0x1][0x2],[CI]
         master 1 pid 103858 bast 0 rseq 4090 mseq 0 history 0x49a5149a
         convert opt KJUSERNODEADLOCKWAIT KJUSERNODEADLOCKBLOCK
         ----------------------------------------
```

从上面的信息可以看到，资源的 GRANTED_Q 队列中有进程 2508016。这个进程正是 CKPT 进程，即 CKPT 进程以 PROTECT READ 模式持有资源，但是 SMON 进程需要以 EXCLUSIVE 模式来持有资源。这两种模式是不兼容的，因此 SMON 进程就被阻塞。

到目前为止，可以确定是由于异常的进程（orapid=207）长时间持有 cache buffers chains child latch，使得 CKPT 被长时间阻塞，导致不能完成检查点，最终导致不能进行任何数据更改和不能连接数据库。

导致系统故障的异常进程，其异常的原因，通常有：

（1）操作系统 Bug 导致，使得异常的进程不能调度。在 trace 文件中看到，使用 procstack 获取进程 call stack 失败，说明进程在操作系统一级异常，而不是在 oracle 内部存在 spin 或 stuck hang 的情况。

（2）Oracle 的 Bug，通过在 MOS 上查找，我们发现有如下 Bug：Bug 6859515　Diagnostic collection may hang or crash the instance，可能会有影响。

如何认定一个 Bug 与故障是匹配的？我们需要从大量的日志和 trace 中去寻找。一般会去从 diag 进程的 trace 和 system state dump 中寻找：

在 diag 进程的 trace 文件中，我们发现现象与 Bug 6859515 比较匹配。

```
*** 2014-01-07 07:24:56.655
Dump requested by process [orapid=46]
REQUEST:custom dump [2] with parameters [46][207][2][2]
Dumping process info of pid[207.5297410] requested by pid[46.1008052]
Dumping process 207.5297410 info:
*** 2014-01-07 07:24:56.655
Dumping diagnostic information for ospid 5297410:
OS pid = 5297410
loadavg : 4.78 4.81 4.52
swap info: free_mem = 77226.37M rsv = 228.00M
alloc = 636.91M avail = 58368.00M swap_free = 57731.09M
    F S      UID     PID    PPID   C PRI NI ADDR        SZ    WCHAN    STIME    TTY  TIME CMD
 240001 A  oracle 5297410       1 120 120 20 3acdbd4590 126808          23:46:25      - 27:22 oracleorcldb1 (DESCRIPTION=(LOCAL=YES)(ADDRESS=(PROTOCOL=beq)))
    open: The file access permissions do not allow the specified action.
procstack: write(/proc/5297410/ctl): The requested resource is busy.
5297410: oracleorcldb1 (DESCRIPTION=(LOCAL=YES)(ADDRESS=(PROTOCOL=beq)))
0x0000000000000000  ????????() + ??
```

从上面的数据可以看到，ospid=5297410 这个进程正是异常的持有 latch 的进程。这个进程在 2014 年 1 月 7 日 07:24:56 被请求转储,从时间上看这与该进程异常的时间点在 2014 年 1 月 7 日 07:23:29 至 2014 年 1 月 7 日 07:28:29 之间匹配。诊断转储是由 orapid=46,ospid=1008052 这个进程请求 diag 进程发起的。

"procstack: write(/proc/5297410/ctl):The requested resource is busy." 表明进程出现了异常。

```
PROCESS 46: ----------------------------------------
SO: 7000028079a4638, type: 2, owner: 0, flag: INIT/-/-/0x00
(process) Oracle pid=46, calls cur/top: 70000275b01c398/70000275b01c398, flag: (0) -
int error: 0, call error: 0, sess error: 0, txn error 0 (post info) last post received: 0 0 55
last post received-location: kjata: wake up enqueue owner
last process to post me: 7000028059c4488 1 6  last post sent: 0 0 90
last post sent-location: KJCS Post snd proxy to flush msg
last process posted by me: 7000028059c4488 1 6
(latch info) wait_event=0 bits=0
Process Group: DEFAULT, pseudo proc: 70000280fb3a820
O/S info: user: oracle, term: pts/3, ospid: 1008052
OSD pid info: Unix process pid: 1008052, image: oracle@kjorcldb1 (TNS V1-V3)
SO: 70000280ffd40b0, type: 4, owner: 7000028079a4638, flag: INIT/-/-/0x00
(session) sid: 6122 trans: 70000276e23b848, creator: 7000028079a4638, flag: (41) USR/-
BSY/-/-/-/-/-
  DID: 0001-002E-001DD0F9, short-term DID: 0001-002E-001DD0FA  txn branch: 0
  oct: 6, prv: 0, sql: 7000026e943b618, psql: 70000268a5dd9d0, user: 49/ORCLADM
  service name: SYS$USERS
  O/S info: user: oracle, term: pts/3, ospid: 3371970, machine: kjorcldb1
    program: SQLplus@kjorcldb1 (TNS V1-V3)
    application name: SQL*Plus, hash value=3669949024
    waiting for 'enq: TX - row lock contention' blocking sess=0x0 seq=4560 wait_time=0 seconds
since wait started=5330 name|mode=54580006, usn<<16 | slot=2b70003, sequence=5324
    Dumping Session Wait History
      for 'enq: TX - row lock contention' count=1 wait_time=488292  name|mode=54580006, usn<<16
| slot=2b70003, sequence=5324
      for 'enq: TX - row lock contention' count=1 wait_time=488292 name|mode=54580006, usn<<16
| slot=2b70003, sequence=5324
```

从上面的数据可以看到,orapid=46,ospid=1008052 这个进程一直在等待 TX 事务行锁,这个进程同样是 SQLplus 程序连接数据库产生的进程,执行的 SQL 语句也与异常进程一样。这个进程被异常进程(即 orapid=207)阻塞,阻塞大约是在 08:37:45 的 5330 秒之前,即 07:08:55 左右。

```
SYS@xj11g> select to_char(sample_time,'yyyy-mm-dd hh24:mi:ss') sample_time,session_id,
event,program,blocking_session
  2    from dba_hist_active_sess_history where instance_number=1 and dbid=3516340257
  3    and sample_id=144574100
  4    order by sample_time,session_id;

SAMPLE_TIME          SESSION_ID EVENT          PROGRAM       BLOCKING_SESSION
-------------------- ---------- ------         ----------    ----------------
2014-01-07 07:15:32        3532                DFGD_LONG
2014-01-07 07:15:32        4323                jlzztb.d
2014-01-07 07:15:32        4339                SQLplus
```

12.5 RAC 数据库频繁 hang 问题诊断案例

```
2014-01-07 07:15:32        4739                                    SQLplus
2014-01-07 07:15:32        5009 enq: TX - row lock contention      DFGD_LONG    3532
2014-01-07 07:15:32        6122 enq: TX - row lock contention      SQLplus      4339
```

会话 6122 在等待事务锁，阻塞的会话是 4339，即 orapid=207 的会话进程。由于 SQL 执行性能的原因，数据锁时间过长，因此会话 6122 即 orapid=46 的进程向 diag 进程申请发起一个对 orapid=207 进程的转储操作，这样就触发了 Bug，导致 orapid=207 的进程异常。

2014 年 1 月 9 日 12:30 左右，数据库节点 2 的实例再次发生 hang 故障。从 diag 的信息来看，同样是因为 diag 触发 Bug 导致进程异常。异常进程持有 redo copy latch 使得其他所有进程，包括 LGWR 进程不能进行日志生成和写出，不能进行任何数据修改，短时间内堵塞了大量会话，同时数据库实例完全挂起。

```
*** 2014-01-09 12:25:06.473
Dumping diagnostic information for ospid 385380:
OS pid = 385380 loadavg : 12.82 13.52 16.37
swap info: free_mem = 31221.03M rsv = 256.00M alloc = 674.18M avail = 65536.00M swap_free
 = 64861.82M
     F S      UID      PID   PPID   C PRI NI  ADDR      SZ    WCHAN    STIME    TTY  TIME CMD
  240001 A   oracle 385380          1  56 88 20 1ea0cf7590 97272        19:08:20    - 39:10
oracleorcldb2 (LOCAL=NO)
open: The file access permissions do not allow the specified action.
procstack: write(/proc/385380/ctl): The requested resource is busy.
385380: oracleorcldb2 (LOCAL=NO)
0x0000000000000000  ????????() + ??

*** 2014-01-09 12:25:06.624
----------------------------------------
SO: 7000028079acc18, type: 2, owner: 0, flag: INIT/-/-/0x00
(process) Oracle pid=165, calls cur/top: 700002641f86eb0/700002641f86eb0, flag: (0) -
int error: 0, call error: 0, sess error: 0, txn error 0
(post info) last post received: 0 0 136
last post received-location: kclrcvt
last process to post me: 7000028059c5448 1 6
last post sent: 0 0 117
last post sent-location: kcbzww
last process posted by me: 7000028049b5968 7 0
(latch info) wait_event=0 bits=0
Process Group: DEFAULT, pseudo proc: 70000280fb3a820
O/S info: user: oracle, term: UNKNOWN, ospid: 385380
OSD pid info: Unix process pid: 385380, image: oracle@kjorcldb2
----------------------------------------
SO: 70000280fdaf820, type: 4, owner: 7000028079acc18, flag: INIT/-/-/0x00
(session) sid: 3252 trans: 700002775b06728, creator: 7000028079acc18, flag: (100141) USR/-
BSY/-/-/-/-/-
DID: 0002-00A5-00000036, short-term DID: 0002-00A5-00000037
txn branch: 0
oct: 0, prv: 0, SQL: 0, pSQL: 70000262f26cbd8, user: 49/ORCLADM
O/S info: user: yxmis, term: , ospid: 25364, machine: kj_orcl_app1
program: DFGD_LONG@kj_orcl_app1 (TNS V1-V3)
application name: DFGD_LONG@kj_orcl_app1 (TNS V1-V3), hash value=3372202780
```

```
last wait for 'gc current request' blocking sess=0x0 seq=8372 wait_time=1086 seconds since
wait started=0
    file#=1f9, block#=2c0a, id#=2000008
```

在 hang analyze 的结果中可以看到，正是 3252 这个会话产生了阻塞。

```
*** 2014-01-09 12:33:12.692
===============
HANG ANALYSIS:
===============
Found 416 objects waiting for <cnode/sid/sess_srno/proc_ptr/ospid/wait_event>
    <1/6568/1/0x69d2bf8/4429090/latch: redo allocation>
Found 13 objects waiting for <cnode/sid/sess_srno/proc_ptr/ospid/wait_event>
    <1/4035/5/0x7a51a38/1242406/log file sync>
Open chains found:
Chain 1 : <cnode/sid/sess_srno/proc_ptr/ospid/wait_event> :
    <1/3252/61/0x79acc18/385380/No Wait>
 -- <1/6568/1/0x69d2bf8/4429090/latch: redo allocation>
 -- <1/2590/3/0x5a263a8/2778422/log file sync>
```

有 416 个会话在等待 6568 这个会话，即 6568 会话阻塞了 416 个进程，而 6568 会话在等待 latch: redo allocation，而 6568 会话正是被 3252 会话阻塞。

6568 会话是 LGWR 进程。毫无疑问，LGWR 进程不能持有 redo allocation latch，将不能写 redo 到日志文件，也就不能完成事务提交，所以在短时间内会产生大量的进程在等待 log file sync，即等待 LGWR 完成日志写入。

接下来再看看 2014 年 1 月 8 日 10:13 左右开始，节点 2 数据库实例很慢然后发生 hang 故障，这同样是 diag 触发 Bug 引起的。

```
*** 2014-01-08 10:13:21.472
Dumping diagnostic information for ospid 652210:
OS pid = 652210
loadavg : 31.73 28.30 25.81
swap info: free_mem = 96209.99M rsv = 256.00M
alloc = 608.77M avail = 65536.00M swap_free = 64927.23M
     F S     UID     PID    PPID   C PRI NI ADDR        SZ WCHAN        STIME    TTY   TIME CMD
240001 A  oracle  652210         1  68 94 20 3aff9d6590 97376             23:03:53       - 40:07
oracleorcldb2 (LOCAL=NO)
    open: The file access permissions do not allow the specified action.
    procstack: write(/proc/652210/ctl): The requested resource is busy.
    652210: oracleorcldb2 (LOCAL=NO)

*** 2014-01-08 10:13:21.611
----------------------------------------
SO: 7000028069f3b98, type: 2, owner: 0, flag: INIT/-/-/0x00
(process) Oracle pid=509, calls cur/top: 70000261e0f2a90/70000261e0f2a90, flag: (0) -
           int error: 0, call error: 0, sess error: 0, txn error 0
 (post info) last post received: 0 0 55
             last post received-location: kjata: wake up enqueue owner
```

12.5 RAC 数据库频繁 hang 问题诊断案例

```
    last process to post me: 7000028059c4488 1 6
    last post sent: 0 0 117
    last post sent-location: kcbzww
    last process posted by me: 7000028059e7b88 1 0
    (latch info) wait_event=0 bits=30
    Process Group: DEFAULT, pseudo proc: 70000280fb3a820
    O/S info: user: oracle, term: UNKNOWN, ospid: 652210
    OSD pid info: Unix process pid: 652210, image: oracle@kjorcldb2

    SO: 700002805ff0c40, type: 4, owner: 7000028069f3b98, flag: INIT/-/-/0x00
    (session) sid: 6064 trans: 70000276e1c5678, creator: 7000028069f3b98, flag: (100141) USR/-
BSY/-/-/-/-/-
    DID: 0002-01FD-00000001, short-term DID: 0002-01FD-00000002
    txn branch: 0
    oct: 0, prv: 0, SQL: 0, pSQL: 70000280daaa140, user: 49/ORCLADM
    O/S info: user: yxmis, term: , ospid: 25365, machine: kj_orcl_app1
    program: DFGD_LONG@kj_orcl_app1 (TNS V1-V3)
    application name: DFGD_LONG@kj_orcl_app1 (TNS V1-V3), hash value=3372202780
    last wait for 'gc current request' blocking sess=0x0 seq=22792 wait_time=315 seconds since
wait started=0
    file#=147, block#=11c2b2, id#=2000001
```

查看 ASH 数据。

```
SYS@xj11g> col program for a40
SYS@xj11g> select to_char(sample_time,'yyyy-mm-dd hh24:mi:ss') sample_time,session_id,
event,program,blocking_session
  2  from dba_hist_active_sess_history where sample_time>=sysdate-3
  3  and instance_number=2 and dbid=3516340257
  4  and sample_id=144689900
  5  order by sample_time,session_id;

SAMPLE_TIME          SESSION_ID    EVENT                          PROGRAM            BLOCKING_SESSION
-------------------  ----------    -----------------------------  -----------------  ----------------
2014-01-08 10:13:15  4218                                         KHFW_LONG
2014-01-08 10:13:15  4541          db file sequential read        BBGL_LONG_D
2014-01-08 10:13:15  4828                                         ZHCX_LONG_C
2014-01-08 10:13:15  5002          db file sequential read        ORCL_TDGL_LONG
2014-01-08 10:13:15  5022          db file parallel read          BBGL_LONG_A
2014-01-08 10:13:15  5237          gc cr multi block request      JL_JLZC_LONG
2014-01-08 10:13:15  5417          db file sequential read        ZHCX_LONG_C
2014-01-08 10:13:15  5440                                         ORCL_ZHCX_LONG
2014-01-08 10:13:15  6064          enq: TX - row lock contention  DFGD_LONG
2014-01-08 10:13:15  6083          enq: TX - row lock contention  DFGD_LONG
2014-01-08 10:13:15  6181          db file sequential read        BBGL_LONG_B
2014-01-08 10:13:15  6205          enq: TX - row lock contention  DFGD_LONG
2014-01-08 10:13:15  6595                                         oracle
-------------------------------------------------------------------------------------
SYS@xj11g> select to_char(sample_time,'yyyy-mm-dd hh24:mi:ss') sample_time,session_id,
event,program,blocking_session
  2  from dba_hist_active_sess_history where sample_time>=sysdate-3
  3  and instance_number=2 and dbid=3516340257
  4  and sample_id=144689910
```

```
    5  order by sample_time,session_id;

SAMPLE_TIME         SESSION_ID EVENT                   PROGRAM       BLOCKING_SESSION
---------------     ---------- ----------------------- ------------- ----------------
2014-01-08 10:13:25       3932 log file sync           DFFX_LONG                 6568
2014-01-08 10:13:25       3942 log file sync           DFYW_SHORT                6568
2014-01-08 10:13:25       4056 db file scattered read  plSQLdev.exe
2014-01-08 10:13:25       4061 log file sync           process.d                 6568
2014-01-08 10:13:25       4134 log file sync           DFYW_LONG_D               6568
…为节省篇幅，去掉大量类似数据…
2014-01-08 10:13:25       6460                         YZX_LONG
2014-01-08 10:13:25       6461 log file sync           DFFX_SHORT                6568
2014-01-08 10:13:25       6462 log file sync           QXGL_SHOR                 6568
2014-01-08 10:13:25       6470 log file sync           QXGL_SHORT                6568
2014-01-08 10:13:25       6525 log file sync           KHFW_SHORT                6568
2014-01-08 10:13:25       6582 gcs log flush sync      oracle                    6568
2014-01-08 10:13:25       6596 gcs log flush sync      oracle                    6568
```

从 ASH 数据可以看到，10:13:15 系统还处于正常状态，仅仅在 10 秒之后（在 AWR 中的 ASH 数据时间间隔为 10 秒），即 10:13:25 系统就产生了大量的阻塞。diag 导致 Bug 被触发是在 10:13:21 这个时间。

12.5.3 案例总结

经过上述的深入分析，我们判断近段时间故障频发的主要原因是触发了 Oracle 的 Bug 6859515，这个 Bug 为：Diagnostic collection may hang or crash the instance。具体如下所示。

（1）当 Oracle 的一个进程在等待某个资源（最常见的是事务锁，即 TX 锁）时间过长时，会向 diag 进程发起转储请求。diag 进程对引起堵塞的进程发起转储（dump）操作，主要包括操作系统层面进程的信息、进程的 call stack、在 Oracle 内部进程的状态和会话的状态等非常详细的数据，用于阻塞问题的诊断。

（2）diag 进程会通过操作系统级的命令来收集被诊断进程的 call stack 信息，包括 gdb 调试工具或 procstack 等（不同的操作系统有所不同）。

gdb 或 procstack 工具在取进程的 call stack 信息时是不安全的，可能会造成进程异常或挂起。如果此进程当时持有 latch 等重要资源，这些重要资源将不会被释放。因为进程只是异常或挂起，没有消失，因此 Oracle 的 pmon 进程不会清理和释放这些被一直占有的重要资源。

latch 是 Oracle 内部一种低级的内存锁，使用非常频繁。通常 latch 的持有时间在毫秒级以下，如果进程长时间持有 latch，将会使其他请求 latch 的进程也会一直挂起。最终关键的后台进程，如 CKPT、LGWR 等进程也会由于请求此 latch 获取无法得到响应而挂起，关键是后台进程的挂起会导致数据库挂起。

（3）异常进程持有的 latch 不同，系统的故障现象有所不同。

1）如果是持有 cache buffer chains child latch，由于这种 latch 的子 latch 数量很多，每个子 latch 只是保护部分数据块头，所以只有在访问这个被持有的子 latch 保护的数据块时才会被阻塞（挂起）。进程一直会等待 latch: cache buffer chains，随后会出现 buffer busy waits、gc buffer busy 等与数据块访问相关的等待。被挂起的进程如果处于事务之中，又会出现 TX 锁等待。如果这个 latch 保护的数据块是脏块，会导致 DBWn（数据库写进程）进程等待这个 latch 而挂起。CKPT 进程如果需要访问这个 latch 也会挂起，最终结果会导致不能完成检查点，日志不能切换，数据库挂起。cache buffer chains child latch 数量众多，如果是由于一个 child latch 被异常进程持有一直到数据库挂起，历时时间较长，甚至可能达到数小时。

2）如果是持有 redo copy latch、redo allocation latch 等与 redo 相关的 latch，由于这种 latch 只有一个或者是子 latch 数量极少，同时使用极为频繁，因此在短时间内 LGWR 这样的关键后台进程会挂起，大量进程出现 log file sync 等待，LMS 等关键进程出现 gcs log flush sync 等待而挂起。其最终结果是短时间内出现大量进程挂起和系统挂起。

3）如果是持有 shared pool latch、library cache latch 等与 SQL 解析相关的 latch，同样会在短时间内出现大量与解析相关的等待，导致大量进程挂起和系统挂起。

（4）在 gdb 或 procstack 导致进程异常的情况下，进程可能会被 KILL 命令杀掉，也可能不能被杀掉。如果进程能够被杀掉，那么数据库 pmon 进程会自动清理异常进程占用的资源，系统能够恢复正常。如果进程不能被杀掉，或者是 pmon 进程本身已经被阻塞而挂起，这样杀进程也不能使系统恢复正常，需要强制重启数据库才能使系统恢复运行。

要解决此问题，可以安装数据库补丁，补丁为 8929701：TRACKING Bug FOR 6859515 ON AIX。安装这个补丁后，diag 进程不再通过操作系统的 gdb 或 procstack 等工具获取被诊断进程的 call stack 等信息，而是改为使用 Oracle 内部的 oradeBug short_stack 命令来获取。如果不能打补丁，可以将隐含参数"_ksb_disable_diagpid"设置为 TRUE 作为临时解决方案。注意，这种设置之后，将不能自动收集进程诊断信息，这样当出现一些其他问题时可能会缺乏一些有效的诊断数据。

这个案例最终通过打补丁解决了问题。

通过这个案例，我想有如下的两点，对我们有些帮助。

1）很多朋友在安装数据库时，没有安装任何补丁。有些有经验的 DBA，也只是安装了最新的 PSU 补丁，实际上有些对系统影响大的补丁并不在 PSU 中。比如在 MOS 文档"11.2.0.4 Patch Set - Availability and Known Issues（文档 ID 1562139.1）"中就列出了 11.2.0.4 这个版本中发现的出现频繁比较高的 Bug。在可能的情况下，要把提到的 Bug 的补丁都打上。

2）在分析问题时，alert 日志、system state dump、hang analyze trace、diagnostic 进程 trace 和 ASH（active session history）数据都是很有用的信息来源。多种数据交叉参考，可以避免单

一数据形成的信息量不足没有思路，或者单一数据形成错误结论。

12.6 Exadata 环境下 SQL 性能问题诊断案例

一位客户的 Exadata 数据库在大量数据处理期间经常出现部分 SQL 性能急剧下降的故障。正常执行只需要几分钟，但在故障时需要执行 1 小时至数小时。SQL 异常会导致数据不能及时处理，严重影响了数据统计的及时性。

出现性能问题的 SQL 语句具体如下（由于 SQL 语句较长，仅列出了关键代码）。

```
create table TBAS.tmp_2_o_user_2_xxxx compress for query high nologging parallel as select
/*+parallel (8) */ t1.ACC_NBR, ….
    case when traffic_wlan/1024 > 20480 then 32 when t42.traffic_fund_type_id is not null then
t42.traffic_fund_type_id else (select traffic_fund_type_id from pu_meta_dim.d_traffic_fund_type
t where state = 1 andtraffic_wlan/1024 < t.max_value and traffic_wlan/1024 > t.min_value) end
MBL_wlan_FUND_TYPE_ID, ….
    from TBAS.tmp_1_o_user_2_xxx t1…
```

where 关键字后面的代码通常很简单，甚至没有 where 条件。

分析 SQL 问题一般会从 AWR 报告、SQL 的执行统计信息和 SQL 执行计划着手。

12.6.1 AWR 报告

下列是检查故障时间段（比如：2014 年 10 月 15 日 11:00 至 12:00）的 AWR 报告。

```
              Snap Id    Snap Time          Sessions Curs/Sess
            ---------  ------------------- --------- ---------
Begin Snap:    19332  05-Oct-14 11:00:18     257       1.8
End Snap:      19333  05-Oct-14 12:00:18     254       1.9
Elapsed:              59.99 (mins)
DB Time:             943.03 (mins)

Cache Sizes                       Begin        End
~~~~~~~~~~~                   ---------- ----------
Buffer Cache:      19,456M    19,456M  Std Block Size:         8K
Shared Pool Size:  11,264M    11,264M  Log Buffer:       227,180K

Load Profile              Per Second    Per Transaction  Per Exec   Per Call
~~~~~~~~~~~~            --------------- --------------- ---------- ---------
DB Time(s):                  15.7           4.6           0.01       0.00
DB CPU(s):                    6.6           1.9           0.00       0.00
Redo size:              441,825.6       127,927.9
Logical reads:           77,377.6        22,404.2
Block changes:              419.4           121.4
Physical reads:           9,898.7         2,866.1
Physical writes:          2,049.4           593.4
User calls:               9,804.6         2,838.9
Parses:                   1,970.4           570.5
```

```
Hard parses:           0.7          0.2
W/A MB processed:      9.0          2.6
Logons:            1,959.8        567.4
Executes:          1,969.7        570.3
Rollbacks:             0.0          0.0
Transactions:          3.5
```

从上面的数据可以看到，平均每秒 Logons 数达到 1959 次。这对于任何一个 Oracle 数据库来说，都是异常高的值，正常数据库的每秒 Logon 数都在 30 以下。出现这种情况的最可能的原因是空闲的并行进程加入到了会话当中，即数据库中频繁地发生大量的并发会话的启动和释放。

同时平均每秒 1970 次解析（parse）对数据仓库类应用来说也显得异常得高。

12.6.2 生成 SQL 报告

下列是生成故障 SQL 的 SQL 报告。

```
SQL ID: fym9dtwdyjpna          DB/Inst: ORCLNEW/orclnew2  Snaps: 19321-19337
 -> 1st Capture and Last Capture Snap IDs
    refer to Snapshot IDs witin the snapshot range
 -> create table TBAS.tmp_2_o_user_2_xxxx compress for query high nologgin...

Plan Hash      Total Elapsed                 1st Capture   Last Capture
 #  Value       Time(ms)    Executions       Snap ID       Snap ID
--- ----------  ------------ -----------    ----------    ----------
 1  3675582523  12,699,298   0               19332         19337
-------------------------------------------------------------------

Plan 1(PHV: 3675582523)
-----------------------

Plan Statistics   DB/Inst: ORCLNEW/orclnew2  Snaps: 19321-19337
 -> % Total DB Time is the Elapsed Time of the SQL statement divided
into the Total Database Time multiplied by 100

Stat Name                      Statement   Per Execution % Snap
------------------------------ ----------- --------------- -------
Elapsed Time (ms)              1.2699E+07    N/A           0.9
CPU Time (ms)                  6,176,679     N/A           0.9
Executions                     0             N/A           N/A
Buffer Gets                    1.6908E+07    N/A           0.2
Disk Reads                     1,378,166     N/A           0.1
Parse Calls                    9,875,656     N/A          23.8
Rows                           0             N/A           N/A
User I/O Wait Time (ms)        1,707,961     N/A           N/A
Cluster Wait Time (ms)         55            N/A           N/A
Application Wait Time (ms)     840,484       N/A           N/A
Concurrency Wait Time (ms)     746,191       N/A           N/A
Invalidations                  1             N/A           N/A
Version Count                  57            N/A           N/A
```

```
Sharable Mem(KB)              28,536          N/A     N/A
----------------------------------------------------------

Execution Plan
------------------------------------------------------------------------------------
| Id  | Operation                          | Name          | Rows  | Bytes | Cost (%CPU)| Time     | Pstart|
------------------------------------------------------------------------------------
|  0  | CREATE TABLE STATEMENT             |               |       |17915 (100)|       |          |       |
|  1  |  PX COORDINATOR                    |               |       |       |       |          |          |       |
|  2  |   PX SEND QC (RANDOM)              | :TQ10000      |    1  |   52  |    2 (0)| 00:00:01 |       |
|  3  |    PX BLOCK ITERATOR               |               |    1  |   52  |    2 (0)| 00:00:01 |       |
|  4  |     TABLE ACCESS STORAGE FULL FIRST ROWS | D_CHARGE_FUND_TYPE | 1 | 52 | 2 (0)| 00:00:01 |
|  5  |  PX COORDINATOR                    |               |       |       |       |          |          |       |
|  6  |   PX SEND QC (RANDOM)              | :TQ20000      |    1  |   52  |    2 (0)| 00:00:01 |       |
|  7  |    PX BLOCK ITERATOR               |               |    1  |   52  |    2 (0)| 00:00:01 |       |
|  8  |     TABLE ACCESS STORAGE FULL FIRST ROWS | D_CHARGE_FUND_TYPE | 1 | 52 | 2 (0)| 00:00:01 |
|  9  |  PX COORDINATOR                    |               |       |       |       |          |          |       |
| 10  |   PX SEND QC (RANDOM)              | :TQ30000      |    1  |   52  |    2 (0)| 00:00:01 |       |
| 11  |    PX BLOCK ITERATOR               |               |    1  |   52  |    2 (0)| 00:00:01 |       |
| 12  |     TABLE ACCESS STORAGE FULL FIRST ROWS | D_CHARGE_FUND_TYPE | 1 | 52 | 2 (0)| 00:00:01 |
| 13  |  PX COORDINATOR                    |               |       |       |       |          |          |       |
| 14  |   PX SEND QC (RANDOM)              | :TQ40000      |    1  |   52  |    2 (0)| 00:00:01 |       |
| 15  |    PX BLOCK ITERATOR               |               |    1  |   52  |    2 (0)| 00:00:01 |       |
| 16  |     TABLE ACCESS STORAGE FULL FIRST ROWS | D_TRAFFIC_FUND_TYPE | 1 | 52 | 2 (0)| 00:00:01 |
| 17  |  PX COORDINATOR                    |               |       |       |       |          |          |       |
| 18  |   PX SEND QC (RANDOM)              | :TQ50000      |    1  |   52  |    2 (0)| 00:00:01 |       |
| 19  |    PX BLOCK ITERATOR               |               |    1  |   52  |    2 (0)| 00:00:01 |       |
| 20  |     TABLE ACCESS STORAGE FULL FIRST ROWS | D_TRAFFIC_FUND_TYPE | 1 | 52 | 2 (0)| 00:00:01 |
| 21  |  PX COORDINATOR                    |               |       |       |       |          |          |       |
| 22  |   PX SEND QC (RANDOM)              | :TQ60000      |    1  |   52  |    2 (0)| 00:00:01 |       |
| 23  |    PX BLOCK ITERATOR               |               |    1  |   52  |    2 (0)| 00:00:01 |       |
| 24  |     TABLE ACCESS STORAGE FULL FIRST ROWS | D_TRAFFIC_FUND_TYPE | 1 | 52 | 2 (0)| 00:00:01 |
| 25  |  PX COORDINATOR                    |               |       |       |       |          |          |       |
| 26  |   PX SEND QC (RANDOM)              | :TQ70000      |    1  |   52  |    2 (0)| 00:00:01 |       |
| 27  |    PX BLOCK ITERATOR               |               |    1  |   52  |    2 (0)| 00:00:01 |       |
| 28  |     TABLE ACCESS STORAGE FULL FIRST ROWS | D_TRAFFIC_FUND_TYPE | 1 | 52 | 2 (0)| 00:00:01 |
| 29  | PX COORDINATOR                     |               |       |       |       |          |          |       |
| 30  |  PX SEND QC (RANDOM)               | :TQ80000      |    1  |   52  |    2 (0) | 00:00:01 |       |
| 31  |   PX BLOCK ITERATOR                |               |    1  |   52  |    2 (0) | 00:00:01 |       |
| 32  |    TABLE ACCESS STORAGE FULL FIRST ROWS | D_TRAFFIC_FUND_TYPE | 1 | 52 | 2 (0)| 00:00:01 |
| 33  |  PX COORDINATOR                    |               |       |       |       |          |          |       |
| 34  |   PX SEND QC (RANDOM)              | :TQ90009      | 2088K | 4519M |  2839 (2) | 00:00:35 |       |
```

```
|35|   LOAD AS SELECT          |          |       |       |        |          |       |
|36|    HASH JOIN RIGHT OUTER|         | 2088K  |4519M|  2839   (2)| 00:00:35 |       |
|37|     PX RECEIVE           |         | 44400  | 1127K|   2  (0)| 00:00:01 |       |
|38|      PX SEND BROADCAST| :TQ90000| 44400  | 1127K|   2  (0)| 00:00:01 |       |
|39|       PX BLOCK ITERATOR |         | 44400  | 1127K|   2  (0) | 00:00:01 |  34  |
|40|        TABLE ACCESS STORAGE FULL | WT_FINANCIAL_CARD | 44400 | 1127K|   2  (0)| 00:00:01 |  34 |
|41|     HASH JOIN RIGHT OUTER|         | 2088K|  4467M  2835   (2)| 00:00:35 |       |
|42|      PX RECEIVE           |        |   25  |  975  |   2  (0) | 00:00:01 |       |
|43|       PX SEND BROADCAST| :TQ90001|  25  |  975  |   2  (0)  | 00:00:01 |       |
|44|        PX BLOCK ITERATOR|         |   25  |  975  |   2  (0)  | 00:00:01 |       |
|45|         TABLE ACCESS STORAGE FULL| D_CHARGE_FUND_TYPE| 25 | 975 |2 (0)| 00:00:01 |       |
|46|      HASH JOIN RIGHT OUTER|         | 2088K|  4389M   2832   (2) | 00:00:34 |       |
|47|       PX RECEIVE           |        |   25  |  975  |   2  (0) | 00:00:01 |       |
|48|        PX SEND BROADCAST  |:TQ90002|  25  |  975  |   2  (0)  | 00:00:01 |       |
|49|         PX BLOCK ITERATOR  |        |   25  |  975  |   2  (0)  | 00:00:01 |       |
|50|          TABLE ACCESS STORAGE FULL    | D_CHARGE_FUND_TYPE |  25 |  975  |   2  (0)| 00:00:01 |       |
|51|       HASH JOIN RIGHT OUTER|         | 2088K|  4312M   2829   (2)  | 00:00:34 |       |
|52|        PX RECEIVE           |        |   25  |  975  |   2  (0) | 00:00:01 |       |
|53|         PX SEND BROADCAST|:TQ90003|   25|  975  |   2  (0)  | 00:00:01 |       |
|54|          PX BLOCK ITERATOR|         |   25  |  975  |   2  (0)  | 00:00:01 |       |
|55|           TABLE ACCESS STORAGE FULL | D_CHARGE_FUND_TYPE | 25 | 975 |2(0)| 00:00:01 |       |
|56|        HASH JOIN RIGHT OUTER|         | 2088K|  4234M   2826   (2)| 00:00:34 |       |
|57|         PX RECEIVE           |        |   16  |  832  |   2  (0)| 00:00:01 |       |
|58|          PX SEND BROADCAST|:TQ90004|   16  |  832  |   2  (0) | 00:00:01 |       |
|59|           PX BLOCK ITERATOR|         |   16  |  832  |   2  (0)| 00:00:01 |       |
|60|            TABLE ACCESS STORAGE FULL      | D_TRAFFIC_FUND_TYPE |  16 |  832  |   2  (0)| 00:00:01 |       |
|61|         HASH JOIN RIGHT OUTER|         | 2088K|  4131M   2822   (1)| 00:00:34 |       |
|62|          PX RECEIVE           |        |   16  |  832  |   2  (0)| 00:00:01 |       |
|63|           PX SEND BROADCAST|:TQ90005|   16  |  832  | 2  (0)| 00:00:01 |       |
|64|            PX BLOCK ITERATOR|   16  |  832  |   2  (0)| 00:00:01 |       |
|65|             TABLE ACCESS STORAGE FULL| D_TRAFFIC_FUND_TYPE | 16 |  832 | 2 (0)| 00:00:01 |       |
|66|          HASH JOIN RIGHT OUTER|         | 2088K|  4027M   2819   (1)| 00:00:34 |       |
|67|           PX RECEIVE |         |   16  |  832  |   2  (0)| 00:00:01 |       |
|68|            PX SEND BROADCAST| :TQ90006|   16  |  832  |   2  (0)| 00:00:01 |       |
|69|             PX BLOCK ITERATOR|   16  |  832  |   2  (0)| 00:00:01 |       |
|70|              TABLE ACCESS STORAGE FULL  | D_TRAFFIC_FUND_TYPE |  16 |  832  |   2  (0)| 00:00:01 |       |
|71|           HASH JOIN RIGHT OUTER |         | 2088K|  3923M   2816   (1)| 00:00:34 |       |
|72|            PX RECEIVE           |        |   16  |  832  |   2  (0)| 00:00:01 |       |
|73|             PX SEND BROADCAST   |:TQ90007|  16  |  832  |   2   (0)| 00:00:01 |       |
|74|              PX BLOCK ITERATOR  |        |   16  |  832  |   2  (0)| 00:00:01 |       |
|75|               TABLE ACCESS STORAGE FULL | D_TRAFFIC_FUND_TYPE | 16 | 832 |2  (0)| 00:00:01 |       |
|76|            HASH JOIN RIGHT OUTER  |        | 2088K|  3820M   2813   (1)| 00:00:34 |       |
|77|             PX RECEIVE   |        |   16  |  832  |   2  (0)| 00:00:01 |       |
|78|              PX SEND BROADCAST              |:TQ90008           |  16 |  832  |   2   (0)| 00:00:01 |       |
|79 |              PX BLOCK ITERATOR |  |   16  |  832  |   2  (0)| 00:00:01 |       |
|80 |               TABLE ACCESS STORAGE FULL| D_TRAFFIC_FUND_TYPE |  16 |  832  |   2  (0)| 00:00:01 |       |
```

```
|81|   PX BLOCK ITERATOR      |  | 2088K| 3716M| 2809   (1)| 00:00:34 | |
|82|    TABLE ACCESS STORAGE FULL | TMP_1_0_USER_2_0817 | 2088K|  3716M|  2809   (1)|
00:00:34 |             |
---------------------------------------------------------------------------

Note
-----
 - dynamic sampling used for this statement (level=4)
 - automatic DOP: Computed Degree of Parallelism is 8
```

可以看到，这条 SQL 语句执行了 12699 秒（即 3 个半小时）还没有执行完成，这条 SQL 语句产生的 parse calls 高达 987 万以上。

从 SQL 执行时间上看，CPU 时间、IO 时间、集群等待时间和并行争用等待时间的合计远小于 SQL 执行总时间。这种情况下通常是由于 CPU 利用率过高，执行 SQL 的进程不能得到足够的 SQL，或者是由于执行 SQL 的进程在等待"空闲"事件，没有计入到等待时间中。

从执行计划上看，SQL 是并行执行的执行计划，同时在 SELECT 字段列表部分有子查询存在，并且子查询同样是并行执行计划。

12.6.3 检查历史数据

检查 SQL 执行的历史数据的代码如下所示。

```
SYS@orclnew2>   select    sample_id,to_char(min(sample_time),'yyyy-mm-dd  hh24:mi:ss')
sample_time,instance_number,QC_SESSION_ID,event,count(*)
  2  from dba_hist_active_sess_history a
  3  where a.sample_time>=to_date('2014-10-05 09:00','yyyy-mm-dd hh24:mi:ss')
  4  and SQL_id='fym9dtwdyjpna'
  5  group by sample_id,event,instance_number,QC_SESSION_ID
  6  order by 1 ;

 SAMPLE_ID SAMPLE_TIME         INSTANCE_NUMBER QC_SESSION_ID EVENT                COUNT(*)
---------- ------------------- --------------- ------------- -------------------- --------
  69710880 2014-10-05 10:24:31               2           406                             1
  69710890 2014-10-05 10:24:41               2           406                             1
  69710910 2014-10-05 10:25:01               2           406                             5
  69710920 2014-10-05 10:25:11               2           406                             2
  69710930 2014-10-05 10:25:21               2           406                             1
  69710940 2014-10-05 10:25:31               2           406 PX Deq: Signal ACK RSG      1
  69710940 2014-10-05 10:25:31               2           406                             2
  69710950 2014-10-05 10:25:41               2           406                             6
  69710960 2014-10-05 10:25:51               2           406 PX Deq: Signal ACK RSG      1
  69710970 2014-10-05 10:26:01               2           406 PX Deq: Signal ACK RSG      1
  69710970 2014-10-05 10:26:01               2           406                             1
  69710980 2014-10-05 10:26:11               2           406 PX Deq: Signal ACK RSG      1
  69710980 2014-10-05 10:26:11               2           406                             2
  69710990 2014-10-05 10:26:21               2           406 PX Deq: Signal ACK RSG      1
```

69710990	2014-10-05 10:26:21	2	406		3
69711000	2014-10-05 10:26:31	2	406	PX Deq: Slave Session Stats	1
69711010	2014-10-05 10:26:41	2	406	PX Deq: Signal ACK RSG	1
69711010	2014-10-05 10:26:41	2	406		3
69711020	2014-10-05 10:26:51	2	406	PX Deq: Signal ACK RSG	1
69711020	2014-10-05 10:26:51	2	406		3
69711030	2014-10-05 10:27:01	2	406		1
69711040	2014-10-05 10:27:11	2	406	PX Deq: Signal ACK RSG	1
69711040	2014-10-05 10:27:11	2	406		1
69711050	2014-10-05 10:27:21	2	406		3
69711060	2014-10-05 10:27:31	2	406	PX Deq: Signal ACK EXT	1
69711080	2014-10-05 10:27:51	2	406		1
69711090	2014-10-05 10:28:01	2	406	enq: KO - fast object checkpoint	1
69711100	2014-10-05 10:28:11	2	406	cell smart table scan	1
69711110	2014-10-05 10:28:21	2	406	PX Deq: Slave Session Stats	1
69711120	2014-10-05 10:28:31	2	406		1
69711130	2014-10-05 10:28:41	2	406		1
69711140	2014-10-05 10:28:51	2	406		1
69711150	2014-10-05 10:29:01	2	406	PX Deq: Signal ACK RSG	1
69711150	2014-10-05 10:29:01	2	406		1
69711160	2014-10-05 10:29:11	2	406	PX Deq: Signal ACK RSG	1
69711160	2014-10-05 10:29:11	2	406		3
69711170	2014-10-05 10:29:21	2	406	cell smart table scan	1
69711180	2014-10-05 10:29:31	2	406	cell smart table scan	1
69711190	2014-10-05 10:29:41	2	406		1
69711200	2014-10-05 10:29:51	2	406		1
69711210	2014-10-05 10:30:01	2	406	cell smart table scan	1
69711220	2014-10-05 10:30:12	2	406		3
69711230	2014-10-05 10:30:22	2	406	enq: KO - fast object checkpoint	1
69711240	2014-10-05 10:30:32	2	406		1
69711250	2014-10-05 10:30:42	2	406		1
69711260	2014-10-05 10:30:52	2	406	cell smart table scan	1
69711270	2014-10-05 10:31:02	2	406	cell smart table scan	1
69711280	2014-10-05 10:31:12	2	406		1
69711290	2014-10-05 10:31:22	2	406	reliable message	1
69711300	2014-10-05 10:31:32	2	406		1
69711310	2014-10-05 10:31:42	2	406	PX Deq: Signal ACK RSG	1
69711310	2014-10-05 10:31:42	2	406		1
69711320	2014-10-05 10:31:52	2	406		1
69711330	2014-10-05 10:32:02	2	406	cell smart table scan	1
69711350	2014-10-05 10:32:22	2	406	reliable message	1
69711370	2014-10-05 10:32:42	2	406	cell smart table scan	1
69711380	2014-10-05 10:32:52	2	406	cell smart table scan	1
69711390	2014-10-05 10:33:02	2	406	PX Deq: Signal ACK RSG	1
69711390	2014-10-05 10:33:02	2	406		1

从会话的历史数据来看，SQL 语句的确是并行执行，但是各个时刻并行进程的数量各不相同。

12.6.4 判断问题产生的流程

由于此 SQL 语句产生了大量的 parse calls,并且在 SELECT 字段列表部分有子查询,结合 SQL 执行的会话历史数据,初步判断问题的产生是这样一个流程。

(1) SQL 语句执行主体查询部分。

(2) 根据主体查询部分返回的数据,执行 SELECT 字段列表中的标量子查询。

(3) 每次执行子查询都进行并行查询。并行查询每次都要启动并行进程、查询数据、释放并行进程这样的过程,这样大量的动作序列使得 SQL 执行非常缓慢。

(4) 新启并行进程会进行 SQL 的解析,所以产生大量的 parse calls。

并行查询表时,执行计划中有提示 TABLE ACCESS STORAGE FULL FIRST ROWS,实际执行时也存在 cell smart table scan。这表明由存储节点进行数据扫描,在扫描数据之前,需要对表做 object checkpoint。会话历史中的 enq: KO - fast object checkpoint 等待事件表示该表正在进行 object checkpoint。

因此,针对此故障,初步判断是语句的 SELECT 字段列表部分有子查询,在子查询执行时出现存在问题。那么需要验证以下两点。

(1) 所有出现此类故障的 SQL 语句是否都是这一类 SQL,即 SELECT 字段列表部分有标量子查询的 SQL。

(2) 同一条 SQL 在故障期间执行时与正常执行时的差别。

12.6.5 查询历史数据

由于故障 SQL 的 parse calls 较高,因此可以通过下列代码来查询历史数据。

```
select snap_id,parse_calls_delta,SQL_id,elapsed_time_delta/1000000 ela,executions_delta exec
from dba_hist_SQLstat
where parse_calls_delta>=10000
and parse_calls_delta/decode(executions_delta,0,1,null,1,executions_delta)>10
and elapsed_time_delta/1000000>=500
order by 1;
```

通过这个 SQL 语句查询出来的结果,可以证实是同一类 SQL 出现了问题。

12.6.6 并列执行的序列过程

对异常 SQL 的会话进行 trace,会重复出现下列的等待事件。

12.6　Exadata 环境下 SQL 性能问题诊断案例

```
    WAIT nam='PX Deq: Join ACK' ela= 643 sleeptime/senderid=268566613 passes=2 p3=36959538792
obj#=13123575
    WAIT nam='PX Deq: Join ACK' ela= 512 sleeptime/senderid=268566619 passes=5 p3=35004876592
obj#=13123575
    WAIT nam='PX Deq: Join ACK' ela= 1 sleeptime/senderid=0 passes=0 p3=0 obj#=13123575
tim=1412491416551083
    WAIT nam='PX Deq: Join ACK' ela= 1 sleeptime/senderid=0 passes=0 p3=0 obj#=13123575
tim=1412491416551095
    WAIT nam='PX Deq: Join ACK' ela= 0 sleeptime/senderid=0 passes=0 p3=0 obj#=13123575
tim=1412491416551105
    WAIT nam='PX Deq: Join ACK' ela= 1 sleeptime/senderid=0 passes=0 p3=0 obj#=13123575
tim=1412491416551116
    WAIT nam='PX Deq: Join ACK' ela= 0 sleeptime/senderid=0 passes=0 p3=0 obj#=13123575
tim=1412491416551126
    WAIT nam='PX Deq: Join ACK' ela= 0 sleeptime/senderid=0 passes=0 p3=0 obj#=13123575
tim=1412491416551136
    WAIT nam='PX Deq: Parse Reply' ela= 3801 sleeptime/senderid=200 passes=1 p3=0 obj#=13123575
    WAIT nam='PX Deq: Parse Reply' ela= 2329 sleeptime/senderid=200 passes=1 p3=0 obj#=13123575
    WAIT nam='PX Deq: Parse Reply' ela= 1 sleeptime/senderid=0 passes=0 p3=0 obj#=13123575
    WAIT nam='PX Deq: Parse Reply' ela= 514 sleeptime/senderid=200 passes=2 p3=0 obj#=13123575
    WAIT nam='PX Deq: Parse Reply' ela= 196 sleeptime/senderid=200 passes=1 p3=0 obj#=13123575
    WAIT nam='PX Deq: Parse Reply' ela= 1 sleeptime/senderid=0 passes=0 p3=0 obj#=13123575
    WAIT nam='PX Deq: Parse Reply' ela= 472 sleeptime/senderid=200 passes=2 p3=0 obj#=13123575
    WAIT nam='PX Deq: Parse Reply' ela= 1 sleeptime/senderid=0 passes=0 p3=0 obj#=13123575
    WAIT nam='PX Deq: Execute Reply' ela= 2 sleeptime/senderid=0 passes=0 p3=0 obj#=13123575
    WAIT nam='PX Deq: Execute Reply' ela= 36 sleeptime/senderid=200 passes=3 p3=0 obj#=13123575
    WAIT nam='PX Deq: Signal ACK RSG' ela= 24 sleeptime/senderid=10 passes=2 p3=0 obj#=13123575
    WAIT nam='PX Deq: Signal ACK RSG' ela= 1 sleeptime/senderid=0 passes=0 p3=0 obj#=13123575
    WAIT nam='PX Deq: Signal ACK RSG' ela= 1 sleeptime/senderid=0 passes=0 p3=0 obj#=13123575
    WAIT nam='PX Deq: Signal ACK RSG' ela= 0 sleeptime/senderid=0 passes=0 p3=0 obj#=13123575
    WAIT nam='PX Deq: Signal ACK RSG' ela= 2 sleeptime/senderid=0 passes=0 p3=0 obj#=13123575
    WAIT nam='PX Deq: Signal ACK RSG' ela= 1 sleeptime/senderid=0 passes=0 p3=0 obj#=13123575
    WAIT nam='PX Deq: Signal ACK RSG' ela= 203 sleeptime/senderid=10 passes=3 p3=0 obj#=13123575
    WAIT nam='PX Deq: Signal ACK RSG' ela= 196 sleeptime/senderid=10 passes=2 p3=0 obj#=13123575
    WAIT    nam='reliable    message'    ela=    365    channel    context=38950850616    channel
handle=38549009072 broadcast message=39084394200 obj#=13123575
    WAIT nam='enq: KO - fast object checkpoint' ela= 95 name|mode=1263468550 2=131327 0=1
obj#=13123575
    WAIT nam='enq: KO - fast object checkpoint' ela= 102 name|mode=1263468545 2=131327 0=2
obj#=13123575
    WAIT nam='cell smart table scan' ela= 447 cellhash#=88802347 p2=0 p3=0 obj#=13123562
tim=1412491416561265
    WAIT nam='cell smart table scan' ela= 498 cellhash#=88802347 p2=0 p3=0 obj#=13123562
tim=1412491416561821
    WAIT    nam='reliable    message'    ela=    503    channel    context=38950850616    channel
handle=38549009072 broadcast message=31940845352 obj#=13123562
    WAIT nam='enq: KO - fast object checkpoint' ela= 174 name|mode=1263468550 2=131327 0=1
obj#=13123562
    WAIT nam='enq: KO - fast object checkpoint' ela= 100 name|mode=1263468545 2=131327 0=2
```

```
obj#=13123562
    WAIT nam='cell smart table scan' ela= 360 cellhash#=88802347 p2=0 p3=0 obj#=13123562
tim=1412491416563438
    WAIT nam='cell smart table scan' ela= 612 cellhash#=88802347 p2=0 p3=0 obj#=13123562
tim=1412491416564115
    WAIT nam='PX Deq: Signal ACK EXT' ela= 247 sleeptime/senderid=200 passes=1 p3=0
obj#=13123562
    WAIT nam='PX Deq: Signal ACK EXT' ela= 256 sleeptime/senderid=200 passes=2 p3=0
obj#=13123562
    WAIT nam='PX Deq: Signal ACK EXT' ela= 2 sleeptime/senderid=0 passes=0 p3=0 obj#=13123562
    WAIT nam='PX Deq: Slave Session Stats' ela= 3 sleeptime/senderid=0 passes=0 p3=0
obj#=13123562
    WAIT nam='PX Deq: Slave Session Stats' ela= 1 sleeptime/senderid=0 passes=0 p3=0
obj#=13123562
    WAIT nam='PX Deq: Slave Session Stats' ela= 2 sleeptime/senderid=0 passes=0 p3=0
obj#=13123562
    WAIT nam='PX Deq: Slave Session Stats' ela= 2 sleeptime/senderid=0 passes=0 p3=0
obj#=13123562
    WAIT nam='PX Deq: Slave Session Stats' ela= 1 sleeptime/senderid=0 passes=0 p3=0
obj#=13123562
    WAIT nam='PX Deq: Slave Session Stats' ela= 2 sleeptime/senderid=0 passes=0 p3=0
obj#=13123562
    WAIT nam='PX Deq: Slave Session Stats' ela= 1 sleeptime/senderid=0 passes=0 p3=0
obj#=13123562
    WAIT nam='PX Deq: Slave Session Stats' ela= 139 sleeptime/senderid=200 passes=3 p3=0
obj#=13123562
```

一个并行执行的序列过程如下。

（1）启动并行进程。

（2）QC 进程（执行 SQL 的发起进程）向所有并行进程发送 JOIN 消息，使并行进程加入到 Server Group（并行执行同一条 SQL 的所有并行查询进程加入到 Server Group 中）。这个时候 QC 进程会出现 PX Deq: Join ACK 等待，表示 QC 进程在等待并行查询进程加入到 Server Group。

（3）QC 进程向每个查询进程发送 SQL 语句，让每个并行查询进程解析 SQL。这时 QC 进程出现 PX Deq: Parse Reply，等待并行进程确认解析完毕。

（4）QC 进程发送要执行的任务给每个并行进程，此时 QC 进程出现 PX Deq: Execute Reply，表示 QC 进程在待待并行进程执行返回结果。

（5）执行完成后到最后并行进程退出 Server Group，并收集并行查询会话的统计信息。此时 QC 进程出现的等待事件是 PX Deq: Sinal ACK EXT 和 PX Deq: Slave Session Stats。

查询并行涉及的对象如下所示。

```
SQL> select owner,object_name,object_type from dba_objects where object_id=13123575;

OWNER                OBJECT_NAME                  OBJECT_TYPE
-------------------  ---------------------------  -------------------
PU_META_DIM          D_CHARGE_FUND_TYPE           TABLE
SQL> select owner,object_name,object_type from dba_objects where object_id=13123562;
OWNER                OBJECT_NAME                  OBJECT_TYPE
-------------------  ---------------------------  -------------------
PU_META_DIM          D_TRAFFIC_FUND_TYPE          TABLE
```

可以看出这些表正是 SELECT 字段列表中的标量子查询中的表。

检查 call stack

SQL 在故障期间执行时 QC 进程的 call stack（通过 oradeBug）：

```
w2> oradeBug short_stack
ksedsts()+461<-ksdxfstk()+32<-ksdxcb()+1876<-sspuser()+112<-__sighandler()<-sendmsg()
+16<-sskgxp_sndmsg()+444<-skgxpipost()+979<-skgxppost()+144<-ksxppst_real()+289<-ksxppst_o
ne()+34<-kslpsprns()+521<-kskpthr()+112<-kslpspr()+9<-kxfprienq()+1926<-kxfpqrenq()+235<-k
xfxpf()+145<-kxfxcp1()+4146<-kxfxcp()+491<-qerpxSendParse()+1059<-kxfpValidateSlaveGroup()
+192<-kxfpgsg()+4707<-kxfrAllocSlaves()+483<-kxfrialo()+2589<-kxfralo()+318<-qerpx_rowsrc_
start()+1379<-qerpxStart()+578<-subex1()+138<-subsr3()+183<-evaopn3()+2537<-expepr()+576<-
evaand()+59<-expepr()+47<-evacssr()+128<-qerlt_snv()+85<-qerltRop()+320<-kdstf11001010011k
m()+3284<-kdsttgr()+133176<-qertbFetch()+2525<-qergiFetch()+317<-rwsfcd()+103<-qerltFetch(
)+599<-qertqoFetch()+952<-qerpxFetch()+11817<-ctcdrv()+13941<-opiexe()+21743<-opiosq0()+3932<-
opipls()+11479<-opiodr()+916<-rpidrus()+211<-skgmstack()+148<-rpiswu2()+638<-rpidrv()+1384<-ps
ddr0()+473<-psdnal()+457<-pevm_EXIM()+308<-pfrinstr_EXIM()+53<-pfrrun_no_tool()+63<-pfrrun()+6
27<-plSQL_run()+649<-peicnt()+301<-kkxexe()+525<-opiexe()+17667<-kpoal8()+2124<-opiodr()+916<-
ttcpip()+2242<-opitsk()+1673<-opiino()+966<-opiodr()+916<-opidrv()+570<-sou2o()+103<-opimai_re
al()+133<-ssthrdmain()+252<-main()+201<-__libc_start_main()+244<-_start()+36
```

call stack 中，函数的执行顺序是从后往前看。plSQL_run 表明 SQL 是 PL/SQL 存储过程中的 SQL 语句，qerpxFetch 表明是并行执行（此处要注意的是，Oracle 根据执行计划以并行方式开始执行，但是可能会发生执行时没有足够的并行进程，结果降级为串行方式执行的情况）subex1 表示在执行子查询，qerpx_rowsrc_start 表示这个是真正的并行开始，kxfrAllocSlaves 表示分配并行进程，qerpxSendParse 表示发送 SQL 解析请求给并行进程。

call stack 跟之前的现象是相符合的，的确是在子查询部分新启动了并行。如果是事先就分配好的并行进程，则不会出现 kxfrAllocSlaves 这样的函数。反复多次查看 QC 进程的 call stack，都能确认是相同的结果。

12.6.8 检查并行会话

检查 SQL 异常执行时的并行会话具体如下。

```
SYS@orclnew2> select sid from v$px_session where qcsid=1044;

       SID
----------
       851
       918
       982
       280
      1044

SYS@orclnew2> /

       SID
----------
       851
       923
       982
       280
      1044

SYS@orclnew2> /

       SID
----------
       867
       919
       982
       280
      1044
```

可以看到，同一个 QC 会话对应的并行进程其会话 SID 会不停地发生变化，这表明并行进程不停地启用和释放，这跟之前的现象、描述以及结论是一致的。这也和 AWR 报告中的 Logons 非常高是一致的。

确认了子查询在不停地重启并行进程的问题后，还需要确认同一条 SQL 语句为什么大多数时候能正常？并且在故障后中止 SQL 运行再重新执行又能正常运行？

在故障期间，运行下面的测试代码。

```
alter session set events 'trace[SQL_Compiler.*] disk=highest';
alter session set events 'trace[SQL_Execution.*] disk=highest';

create table tu_tbas.WT_P_OFFER_2014101802812_XJ compress for query high nologging parallel 64
    as SELECT
        SERV_ID….
```

检查生成的 trace 文件，在 SQL 开始尝试并行执行时，信息如下所示。

```
2014-10-22 04:58:23.701176*:PX_Messaging:kxfp.c@3325:kxfpqsod():
Query end, buffer cache support for numa enabled: YES
2014-10-22 04:58:23.701176*:PX_Messaging:kxfp.c@3329:kxfpqsod(end):
serial - no slave alloc'd for 1st server set
2014-10-22 04:58:23.701176*:PX_Messaging:kxfp.c@10797:kxfpgsg(end):
slave group q=(nil)
kxfpallocslaves                                           [    30/    10]
actual num slaves alloc'd = 0 (kxfpqcthr)
kxfrialo                                                  [    30/     0]
Finish: allocated actual 0 slaves for non-GV query
```

可以看到，在分配第一组并行进程（1st server set）时，没有并行进程，只能以串行方式运行。

```
qerpxStart                                                [    30/     0]
        rwsrid:11 pxid:3 qbas:2339749:err:0
        pgakid:1 pgadep:0

*** 2014-10-22 04:58:23.702
        START no parallel resources
        ksdftm:958862290
qertqoStart                                               [    30/     0]
```

虽然没有分配到并行进程，但还是要以并行模式来执行 SQL。因为 SQL 执行的入口已经是并行模式，只是在实际执行过程中没有并行进程，只能串行执行了。

在 SQL 执行过程中，会出现大量的类似下面的 trace 信息。

```
2014-10-22 04:58:23.772279*:PX_Granule:kxfr.c@2291:kxfrialo(): Start: building object table (1)
Start: building obj table for obj# 0
kxfrExtInfo                                               [   100/     0]

prefetch part:0 objn:13123568 objd:13123568 seghdr tsn:11 rdba:0x0a80d47a
kxfrExtInfo                                               [   100/     0]

kxfrFileAdd                                               [   100/     0]
filenum:42 #nodes:0 arch:255 test hard affinity:0
dumping object objn:13123568 PU_META_DIM.D_STATE
so:0x838788d70 mo:0x2b5257602150
flg:4201 (KXFRO_PSC/KXFRO_RID/KXFRO_HINT/)
nbparts:1 nfiles:1 nnodes:0 ninst:0
ecnt:1 size(blocks):1 mtfl:0

files for object 1:
kfil:42 size:1 nnodes:0 nodes:(nil) naff:65535

partitions of object 1:
part:0 (abs:-1 pnum:-1 spnum:-1 objn:-1 objd:13123568 tsn:11)
ecnt:1 size:1 nfiles:1 naff:65535  flags: BIT
Files in partition 0:
kfil:42 size:1
Extent map of partition 0:
afn:42  blk:54395    size:1    coalescable:1
```

```
No nodes information.
kxfrialo                                                  [    100/     0]
DOP trace -- requested thread from best ref obj = 16 (from kxfrIsBestR
ef())
……
kxfrialo                                                  [    100/     0]
best object 0x838788d70
hgt:0 blks:1 acp:0 nds:1 thr:16
kxfrialo                                                  [    100/     0]
threads requested = 16 (from kxfrComputeThread())
kxfrialo                                                  [    100/     0]
adjusted no. threads = 16 (from kxfrAdjustDOP())
kxfrAllocSlaves                                           [    100/     0]
DOP trace -- call kxfpgsg to get 16 slaves
2014-10-22 04:58:23.773172*:PX_Messaging:kxfp.c@9868:kxfpgsg(begin):
reqthreads=16 height=0 lsize=0 alloc_flg=0x230
2014-10-22 04:58:23.773172*:PX_Messaging:kxfp.c@9941:kxfpgsg():
reqthreads=16 KXFPLDBL/KXFPADPT/ load balancing=on adaptive=off
Getting instance info for open group
……
2014-10-22 04:58:24.436235*:PX_Messaging:kxfp.c@11465:kxfpg1sg():
got 4 servers (sync), errors=0x0 returning
Acquired 4 slaves on 1 instances avg height=4 #set=1 qser=2045742594
       P011 inst 2 spid 16249
       P016 inst 2 spid 16265
       P017 inst 2 spid 16267
       P018 inst 2 spid 12448
2014-10-22 04:58:24.436235*:PX_Messaging:kxfp.c@10671:kxfpgsg():
Instance(servers):
inst=2 #slvs=4
```

上述 trace 信息表明，在执行子查询时，在不停地启用并行进程并释放并行进程。并且随着系统负载的不同，每次获得的并行进程数不一样，这跟会话历史信息中检查发现的不同时刻并行进程数不同的现象是一致的。

至此，故障原因已经明确，是由于在 SELECT 字段列表有子查询的 SQL 语句，在主体查询以串行方式执行时，子查询 SQL 以并行方式查询所引起的。

在此，我们小结一下这个问题，这个故障的出现需要同时满足以下两个条件。

（1）SQL 语句的 SELECT 字段列表部分有子查询与下列代码相似。

```
create table TBAS.tmp_2_o_user_2_xxxx compress for query high nologging parallel as select
/*+parallel (8) */ t1.ACC_NBR, ...
    case when traffic_wlan/1024 > 20480 then 32 when t42.traffic_fund_type_id is not null then
t42.traffic_fund_type_id else (select traffic_fund_type_id from pu_meta_dim.d_traffic_fund_type
t where state = 1 andtraffic_wlan/1024 < t.max_value and traffic_wlan/1024 > t.min_value) end
MBL_wlan_FUND_TYPE_ID, ….
    from TBAS.tmp_1_o_user_2_xxxx t1…
```

（2）实例上的并行进程用完，即所有的并行进程全部处于活动状态。

当 SELECT 字段列表部分带有子查询的 SQL 执行时,正常情况下,Oracle 会发起两组并行进程(parallel slave sets)。一组进程为数据的生产者(比如从表读数据),一组进程为数据的消费者(比如对"生产者"进程传过来的表数据进行汇总排序等)。在有并行进程时,SQL 语句运行正常。

在实例的并行进程用完后,由于没有并行进程,因此 SQL 语句的主体部分就会降级到串行方式运行。但是在提取 SELECT 字段列表中的子查询数据时,子查询的表需要使用并行进程。这个时候 Oracle 会尝试以并行方式运行子查询。Oracle 对主查询中返回的每一行数据,都要经过"启动并行进程,查询子查询数据,释放并行进程"的过程。这个过程效率非常低,导致 SQL 语句执行需要非常长的时间。

下面是另一条出现故障的 SQL 语句的执行计划(为了便于排版,对执行计划做了剪裁)。

```
-------------------------------------------------------------------------------
| Id  | Operation                            | Name           | Rows  | Bytes |
-------------------------------------------------------------------------------
|   0 | CREATE TABLE STATEMENT               |                |       |       |
|   1 |  PX COORDINATOR                      |                |       |       |
|   2 |   PX SEND QC (RANDOM)                | :TQ10000       |    1  |    19 |
|   3 |    PX BLOCK ITERATOR                 |                |    1  |    19 |
|*  4 |     TABLE ACCESS STORAGE FULL FIRST ROWS | D_STATE    |    1  |    19 |
|*  5 | COUNT STOPKEY                        |                |       |       |
|   6 |  PX COORDINATOR                      |                |       |       |
|   7 |   PX SEND QC (RANDOM)                | :TQ20000       |   11  |   209 |
|*  8 |    COUNT STOPKEY                     |                |       |       |
|   9 |     PX BLOCK ITERATOR                |                |   11  |   209 |
|* 10 |      TABLE ACCESS STORAGE FULL FIRST ROWS| D_STATE    |   11  |   209 |
|  11 | PX COORDINATOR                       |                |       |       |
|  12 |  PX SEND QC (RANDOM)                 | :TQ30000       |   14M |  3644M|
|  13 |   LOAD AS SELECT                     |                |       |       |
|  14 |    PX BLOCK ITERATOR                 |                |   14M |  3644M|
|* 15 |     TABLE ACCESS STORAGE FULL        | F_2_OFFER_SERV_D|  14M |  3644M|
-------------------------------------------------------------------------------
```

上面的执行计划中,第 11~15 步是 SQL 主体查询部分,第 2~4 步是 SELECT 字段列表中的子查询执行部分。正常情况下,是由两组并行进程负责,一组负责主体查询,然后将结果发送到另一组进程来进行子查询。但是在主体查询部分以串行方式进行时,SQL 语句的主进程即 QC 进程,对 F_2_OFFER_SERV_ID 返回的每一行数据都要以并行方式执行一次子查询。这种方式的效率极为低下。

针对这样的问题,解决方案有如下几种。

(1)避免 SELECT 字段列表带有子查询的 SQL 以串行方式执行。在数据仓库应用中,串行执行的 SQL 语句,不管何种情况性能都比较差,因此要避免此种情况的产生。具体方法是在并行进程用光时,对并行执行的 SQL 进行排队。有以下几种方法。

1)将 PARALLEL_DEGREE_POLICY 参数设置为 AUTO,开启自动并行。在系统级设置此参数在没有经过充分测试的情况下有比较大的风险，但是可以考虑在会话级设置此参数。在调度具有故障类 SQL 语句的脚本中设置会话级参数。

2）在故障类 SQL 中加上 Hint "parallel(auto)"。

3）在故障类 SQL 中加上 Hint "STATEMENT_QUEUING"。

（2）避免 SELECT 字段列表中使用子查询，尽量改为表关联。或者是针对此类子查询使用 Hint no_parallel(表别名)"，但是需要测试此种更改对性能的影响。

在数据仓库应用中，需要控制并行度来避免单条 SQL 占用过多的并行进程。控制 SQL 并行度的方法有以下两种（以下两个方法都要同时采用）。

1）设置自动并行时的并行度限制，这是通过设置参数 PARALLEL_DEGREE_LIMIT 来实现，比如将此参数设置为 8。

2）通过资源管理限制来限制并行度。

```
exec dbms_resource_manager.clear_pending_area();
exec dbms_resource_manager.create_pending_area();
exec dbms_resource_manager.create_plan( plan =>'LIMIT_DOP', comment => 'Limit Degree of Parallelism');
exec dbms_resource_manager.create_plan_directive(plan=> 'LIMIT_DOP', group_or_subplan => 'OTHER_GROUPS' , comment => 'limits the parallelism', parallel_degree_limit_p1=> 8);
exec dbms_resource_manager.validate_pending_area();
exec dbms_resource_manager.submit_pending_area();
```

调整前建议进行充分测试，调整后需要密切关注系统的运行。

此外，此类的 SQL 应该比较多，我们可以通过下列代码查询出现过此类故障的 SQL。

```
select   snap_id,parse_calls_delta,SQL_id,elapsed_time_delta/1000000   ela,executions_delta exec from dba_hist_SQLstat
where parse_calls_delta>=10000
and parse_calls_delta/decode(executions_delta,0,1,null,1,executions_delta)>10
and elapsed_time_delta/1000000>=500
order by 1;
```

12.7　关于 RAC 数据库 load balance 案例分析

在这一节我将通过两个案例入手，来讲一下 RAC 数据库 load balance 相关的问题。两个案例会以比较直接简单的方式进行描述，对于案例中涉及的深入的知识将在后面展开。

案例一：一套 2 节点的 Oracle 10g RAC 数据库。系统维护人员报告称不能实现负载均衡，大部分的会话都连接到了第二个节点上。 环境是 Oracle 10.2.0.3 for AIX。

分析过程如下所示。

12.7 关于 RAC 数据库 load balance 案例分析

（1）连接到数据库，检查数据库连接，发现实例 resrac2 的连接数超过 2000，而实例 resrac1 的连接数不到 300。相差非常大。

（2）检查 2 个节点的监听状态，没有异常。

（3）查看应用的 TNS 配置，没有发现异常。

（4）仔细检查 2 个节点的连接，发现几乎所有的 JAVA 应用都连接到了 resrac2，而其他的非 JAVA 应用，大部分也同样是连接到了 resrac2。

（5）使用 SQL*Plus 进行连接测试，发现所有的连接都被重定向了 resrac2。但是修改 TNSNAME，指定 INSTANCE_NAME 为 resrac1 时，能够正常连接 resrac1，这表明实例 resrac1 本身连接接受连接。

（6）检查 listener.log 日志，发现 PMON 能够正常进行监听注册和服务更新（Service Update）。

（7）检查发现 ONS 没有启动，手工启动 ONS 后，问题依然存在。

（8）重启监听，问题依然存在。

（9）检查服务状态，具体如下所示。

```
SQL> select name,goal,clb_goal from dba_services;

NAME                       GOAL         CLB_G
-------------------------- ------------ -----
SYS$BACKGROUND             NONE         SHORT
SYS$USERS                  NONE         SHORT
seeddataXDB                             LONG
seeddata                                LONG
resracXDB                               LONG
resrac                     NONE         LONG
```

可以看到 resrac 服务配置正常。

（10）检查 Service Metric 数据，具体如下所示。

```
SQL> select inst_id,service_name,goodness,delta from
     gv$servicemetric where service_name='resrac';

   INST_ID SERVICE_NAME      GOODNESS      DELTA
---------- --------------- ---------- ----------
         1 resrac                6999          1
         1 resrac                6999          1
         2 resrac                1976          1
         2 resrac                1976          1
```

从上面的数据可以看出，第 1 个实例 resrac1 存在问题。实例 resrac2 的 goodness 值与以 resrac 服务连接的连接数是接近相等的（这跟更新的周期有关）。但实例 resrac1 的 goodness 值远远大于 resrac1 的实际连接数。

Goodness 值有如下两个特点。

- Goodness 值在 Service 的 CLB_GOAL 为 CLB_GOAL_LONG 时表示实例中该服务的会话连接数。此值越大，连接越多，那么连接时就不会连接到 goodness 值大的实例上，而是连接到值小的实例上。
- Goodness 值由 MMON 进程进行计算，然后定期由 PMON 发布给监听。

检查 MMON 进程没有发现异常。

```
$ truss -p 1467404
thread_wait(3000)                    (sleeping...)
thread_wait(3000)                                  = 1
thread_wait(3000)                                  Err#4  EINTR
    Received signal #14, SIGALRM [caught]
sigprocmask(0, 0x0FFFFFFFFFFFC1E0, 0x0000000000000000) = 0
_poll(0x0FFFFFFFFFFFB250, 1, 0)                    = 0
sigprocmask(0, 0x0FFFFFFFFFFFBEB8, 0x0000000000000000) = 0
incinterval(0, 0x0FFFFFFFFFFFBC50, 0x0FFFFFFFFFFFBC70) = 0
sigprocmask(1, 0x0FFFFFFFFFFFBDF0, 0x0000000000000000) = 0
incinterval(0, 0x0FFFFFFFFFFFC060, 0x0FFFFFFFFFFFC080) = 0
sigprocmask(1, 0x0FFFFFFFFFFFC1E0, 0x0000000000000000) = 0
ksetcontext_sigreturn(0x0FFFFFFFFFFFC320,     0x000000011023F1C8,     0x0000000000000000,
0x800000000000D0B2, 0x0000000000000000, 0x0000000000000134, 0x0000000000000000, 0x0000000000000000)
thread_wait(1230)                                  = 1
times(0x0FFFFFFFFFFFC8B0)                          = 1776218558
times(0x0FFFFFFFFFFFC800)                          = 1776218558
thread_wait(3000)                    (sleeping...)
thread_wait(3000)                                  = 1
thread_wait(3000)                                  Err#4  EINTR
    Received signal #14, SIGALRM [caught]
sigprocmask(0, 0x0FFFFFFFFFFFC1E0, 0x0000000000000000) = 0
```

（11）从上面的分析可以看到，Service Metric 的数据异常，导致了实例 resrac1 不能接受新的连接。由于 resrac1 实例的 resrac 服务的 goodness 值异常得高，怀疑是 Bug 引起。

（12）搜索 Oracle 的 Bug 数据库，发现与下面的 Bug 匹配。

```
Bug 6442431 A node may stop receiving connections with RAC load balancing
Details: In a RAC cluster using server side load balancing, one of the nodes may stop receiving
any new connections even though it is idle.
Fixed-Releases: 10.2.0.4 & 11.1.0.7 & 11.2
```

但是当前的版本没有此 Bug 的补丁，只有升级到 10.2.0.4 才能解决。

（13）解决办法有如下 3 种。

1）重启实例 resrac1，但是以后 Bug 可能被再次触发。

2）将 Oracle 数据库升级到 10.2.0.4。

3）关闭服务端（Server Side）的 Load balancing，修改应用的连接配置，使用客户端连接

12.7 关于 RAC 数据库 load balance 案例分析

时负载均衡（Connection Time Load Balancing）。

案例二：一套 2 节点的 Oracle 11g RAC Active Data Guard，应用通过 SCAN 地址连接到数据库，绝大部分连接都连到了节点 1 上。

环境：某银行重要交易系统，Oracle 11.2.0.4 for Linux，2 节点 RAC 主库+2 节点本地 RAC ADG+2 节点同城异机房 RAC ADG，读写分离架构，ADG 用于只读查询，主库实现事务交易，分布式存储架构，同城异机房实现容灾。SCAN 地址通过域名解析。

问题分析和处理思路与前面的案例大体一致，关键的差异在于如下一些要点。

（1）Active Data Guard 指定与主库不同的 db_unique_name，那么默认注册到监听的服务与 db_unique_name 是一致的。但是这个服务名通过 dba_services 视图查询并没有。

```
INST_ID TYPE        SERVICE_NAME              COUNT(*)
------- ----------  ------------------------  ----------
      1 USER        cusr                            226
      1 USER        SYS$USERS                         4
      2 USER        cusr                            218
      1 BACKGROUND  SYS$BACKGROUND                   50
      2 USER        SYS$USERS                         2
      2 BACKGROUND  SYS$BACKGROUND                   50
```

（2）查询 v$servicemetric 视图也没有这个服务的统计数据。

```
INST_ID SERVICE_NAME              GOODNESS     DELTA
------- ------------------------  ----------  ----------
      1 cusrXDB                           0         1
      1 cusrXDB                           0         1
      1 SYS$BACKGROUND                 2150        50
      1 SYS$BACKGROUND                 2150        50
      1 SYS$USERS                      4100        50
      1 SYS$USERS                      4100        50
      2 cusrXDB                           0         1
      2 cusrXDB                           0         1
      2 SYS$BACKGROUND                 2100        50
      2 SYS$BACKGROUND                 2100        50
      2 SYS$USERS                         0       100
      2 SYS$USERS                         0       100
```

那么，会不会是因为 pmon 不能发布服务的负载信息呢？好在有这样一个事件 10257。

```
SQL> oradeBug setospid <pmon 的 pid>
SQL> oradeBug event 10257 trace name context forever,level 16
```

在 pmon 的 trace 文件中不断重复出现如下信息。

```
err=-300 lbflgs=0x0 tbtime=0 tntime=0 etime=300 srvs=1 nreqs=0 sreqs=0 asrvs=1
error=-300 etime=300 control=0 integral=0 lasterr=-300 lastetm=300
kmmlrl: status: succ=4, wait=0, fail=0
```

我们需要有点耐心，一会儿会出现下示情况。

```
err=-300 lbflgs=0x0 tbtime=0 tntime=0 etime=300 srvs=1 nreqs=0 sreqs=0 asrvs=1
error=-300 etime=300 control=0 integral=0 lasterr=-301 lastetm=301
kmmlrl: status: succ=4, wait=0, fail=0
kmmlrl: update for time delta: 60019
kmmgdnu: cusrXDB
        goodness=0, delta=1,
        flags=0x5:unblocked/not overloaded, update=0x6:G/D/-
kmmlrl: node load 284
kmmlrl: D000 load 0
kmmlrl: nsgr update returned 0
```

如果我们手工执行 alter system register 命令，会产生如下日志。

```
kmmlrl: status: succ=4, wait=0, fail=0
kmmlrl: register now
kmmgdnu: cusrXDB
        goodness=0, delta=1,
        flags=0x5:unblocked/not overloaded, update=0x6:G/D/-
kmmlrl: 556 processes
kmmlrl: node load 266
kmmlrl: instance load 508
kmmlrl: nsgr update returned 0
kmmlrl: nsgr register returned 0
```

应用连的服务名是 cusrdg，很显然这个并不在里面。以上输出信息中，kmmlrl 是 Kernel Message Monitor Listener Register Load 的含义，kmmgdnu 是 KMM GooDNess Update 的含义。

那么，到现在要的方向就是 pmon 没有发布 cusrdg 这个服务，如何来验证这个问题？好在有一套测试环境，是同样的架构。在测试环境中，使用 ADG 的 db_unique_name 指定的服务，同样不能达到负载均衡。但是如果用 dbms_service.add_service 增加一个服务（需要在主库上操作），然后在备库启用这个服务，使用这个服务则可以正常进行负载均衡。

当然也可以用 srvctl add service 来增加服务，但使用 srvctl start service 时，同样会去调用 dbms_service 的 add_service 命令。

使用的测试脚本具体如下所示。

```
#!/bin/sh
SQLplus system/xxxx/@SCAN域名/service_name << EOF
exec dbms_lock.sleep(120);
EOF
```

将上述脚本在后台跑上几十个，检查 gv$session 就可以发现负载均衡是否起作用。我们在测试 ADG 库上看到的 10257 event trace 的输出如下所示。

```
kmmlrl: status: succ=4, wait=0, fail=0
kmmlrl: status: succ=4, wait=0, fail=0
kmmlrl: register now
kmmgdnu: drtestsvc
        goodness=0, delta=1,
        flags=0x4:unblocked/not overloaded, update=0x6:G/D/-
```

12.7 关于 RAC 数据库 load balance 案例分析

```
kmmgdnu: oradbXDB
        goodness=0, delta=1,
        flags=0x5:unblocked/not overloaded, update=0x6:G/D/-
kmmlrl: node load 10
kmmlrl: nsgr update returned 0
kmmlrl: nsgr register returned 0
```

可以看到有 drtestsvc 这个单独建的服务名,这个服务名出现在 pmon 向监听注册的信息中。

要继续解析这个问题,还必须了解 RAC 数据库负载均衡原理。负载均衡分两种,Client Side Load Balance 和 Server Side Load Balance。

客户端的负载均衡通过连接描述符来指定,具体如下所示。

```
(DESCRIPTION =
   (ADDRESS_LIST =
     (ADDRESS = (PROTOCOL = TCP)(HOST = 192.168.22.24)(PORT = 1521))
     (ADDRESS = (PROTOCOL = TCP)(HOST = 192.168.22.44)(PORT = 1521))
   )
   (LOAD_BALANCE = yes)
   (CONNECT_DATA =
     (SERVER = DEDICATED)
     (SERVICE_NAME = resrac)
   )
```

LOAD_BALANCE=yes 表示连接时会随机从 ADRESS LIST 中选择一个地址进行连接。要实现 LOAD BALANCE,那么不能指定 INSTANCE_NAME,也不能指定 SID,同时连接时"面向服务"。客户端连接描述符在指定多个地址还默认打开了 failover,也就是一个地址连接不上就会连接另一个地址。

服务器端的负载均衡,又是怎么一种机制呢? 10g 和 11g 有很大的不同,这里以 11g RAC 为例来解释,如图 12-1 所示。

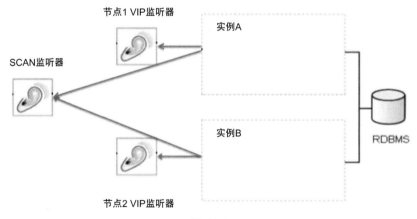

图 12-1

我们配合监听的信息来展示一下（这是在第 2 个节点上获得的信息）。

```
lsnrctl services LISTENER_SCAN1

Connecting to (DESCRIPTION=(ADDRESS=(PROTOCOL=IPC)(KEY=LISTENER_SCAN1)))
Services Summary...
Service "appadb" has 2 instance(s).
  Instance "appadb1", status READY, has 2 handler(s) for this service...
    Handler(s):
      "DEDICATED" established:0 refused:0 state:ready
         REMOTE SERVER
         (DESCRIPTION=(ADDRESS=(PROTOCOL=TCP)(HOST=192.168.233.143)(PORT=1521)))
      "DEDICATED" established:0 refused:0 state:ready
         REMOTE SERVER
         (DESCRIPTION=(ADDRESS=(PROTOCOL=TCP)(HOST=192.168.233.143)(PORT=1522)))
  Instance "appadb2", status READY, has 2 handler(s) for this service...
    Handler(s):
      "DEDICATED" established:0 refused:0 state:ready
         REMOTE SERVER
         (DESCRIPTION=(ADDRESS=(PROTOCOL=TCP)(HOST=192.168.233.145)(PORT=1521)))
      "DEDICATED" established:0 refused:0 state:ready
         REMOTE SERVER
         (DESCRIPTION=(ADDRESS=(PROTOCOL=TCP)(HOST=192.168.233.145)(PORT=1522)))
```

VIP 监听信息如下所示。

```
lsnrctl services LISTENER

LSNRCTL for Linux: Version 11.2.0.3.0 - Production on 17-DEC-2015 17:16:35

Copyright (c) 1991, 2011, Oracle.  All rights reserved.

Connecting to (DESCRIPTION=(ADDRESS=(PROTOCOL=IPC)(KEY=LISTENER)))
Services Summary...
Service "+ASM" has 1 instance(s).
  Instance "+ASM2", status READY, has 1 handler(s) for this service...
    Handler(s):
      "DEDICATED" established:0 refused:0 state:ready
         LOCAL SERVER
Service "appadb" has 1 instance(s).
  Instance "appadb2", status READY, has 1 handler(s) for this service...
    Handler(s):
      "DEDICATED" established:1858112 refused:0 state:ready
         LOCAL SERVER
The command completed successfully
```

打开数据库后，pmon 进程会向监听注册服务。

```
SQL> show parameter listener

NAME                                 TYPE        VALUE
------------------------------------ ----------- ------------------------------
listener_networks                    string
```

12.7 关于 RAC 数据库 load balance 案例分析

```
local_listener                  string    (DESCRIPTION=(ADDRESS_LIST=(AD
                                          DRESS=(PROTOCOL=TCP)(HOST=133.
                                          37.233.145)(PORT=1522))(ADDRES
                                          S=(PROTOCOL=TCP)(HOST=192.168.2
                                          33.145)(PORT=1521))))
remote_listener                 string    appadb-scan:1521
```

remote_listener 指向 SCAN LISTENER，local_listener 指向本节点 VIP 监听。这里有一个 PORT 为 1522 的是一个单独的监听，这里可以先忽略。

我们会看到，VIP 监听上的服务只有一个实例，即本节点的实例的服务。

```
Service "appadb" has 1 instance(s).
  Instance "appadb2", status READY, has 1 handler(s) for this service...
    Handler(s):
      "DEDICATED" established:1858112 refused:0 state:ready
         LOCAL SERVER
```

但是 SCAN 监听上有两个实例。

```
Service "appadb" has 2 instance(s).
  Instance "appadb1", status READY, has 2 handler(s) for this service...
  ….
  Instance "appadb2", status READY, has 2 handler(s) for this service...
    Handler(s):
      "DEDICATED" established:0 refused:0 state:ready
         REMOTE SERVER
         (DESCRIPTION=(ADDRESS=(PROTOCOL=TCP)(HOST=192.169.233.145)(PORT=1521)))
      "DEDICATED" established:0 refused:0 state:ready
         REMOTE SERVER
         (DESCRIPTION=(ADDRESS=(PROTOCOL=TCP)(HOST=192.169.233.145)(PORT=1522)))
```

REMOTE SERVER 表示，这个实例是需要通过另一个监听来跳转的。这个监听不能直接连接到该实例上。对于 SCAN 监听，所有的实例都是远程实例。

那么 REMOTE SERVER 中的信息是从哪里来，实际上是 PMON 进程将 local listener 的信息告诉 SCAN 监听。大家可以看到，local listener 设定的信息跟 SCAN LISTENER 中 Remote Server 的信息是一样的。

了解了这些之后，下面来看一下如何实例负载均衡，客户端与数据库的交互顺序如下所示。

（1）客户端连接到 SCAN 监听。

（2）SCAN 监听根据客户端连接时指定的条件（SERVICE_NAME），选择一个实例。

（3）SCAN 监听把选定的实例的信息，包括地址，端口，协议，实例名发给客户端。

（4）客户端根据这个信息，重新发起一次网络连接，这次连接到实例的 VIP 监听上，VIP 监听的服务是 LOCAL SERVER，所以它会建立真正的连接。

（5）接下来就是登录验证。

这个过程有 2 个关键点。

（1）上述第 2 步的实例选择中，监听是根据从 PMON 得到的实例的负载信息来选择实例的。这一步实现了负载均衡。如果连接 SCAN 时，指定了 INSTANCE NAME 参数，那 SCAN 监听只会重定向到指定参数。

（2）如果 local listener 参数设置不对（正常情况下，11g 不需要显式设置，由 Grid Infrastructure 设置好就行），那么将不能连接，会报错。这个参数正常指向 VIP 地址，如果设置为主机名或域名，但是客户端不能解析，就连接不上。

到这里，会有一个跟网络安全相关的话题。有的企业的应用是在另外的网络，中间有网闸或 NAT，连接 SCAN 监听时，用的一个外网地址，外网地址通过 NAT 转换，得到内网的地址。例如，外网是 10.10.100.144:1521，内网是 192.168.100.144:1521，有一个 NAT 设备进行转换。SCAN 监听会返回一个 VIP 监听地址，不幸的是，它会返回内网 local listener 地址——192.168.100.141:1521，客户端在外网连接这个地址肯定是不通的。那是否可以将数据库的参数设置为外网地址呢？这样的话它就注册不到 vip 监听上了。这个解决方案也有问题。

这样的方式下，用客户端的负载均衡，直接连接 VIP 地址是更好的选择。另外，一些比较老的 JDBC 驱动，不能识别 SCAN 监听发回的要求重定向的协议，这种驱动连接数据库也有问题。

在监听的信息中，会有如下这样的信息。

```
Handler(s):
  "DEDICATED" established:0 refused:0 state:ready
```

这条信息表示接受这个服务的处理器（实例）的状态 ready，已经拒绝多少个连接，建立了多少连接。显然这个应用没有连接 SCAN 监听，因为这个监听的所有服务的连接数都是 0。

只有本地的 VIP 监听才显示有连接。

```
Handler(s):
  "DEDICATED" established:1858112 refused:0 state:ready
    LOCAL SERVER
```

另外，在 VIP 监听日志中会出现如下所示的信息。

```
10-DEC-2015 22:29:01 * (CONNECT_DATA=(CID=(PROGRAM=JDBC Thin
Client)(HOST=__jdbc__)   (USER=app))(SERVER=DEDICATED)(SERVICE_NAME=cusrdg)(FAILOVER_MODE=
(TYPE=session)(METHOD=basic)))  *  (ADDRESS=(PROTOCOL=tcp)(HOST=192.168.13.33)(PORT=44885))  *
establish * cusrdg * 0
```

这表示直接连接或本机 SCAN LISTENER 连接的。如果有 INSTANCE_NAME=这样的字样，则表示是另一个节点转发过来的。我们去检查监听日志，可以查看这个有用的信息。

```
grep "service update" listener.log
grep INSTANCE_NAME= listener.log
```

如果有这样的数据，表明客户端连接指定了 INSTANCE_NAME，这个可能是连接串指定了（当前这种很少）。对于负载均衡来说，更多的是客户端开始连接到另外节点的 SCAN LISTENER 上然后重定向过来的。

12.8 总结

数据库的故障和性能问题千差万别，导致问题的原因也多种多样。所以我个人的建议是，虽然我们可以在不同的案例中总结一些规律，帮助我们处理问题，但生产环境很忌讳经验论。任何具体的问题都需要具体分析，不可以偏概全。希望这一章的内容分享对大家的工作学习有帮助。

第 13 章　反思与总结：轻松从菜鸟到专家

——我的故障分析及处理思路分享（李真旭）

技术的学习是一个不断深入和扩展的过程，在拥有一定的技术基础后，需要不断在应用旧知识的过程中去挖掘新知识，丰富自己对一项技能或者一个领域的认识，主动实现从菜鸟到专家的自我升华。同时，每一个独立的技术领域都包含很多方面。就拿 Oracle 来说，内核研究、架构分析、故障处理、性能优化等各方面都是完整的模块。对于一个做运维的人来说，当然要求每个方面的知识都要有所了解，有全局的概念。但最重要的，是要发现自己的特长，找到最合适的方向深入探究，样样通不如一样精。

我最初在接触 Oracle 的时候，也是全面攻击，但随着我对一些基础知识的掌握，对于整体架构的了解，我开始去摸索一个具体的方向。

作为一个数据运维者，每天都要处理各种各样的故障。最初由于经验不足，虽然感觉各方面的知识都具备，但遇到问题却不会分析。因为任何一个简单的故障，都会涉及运维的很多方面，除了数据库本身，还包含存储或者网络等其他各种因素。

但我渐渐发现，在处理故障的过程中，我会重新梳理自己的知识树。每次完整地处理一个故障后，就会对各方面的知识有更全面的认识。有时候这种完善来自于技术细节，有时候会给我一个更广阔的眼界去认识我所面对的数据，这让我很快成长。而这样的成长过程，除了在故障中去锻炼，其他任何方式都代替不了。

也许有人会问我学习或者分析问题的方法技巧。我要说，学习技术只有一个方法，那就是常反思常总结。也许某些技术是短期不变的，但对于技术的认识一定要随着我们学习的深入不断变化。

接下来，我通过几个真实案例的分享，来跟大家简单介绍我分析、解决问题的思路。

这几个案例包含数据运维的很多方面。其中前三篇关于内存及参数的管理，接下来两篇是有关 SQL 及索引的性能，最后是使用数据泵和 ogg 进行迁移的时候遇到的故障分析。希望这几个案例能够对今后大家处理各类故障有所帮助。

13.1 一波三折：释放内存导致数据库崩溃的案例

这是我一个运营商客户的案例。其现象大致是某天凌晨某 RAC 节点实例被重启了，通过如下的告警日志我们可以发现 RAC 集群的节点 2 实例被强行终止掉了，以下是详细的日志信息。

```
Mon Sep 28 02:00:00 2015
Errors in file /oracle/admin/xxx/bdxxx/xxx2_j000_7604.trc:
ORA-12012: error on auto execute of job 171538
ORA-06550: line ORA-06550: line 4, column 3:
PLS-00905: object XSQD.E_SP_DL_CRM_TERMINAL_MANAGER is invalid
ORA-06550: line 4, column 3:
PL/SQL: Statement ignored, column :
Mon Sep 28 02:03:18 2015
Errors in file /oracle/admin/xxx/udxxx/xxx2_ora_6810.trc:
ORA-00600: internal error code, arguments: [KGHLKREM1], [0x679000020], [], [], [], [], [], []
......
Trace dxxxing is performing id=[cdmp_20150928023925]
Mon Sep 28 02:39:32 2015
Errors in file /oracle/admin/xxx/bdxxx/xxx2_lmd0_24228.trc:
ORA-00600: internal error code, arguments: [KGHLKREM1], [0x679000020], [], [], [], [], [], []
Mon Sep 28 02:39:32 2015
LMD0: terminating instance due to error 482
Mon Sep 28 02:39:33 2015
Shutting down instance (abort)
License high water mark = 145
Mon Sep 28 02:39:37 2015
Instance terminated by LMD0, pid = 24228
Mon Sep 28 02:39:38 2015
```

从上面的数据库告警日志来看，数据库实例 2 从 2:03 就开始报错 ORA-00600 [KGHLKREM1]，一直持续到 2:39，lmd0 进程开始报同样的错误，紧接着 LMD0 进程强行把数据库实例 2 终止掉。如果参照上述 ORA-00600 错误，直接搜索 Oracle MOS，可能会搜到以下结果，Bug 14193240：LMS SIGNALED ORA-600[KGHLKREM1] DURING BEEHIVE LOAD。但是这个 Bug 很容易被排除，根据系统监控就可以直接排除。在故障期间系统负载是非常低的。

这里我们需要注意，从告警日志来看，从 2:03 就开始报错，然而直到 lmd0 进程报错时，实例才被其终止掉。不难看出，数据库节点 2 的 lmd0 报错才是问题的关键。那么我们首先来分析数据库节点 2 的 lmd0 进程的 trace 文件内容。

```
*** 2015-09-28 02:39:24.291
***** Internal heap ERROR KGHLKREM1 addr=0x679000020 ds=0x60000058 *****
```

```
***** Dxxx of memory around addr 0x679000020:
678FFF020 00000000 00000000 00000000 00000000  [................]
......
679000030 00000000 00000000 00000000 00000000  [................]
  Repeat 254 times
Recovery state: ds=0x60000058 rtn=(nil) *rtn=(nil) szo=0 u4o=0 hdo=0 off=0
 Szo:
 UB4o:
 Hdo:
 Off:
 Hla: 0
*********************************************************
HEAP Dxxx heap name="sga heap"  desc=0x60000058
 extent sz=0x47c0 alt=216 het=32767 rec=9 flg=-126 opc=4
 parent=(nil) owner=(nil) nex=(nil) xsz=0x0
 ds for latch 1: 0x60042f70 0x600447c8 0x60046020 0x60047878
......
ksedmp: internal or fatal error
ORA-00600: internal error code, arguments: [KGHLKREM1], [0x679000020], [], [], [], [], [], []
----- Call Stack Trace -----
calling              call      entry           argument values in hex
location             type      point           (? means dubious value)
-------------------- --------- --------------- ------------------------
ksedst()+31          call      ksedst1()       000000000 ? 000000001 ?
                                               7FFF41D71F90 ? 7FFF41D71FF0 ?
                                               7FFF41D71F30 ? 000000000 ?

ksedmp()+610         call      ksedst()        000000000 ? 000000001 ?
                                               7FFF41D71F90 ? 7FFF41D71FF0 ?
                                               7FFF41D71F30 ? 000000000 ?

ksfdmp()+21          call      ksedmp()        000000003 ? 000000001 ?
                                               7FFF41D71F90 ? 7FFF41D71FF0 ?
                                               7FFF41D71F30 ? 000000000 ?

kgerinv()+161        call      ksfdmp()        000000003 ? 000000001 ?
                                               7FFF41D71F90 ? 7FFF41D71FF0 ?
                                               7FFF41D71F30 ? 000000000 ?

kgesinv()+33         call      kgerinv()       0068966E0 ? 2AF92650E2C0 ?
                                               7FFF41D71FF0 ? 7FFF41D71F30 ?
                                               000000000 ? 000000000 ?

kgesin()+143         call      kgesinv()       0068966E0 ? 2AF92650E2C0 ?
                                               7FFF41D71FF0 ? 7FFF41D71F30 ?
                                               000000000 ? 000000000 ?

kghnerror()+342      call      kgesin()        0068966E0 ? 2AF92650E2C0 ?
                                               7FFF41D71FF0 ? 7FFF41D71F30 ?
                                               000000002 ? 679000020 ?

kghadd_reserved_ext  call      kghnerror()     0068966E0 ? 060000058 ?
ent()+1039                                     005AE1C14 ? 679000020 ?
                                               000000002 ? 679000020 ?

kghget_reserved_ext  call      kghadd_reserved_ext 0068966E0 ? 060000058 ?
ent()+250                      ent()           060042F70 ? 060042FB8 ?
                                               000000000 ? 000000000 ?

kghgex()+1622        call      kghget_reserved_ext 0068966E0 ? 060003B98 ?
                               ent()           060042F70 ? 000000B10 ?
                                               000000000 ? 000000000 ?

kghfnd()+660         call      kghgex()        0068966E0 ? 060003B98 ?
```

13.1 一波三折：释放内存导致数据库崩溃的案例

```
                                      060042F70 ? 000002000 ?
                                      000000AC8 ? 06000D600 ?
kghprmalo()+274     call    kghfnd()  0068966E0 ? 060003B98 ?
                                      060042F70 ? 000000AB8 ?
                                      000000AB8 ? 06000D600 ?
```

从上面的信息来看，确实 heap 存在错误的情况。根据这个错误堆栈可以在 MOS 上再次匹配，这时找到文档号 1070812.1 的文章：ORA-600 [KGHLKREM1] On Linux Using Parameter drop_cache On hugepages Configuration，此次故障跟描述基本上一致。

```
***** Internal heap ERROR KGHLKREM1 addr=0x679000020 ds=0x60000058 *****
***** Dxxx of memory around addr 0x679000020:
678FFF020 00000000 00000000 00000000 00000000  [................]
         Repeat 255 times
679000020 60001990 00000000 00000000 00000000  [...`............]
679000030 00000000 00000000 00000000 00000000  [................]
         Repeat 254 times
```

其中地址[0x679000020] 后面的内容也均为 0，跟文档描述一样。其次，文章中提到使用了 linux 内存释放机制以及同时启用了 hugepage 配置。根据文档描述，这应该是 Linux Bug。通过检查对比 2 个节点配置，发现节点 2 的配置确实不同。

```
--节点1
[oracle@xxx-DS01 ~]$ cat /proc/sys/vm/drop_caches
0
--节点2
[oracle@xxx-DS02 ~]$ cat /proc/sys/vm/drop_caches
3
```

当 drop_caches 设置为 3 时，会触发 linux 的内存清理回收机制，可能出现内存错误的情况。然而我们检查配置发现并没有修改。

```
oracle@xxx-DS02 bdxxx]$ cat /etc/sysctl.conf
# Kernel sysctl configuration file for Red Hat Linux
#
# For binary values, 0 is disabled, 1 is enabled.  See sysctl(8) and
# sysctl.conf(5) for more details.
# Controls IP packet forwarding
net.ipv4.ip_forward = 0
# Controls source route verification
net.ipv4.conf.default.rp_filter = 1
# Do not accept source routing
net.ipv4.conf.default.accept_source_route = 0
# Controls the System Request deBugging functionality of the kernel
kernel.sysrq = 0
# Controls whether core dxxxs will append the PID to the core filename
# Useful for deBugging multi-threaded applications
kernel.core_uses_pid = 1
# Controls the use of TCP syncookies
net.ipv4.tcp_syncookies = 1
# Controls the maximum size of a message, in bytes
kernel.msgmnb = 65536
```

```
# Controls the default maxmimum size of a mesage queue
kernel.msgmax = 65536
# Controls the maximum shared segment size, in bytes
kernel.shmmax = 68719476736
# Controls the maximum number of shared memory segments, in pages
kernel.shmall = 4294967296
kernel.shmmni = 4096
kernel.sem = 250 32000 100 128
fs.file-max = 65536
net.ipv4.ip_local_port_range = 1024 65000
net.core.rmem_default = 1048576
net.core.rmem_max = 1048576
net.core.wmem_default = 262144
net.core.wmem_max = 262144
vm.nr_hugepages=15800
```

既然没有修改配置文件，那么为什么会出现这种情况呢？

我怀疑是有人手工执行了 echo 3 > /proc/sys/vm/drop_caches 命令来强制释放内存导致。接下来查看了最近几分钟的操作记录，发现了如下的蛛丝马迹。

```
[root@xxx-DS02 ~]# history|grep echo
   22  2015-09-29 16:12:42 root echo 3 > /proc/sys/vm/drop_caches
   71  2015-09-29 16:12:42 root echo 0 > /proc/sys/vm/drop_caches
   73  2015-09-29 16:12:42 root echo 3 > /proc/sys/vm/drop_caches
   79  2015-09-29 16:12:42 root echo 0 > /proc/sys/vm/drop_caches
  311  2015-09-29 16:12:42 root echo 3 > /proc/sys/vm/drop_caches
  329  2015-09-29 16:12:42 root echo 0 > /proc/sys/vm/drop_caches
 1001  2015-09-29 16:12:49 root history|grep echo
 1005  2015-09-29 16:14:55 root history|grep echo
```

很明显 root 用户确实执行了内存释放的操作。然而运维人员却否认执行过类似操作，这说明事情并不是如此简单。我们进一步检查数据库操作系统日志发现如下信息。

```
Sep 29 00:00:12 xxx-DS02 kernel: Bug: soft lockup - CPU#1 stuck for 10s! [rel_mem.sh:13887]
Sep 29 00:00:12 xxx-DS02 kernel: CPU 1:
......
Sep 29 00:00:12 xxx-DS02 kernel: Pid: 13887, comm: rel_mem.sh Tainted: G      2.6.18-194.el5 #1
Sep 29 00:00:12 xxx-DS02 kernel: RIP: 0010:[<ffffffff800cb229>]  [<ffffffff800cb229>] __invalidate_mapping_pages+0xf3/0x183
Sep 29 00:00:12 xxx-DS02 kernel: RSP: 0018:ffff8112f5f71da8  EFLAGS: 00000207
......
Sep 29 00:00:12 xxx-DS02 kernel:
Sep 29 00:00:12 xxx-DS02 kernel: Call Trace:
Sep 29 00:00:12 xxx-DS02 kernel:  [<ffffffff800cb21a>] __invalidate_mapping_pages+0xe4/0x183
Sep 29 00:00:12 xxx-DS02 kernel:  [<ffffffff800f642e>] drop_pagecache+0xa5/0x13b
Sep 29 00:00:12 xxx-DS02 kernel:  [<ffffffff80096bc1>] do_proc_dointvec_minmax_conv+0x0/0x56
Sep 29 00:00:12 xxx-DS02 kernel:  [<ffffffff800f64de>] drop_caches_sysctl_handler+0x1a/0x2c
Sep 29 00:00:12 xxx-DS02 kernel:  [<ffffffff80096fea>] do_rw_proc+0xcb/0x126
Sep 29 00:00:12 xxx-DS02 kernel:  [<ffffffff80016a49>] vfs_write+0xce/0x174
Sep 29 00:00:12 xxx-DS02 kernel:  [<ffffffff80017316>] sys_write+0x45/0x6e
Sep 29 00:00:12 xxx-DS02 kernel:  [<ffffffff8005e28d>] tracesys+0xd5/0xe0
```

13.1 一波三折：释放内存导致数据库崩溃的案例

我们可以看出，原来是由于调用了 rel_mem.sh 脚本引发了这个问题，这个调用甚至导致 CPU 1 挂起了 10 秒。对于 Oracle RAC 而言，当 CPU 出现挂起，那么极有可能导致 LMS 等进程也挂起，进而引发数据库故障。

既然是调用了 shell 脚本，而客户的运维人员又没有主动做过任何操作，因此我怀疑很可能是之前部署了 crontab 脚本，经过检查发现确实存在相关脚本，如下所示。

```
[root@xxx-DS02 ~]# crontab -l
00 00 * * * /home/oracle/ht/rel_mem.sh
[root@xxx-DS02 ~]# cat /home/oracle/ht//rel_mem.sh
#!/bin/bash
#mkdir /var/log/freemem
time=`date +%Y%m%d`
used=`free -m | awk 'NR==2' | awk '{print $3}'`
free=`free -m | awk 'NR==2' | awk '{print $4}'`
echo "==========================" >> /var/log/freemem/mem$time.log
date >> /var/log/freemem/mem$time.log
echo "Memory usage | [Use:${used}MB][Free:${free}MB]" >> /var/log/freemem/mem$time.log

if [ $free -le 71680 ];then
                sync && echo 3 > /proc/sys/vm/drop_caches
                echo "OK" >> /var/log/freemem/mem$time.log
                free >> /var/log/freemem/mem$time.log
else
                echo "Not required" >> /var/log/freemem/mem$time.log
fi
```

到这里我们已经清楚了原因，到此这个案例已经告一段落了。最后我们还应该深入思考，为什么客户要部署这样一个脚本呢？只有一种解释，说明这个数据库节点之前可能面临内存使用居高不下的问题。既然如此，那么就进一步检查一下目前系统的内存使用情况。

```
[root@xxx-DS02 ~]# free -m
            total     used     free   shared  buffers   cached
Mem:       128976   127808     1168        0      245    88552
-/+ buffers/cache:  39010    89966
Swap:       15999      194    15805
```

我们可以看到操作系统物理内存为 128GB，而其中 cache 内存就占据了 88GB。Linux 文件系统的 cache 分为 2 种：page cache 和 buffer cache。page cache 是用于文件 inode 等操作的，而 buffer cache 是用于块设备的操作。上述 free -m 命令中的 cached 88552 全是 page cache。目前数据库实例的内存分配之后也就 40GB。

```
SQL> show parameter sga
NAME                                 TYPE         VALUE
------------------------------------ ------------ ------------------------------
lock_sga                             boolean      FALSE
pre_page_sga                         boolean      FALSE
sga_max_size                         big integer  30G
sga_target                           big integer  30G
SQL> show parameter pga
```

```
NAME                                  TYPE         VALUE
------------------------------------- ------------ ------------------------------
pga_aggregate_target                  big integer  10G
```

由此可见，操作系统物理内存之所以看上去那么高，并非 Oralce 本身所消耗，大部分为文件系统 cache 所消耗。而对于 Linux 的文件系统缓存，我们是可以通过调整操作系统内核参数来加快回收的，并不需要使用前面提到的强制清理回收内存的暴力解决方式。

根据 Oracle metalink 相关文档建议，通过设置如下参数来避免文本中的问题。

```
sysctl -w vm.min_free_kbytes=1024000
sysctl -w vm.vfs_cache_pressure=200
sysctl -w vm.swappiness=40
```

其中 min_free_kbytes 参数表示操作系统至少保留的空闲物理内存大小，单位是 KB。vm.vfs_cache_pressure 参数用来控制操作系统对内存的回收，默认值为 100，通过增加该值大小，可以加快系统对文件系统 cache 的回收。

而最后的参数 vm.swappiness 则是控制 swap 交换产生的趋势程度，默认值为 100。通过将该值调得更低一些，可以降低物理内存和 disk 之间产生交换的概率。但是 Redhat 官方白皮书中明确提出，不建议将该值设置为 10 或者更低的值。

实际上，通过我们的调整之后，据后续观察，该系统至今未再出现实例宕机的情况。这个案例告诉我们，在一个复杂的系统中，通过单一的手段进行粗暴的问题处理是危险的。正确的途径应该是，找出问题的根源，从源头对症下药，如此才能高枕无忧。

13.2 层层深入：DRM 引发 RAC 的故障分析

在 2015 年 1 月 13 日凌晨 3:44 左右，某客户的一套集群数据库的节点 1 出现 crash。检查该节点的告警日志，我们发现了如下内容。

```
Tue Jan 13 03:44:43 2015
Errors in file /home/oracle/app/admin/XXXX/bdump/XXXX1_lmon_10682988.trc:
ORA-00481: LMON process terminated with error
Tue Jan 13 03:44:43 2015
USER: terminating instance due to error 481
Tue Jan 13 03:44:43 2015
Errors in file /home/oracle/app/admin/XXXX/bdump/XXXX1_lms0_27525728.trc:
ORA-00481: LMON process terminated with error
......
Tue Jan 13 03:44:44 2015
Errors in file /home/oracle/app/admin/XXXX/bdump/XXXX1_lgwr_33489026.trc:
ORA-00481: LMON process terminated with error
Tue Jan 13 03:44:44 2015
Doing block recovery for file 94 block 613368
Tue Jan 13 03:44:45 2015
Shutting down instance (abort)
```

13.2 层层深入：DRM 引发 RAC 的故障分析

```
License high water mark = 1023
Tue Jan 13 03:44:49 2015
Instance terminated by USER, pid = 19333184
Tue Jan 13 03:44:55 2015
Instance terminated by USER, pid = 33554510
Tue Jan 13 03:45:46 2015
Starting ORACLE instance (normal)
sskgpgetexecname failed to get name
```

从上述的告警日志来看，在凌晨 3:44:43 时间点，节点 1 的 LMON 进程出现异常被终止，抛出 ORA-00481 错误。接着节点 1 的数据库实例被强行终止。对于 Oracle RAC 数据库故障，我们在进行分析时，首先要对 Oracle RAC 的一些核心进程原理有所了解才行，这样才能深入本质。这里我们简单描述 Oracle RAC 的几个核心进程的基本原理，例如 LMON、LMS 和 LMD 等。

对于 Oracle 的 LMON 进程，其作用主要是监控 RAC 的 GES（Global Enqueue Service）信息。当然其作用不仅仅局限于此，还负责检查集群中各个 Node 的健康情况。当有节点出现故障的时候，LMON 进程负责进行 reconfig 以及 GRD（global resource Directory）的恢复等。

我们知道 RAC 的脑裂机制，如果 IO fencing 是 Oracle 本身来完成，也就是说由 Clusterware 来完成。那么当 LMON 进程检查到实例级别出现脑裂时，会通知 Clusterware 来进行脑裂操作，当然 LMON 进程其并不会等待 Clusterware 的处理结果。当等待超过一定时间，LMON 进程会自动触发 IMR（instance membership recovery），这实际上也就是我们所说的 Instance membership reconfig。

既然提到 LMON 进程的检测机制，那么 LMON 如何去检查集群中每个节点的健康状态呢？其中主要分为如下几种。

（1）网络心跳 （主要是通过 ping 进行检测）。

（2）控制文件磁盘心跳，其实就是每个节点的 CKPT 进程每三秒更新一次控制文件的机制。

Oracle RAC 核心进程 LMD 主要负责处理全局缓存服务（保持块缓冲区在实例间一致）处理锁管理器服务请求。它向一个队列发出资源请求，这个队列由 LMS 进程处理。同时 LMD 还负责处理全局死锁的检测、解析，并监视全局环境中的锁超时（这就是为什么我们经常看到数据库 alert log 中 LMD 进程发现全局死锁的原因）。

如果是 Oracle 11gR2 rac 环境，那么还存在一个新的核心进程，即 LMBH 进程。该进程是 Oracle 11R2 版本引入的一个进程，该进程的主要作用是负责监控 LMD、LMON、LCK 和 LMS 等核心进程，防止这些 Oracle 核心后台进程 spin（stuck）或被阻塞。该进程会定时地将监控的信息打印输出在相应的 trace 文件中，便于我们进行诊断。这也是 11gR2 一个亮点。当 LMBH 进程发现其他核心进程出现异常时，会尝试发起一些 kill 动作。如果一定时间内仍然无法解决，那么将触发保护，将实例强行终止掉，当然这是为了保证 RAC 节点数据的完整性和一致性。

从上述的日志分析，我们可以看出，节点 1 实例是被 LMON 进程强行终止的，而 LMON 进程由于本身出现异常才采取了这样的措施。那么节点 1 的 LMON 进程为什么会出现异常呢？通过分析节点 1 数据库实例 LMON 进程的 trace 内容，我们可以看到如下内容。

```
*** 2015-01-13 03:44:18.067
kjfcdrmrfg: SYNC TIMEOUT (1295766, 1294865, 900), step 31
Submitting asynchronized dump request [28]
KJC Communication Dump:
 state 0x5  flags 0x0  mode 0x0  inst 0  inc 68
 nrcv 17  nsp 17  nrcvbuf 1000
 reg_msg: sz 456   cur 1235 (s:0 i:1235) max 5251  ini 3750
 big_msg: sz 8240  cur 263  (s:0 i:263)  max 1409  ini 1934
 rsv_msg: sz 8240  cur 0    (s:0 i:0)    max 0     tot 1000
 rcvr: id 1  orapid 7  ospid 27525728
 ……
 send proxy: id 16  ndst 1 (1:16 )
GES resource limits:
 ges resources:  cur 0 max 0 ini 39515
 ges enqueues:   cur 0 max 0 ini 59069
 ges cresources: cur 4235 max 7721
 gcs resources:  cur 4405442 max 5727836 ini 7060267
 gcs shadows:    cur 4934515 max 6358617 ini 7060267
KJCTS state: seq-check:no  timeout:yes  waitticks:0x3  highload no
……
kjctseventdump-end tail 238 heads 0 @ 0 238 @ -744124571
sync() timed out - lmon exiting
kjfsprn: sync status  inst 0  tmout 900 (sec)
kjfsprn: sync propose inc 68  level 85020
kjfsprn: sync inc 68  level 85020
kjfsprn: sync bitmap 0 1
kjfsprn: dmap ver 68 (step 0)
 ……
Dump state for lmd0 (ospid 27198128)
Dump IPC context for lmd0 (ospid 27198128)
Dumping process 6.27198128 info:
```

从上面 LMON 进程的 trace 信息来看，LMON 进程检测到了 DRM 在进行 sync 同步时出现了 timeout，最后 LMON 强制退出了。既然如此，那么我们应该继续分析为什么 DRM 会出现 timeout。另外根据前面讲述的原理，Oracle DRM 的操作主要进程是 LMD 进程来完成，那么我们来分析节点 1 实例的 LMD 进程的 trace 内容，是否能看出蛛丝马迹。

```
*** 2015-01-13 03:44:43.666
lmd abort after exception 481
KJC Communication Dump:
 state 0x5  flags 0x0  mode 0x0  inst 0  inc 68
 nrcv 17  nsp 17  nrcvbuf 1000
 reg_msg: sz 456   cur 1189 (s:0 i:1189) max 5251  ini 3750
 big_msg: sz 8240  cur 261  (s:0 i:261)  max 1409  ini 1934
 rsv_msg: sz 8240  cur 0    (s:0 i:0)    max 0     tot 1000
 rcvr: id 1  orapid 7  ospid 27525728
 ……
```

13.2 层层深入:DRM 引发 RAC 的故障分析

```
  send proxy: id 7  ndst 1 (1:7 )
  send proxy: id 16  ndst 1 (1:16 )
GES resource limits:
 ges resources: cur 0 max 0 ini 39515
 ges enqueues: cur 0 max 0 ini 59069
 ges cresources: cur 4235 max 7721
 gcs resources: cur 4405442 max 5727836 ini 7060267
 gcs shadows: cur 4934515 max 6358617 ini 7060267
KJCTS state: seq-check:no  timeout:yes  waitticks:0x3  highload no
GES destination context:
GES remote instance per receiver context:
GES destination context:
```

我们可以看到,当 LMON 进程遭遇 ORA-00481 错误之后,LMD 进程也被强制 abort 终止掉了。从 LMD 进程的 trace 文件来看,出现了 tickets 等待超时的情况,而且日志中 Oracle 也告诉我们,在该故障时间点,系统负载并不高。

从上述内容来看,我们似乎并没有得到十分有价值信息。由于 LMON 进程被强制终止掉之前,触发了一个 process dump,因此我们进一步来分析进程 dump,继续寻找蛛丝马迹。

```
*** 2015-01-13 03:44:18.114
Dump requested by process [orapid=5]
REQUEST:custom dump [2] with parameters [5][6][0][0]
Dumping process info of pid[6.27198128] requested by pid[5.10682988]
Dumping process 6.27198128 info:
*** 2015-01-13 03:44:18.115
Dumping diagnostic information for ospid 27198128:
OS pid = 27198128
loadavg : 1.71 1.75 2.33
swap info: free_mem = 13497.62M rsv = 96.00M
alloc = 342.91M avail = 24576.00M swap_free = 24233.09M
     F S      UID      PID     PPID   C PRI NI ADDR        SZ    WCHAN    STIME    TTY  TIME CMD
  240001 A oracle 19530440 10682988  10  65 20 16ae3ea590  1916           03:44:18   -   0:00
/usr/bin/procstack 27198128
  242001 T oracle 27198128        1   1   1 60 20 7412f4590 108540        Dec 29     -   -
569:20 ora_lmd0_XXXX1
     procstack: open(/proc/27198128/ctl): Device busy
*** 2015-01-13 03:44:18.420
```

LMD 进程的 processstate dmp 似乎也没有太多有价值的内容。不过我们至少可以肯定的是,系统的资源使用很正常。

通过上述的分析,我们可以看到 ORA-00481 错误的产生是关键,而这个错误是 LMON 进程产生的。ORA-00481 错误在 Oracle MOS 文档(1950963.1)中有非常详细的描述,针对本文 DRM 操作没有结束的情况,一般有如下 2 种原因。

(1)实例无法获得 LE(Lock Elements)锁。

(2)tickets 不足。

根据文档描述,我们从数据库两个节点的 LMS 进程 trace 中没有发现如下的类似关键信息。

```
Start affinity expansion for pkey 81885.0
Expand failed: pkey 81885.0, 229 shadows traversed, 153 replayed 1 retries
```

因此，我们可以排除第一种可能性。我们从 LMD 进程的 trace 文件中可以看到如下类似信息。

```
GES destination context:
Dest 1  rcvr 0  inc 68  state 0x10041  tstate 0x0
 batch-type quick  bmsg 0x0  tmout 0x20f0dd31  msg_in_batch 0
tkt total 1000  avl 743 sp_rsv 242 max_sp_rsv 250
 seq wrp 0  1st 268971339  ack 0  snt 268971336
 sync seq 0.268971339  inc 0  sndq enq seq 0.268971339
 batch snds 546480  tot msgs 5070830  max sz 88  fullload 85  snd seq 546480
 pbatch snds 219682271  tot msgs 267610831
 sndq msg tot 225339578  tm (0 17706)
 sndq msg 0  maxlmt 7060267  maxlen 149  wqlen 225994573
 sndq msg 0  start_tm 0  end_tm 0
```

这里的 tkt total 表示目前数据库实例总的 tickets 数据为 1000，当前可用的 tickets 为 743。因此我们可以排除 ticket 不足的导致 DRM 没有完成的情况。

换句话讲，上述 ORA-00481 错误的产生，本身并不是 Oracle RAC 的配置问题导致。对于 LMON 进程检查到 DRM 操作出现超时，最后导致实例崩溃。超时的原因通常有如下几种。

（1）操作系统 Load 极高，例如 CPU 极度繁忙，导致进程无法获得 CPU 资源。

（2）进程本身处理异常，比如进程挂起。

（3）网络问题，比如数据库节点之间通信出现异常。

（4）Oracle DRM Bug。

从上面的信息来看，系统在出现异常时，操作系统的 Load 是很低的，因此第一点我们可以直接排除。

我们现在的目的是需要分析出 LMON 进程检查到了什么异常，以及为什么会出现异常。LMON 进程在 abort 之前进行了 dump，那么是否能够从相关 dump 中找到一些有价值的线索呢？

```
PROCESS 5:
----------------------------------------
SO: 700001406331850, type: 2, owner: 0, flag: INIT/-/-/0x00
(process) Oracle pid=5, calls cur/top: 7000014054d75e0/7000014054d75e0, flag: (6) SYSTEM
int error: 0, call error: 0, sess error: 0, txn error 0
(post info) last post received: 0 0 24
last post received-location: ksasnd
last process to post me: 700001405330198 1 6
last post sent: 0 0 24
last post sent-location: ksasnd
last process posted by me: 7000014023045d0 1 2
(latch info) wait_event=0 bits=0
Process Group: DEFAULT, pseudo proc: 70000140336dd88
O/S info: user: oracle, term: UNKNOWN, ospid: 10682988
OSD pid info: Unix process pid: 10682988, image: oracle@tpihxdb1 (LMON)
```

```
......
----------------------------------------
SO: 7000014064d88c8, type: 4, owner: 700001406331850, flag: INIT/-/-/0x00
(session) sid: 1652 trans: 0, creator: 700001406331850, flag: (51) USR/- BSY/-/-/-/-/-
DID: 0001-0005-00000006, short-term DID: 0000-0000-00000000
txn branch: 0
oct: 0, prv: 0, SQL: 0, pSQL: 0, user: 0/SYS
service name: SYS$BACKGROUND
last wait for 'ges generic event' blocking sess=0x0 seq=35081 wait_time=158 seconds since wait started=0
   =0, =0, =0
Dumping Session Wait History
for 'ges generic event' count=1 wait_time=158
   =0, =0, =0
......
----------------------------------------
SO: 7000013c2f52018, type: 41, owner: 7000014064d88c8, flag: INIT/-/-/0x00
(dummy) nxc=0, nlb=0
----------------------------------------
SO: 7000014035cb1e8, type: 11, owner: 700001406331850, flag: INIT/-/-/0x00
(broadcast handle) flag: (2) ACTIVE SUBSCRIBER, owner: 700001406331850,
event: 5, last message event: 70,
last message waited event: 70,                       next message: 0(0), messages read: 1
channel: (7000014074b6298) system events broadcast channel
scope: 2, event: 129420, last mesage event: 70,
publishers/subscribers: 0/915,
messages published: 1
----------------------------------------
SO: 7000014054d75e0, type: 3, owner: 700001406331850, flag: INIT/-/-/0x00
(call) sess: cur 7000014064d88c8, rec 0, usr 7000014064d88c8; depth: 0
----------------------------------------
SO: 7000014036e3060, type: 16, owner: 700001406331850, flag: INIT/-/-/0x00
(osp req holder)
```

从上面 LMON 进程本身的 dump 来看，节点 1 实例的 LMON 进程状态是正常的，最后发送消息给 LMON 进程的是 SO: 700001405330198，SO 即为 state obejct。搜索该 SO，我们可以发现为 LMD 进程。

```
PROCESS 6:
----------------------------------------
SO: 700001405330198, type: 2, owner: 0, flag: INIT/-/-/0x00
(process) Oracle pid=6, calls cur/top: 7000014054d78a0/7000014054d78a0, flag: (6) SYSTEM
int error: 0, call error: 0, sess error: 0, txn error 0
(post info) last post received: 0 0 104
last post received-location: kjmpost: post lmd
last process to post me: 700001402307510 1 6
last post sent: 0 0 24
last post sent-location: ksasnd
last process posted by me: 700001407305690 1 6
(latch info) wait_event=0 bits=0
Process Group: DEFAULT, pseudo proc: 70000140336dd88
O/S info: user: oracle, term: UNKNOWN, ospid: 27198128 (DEAD)
OSD pid info: Unix process pid: 27198128, image: oracle@tpihxdb1 (LMD0)
```

```
  ......
  -------------------------------------
  SO: 7000014008a8d00, type: 20, owner: 700001405330198, flag: -/-/-/0x00
  namespace [KSXP] key  = [ 32 31 30 47 45 53 52 30 30 30 00 ]
  -------------------------------------
  SO: 7000014074a73d8, type: 4, owner: 700001405330198, flag: INIT/-/-/0x00
    (session) sid: 1651 trans: 0, creator: 700001405330198, flag: (51) USR/- BSY/-/-/-/-/-
    DID: 0000-0000-00000000, short-term DID: 0000-0000-00000000
    txn branch: 0
    oct: 0, prv: 0, SQL: 0, pSQL: 0, user: 0/SYS
    service name: SYS$BACKGROUND
    last wait for 'ges remote message' blocking sess=0x0 seq=25909 wait_time=163023 seconds
since wait started=62
    waittime=40, loop=0, p3=0
    Dumping Session Wait History
    for 'ges remote message' count=1 wait_time=163023
    waittime=40, loop=0, p3=0
  ......
           ------------process 0x7000014036e3ba0--------------------
    proc version         : 0
    Local node           : 0
    pid                  : 27198128
    lkp_node             : 0
    svr_mode             : 0
    proc state           : KJP_NORMAL
    Last drm hb acked    : 0
    Total accesses       : 31515
    Imm. accesses        : 31478
    Locks on ASTQ        : 0
    Locks Pending AST    : 0
    Granted locks        : 2
```

从 LMON 和 LMD 进程的 process dump 来看，进程本身状态是正常的。因此我们可以排除进程挂起导致出现 Timeout 的可能性。同时我们可以看到 LMD 进程一直在等待 ges remote message，很明显这是和另外一个数据库节点进行通信。因此我们要分析问题的根本原因，还需要分析节点 2 数据库实例的一些信息。

首先我们来分析节点 2 实例的数据库告警日志。

```
Tue Jan 13 03:39:14 2015
Timed out trying to start process PZ96.
Tue Jan 13 03:44:44 2015
Trace dumping is performing id=[cdmp_20150113034443]
Tue Jan 13 03:44:48 2015
Reconfiguration started (old inc 68, new inc 70)
List of nodes:
  1
 Global Resource Directory frozen
 * dead instance detected - domain 0 invalid = TRUE
 Communication channels reestablished
  Master broadcasted resource hash value bitmaps
  Non-local Process blocks cleaned out
```

13.2 层层深入：DRM 引发 RAC 的故障分析

从节点 2 的数据库告警日志来看，在 3:44:48 时间点，实例开始进行 reconfig 操作，这与整个故障的时间点是符合的。告警日志中本身并无太多信息，我们接着分析节点 2 数据库实例的 LMON 进程 trace 信息。

```
*** 2015-01-13 03:18:53.006
Begin DRM(82933)
……
End DRM(82933)
*** 2015-01-13 03:23:55.896
Begin DRM(82934)
……
End DRM(82934)
*** 2015-01-13 03:29:00.374
Begin DRM(82935)
……
sent syncr inc 68 lvl 85020 to 0 (68,0/38/0)
synca inc 68 lvl 85020 rcvd (68.0)
*** 2015-01-13 03:44:45.191
kjxgmpoll reconfig bitmap: 1
*** 2015-01-13 03:44:45.191
kjxgmrcfg: Reconfiguration started, reason 1
kjxgmcs: Setting state to 68 0.
*** 2015-01-13 03:44:45.222
 Name Service frozen
kjxgmcs: Setting state to 68 1.
kjxgfipccb: msg 0x110fffe78, mbo 0x110fffe70, type 22, ack 0, ref 0, stat 34
kjxgfipccb: Send cancelled, stat 34 inst 0, type 22, tkt (1416,80)
……
kjxggpoll: change poll time to 50 ms
* kjfcln: DRM aborted due to CGS rcfg.
* ** 2015-01-13 03:44:45.281
```

从上述 LMON 进程的日志来看，在故障时间点之前，数据库一直存在大量的 DRM 操作。并且注意到节点进行 reconfiguration 时，reason 代码值为 1。关于 reason 值，Oracle Metalink 文档有如下描述。

```
Reason 0 = No reconfiguration
Reason 1 = The Node Monitor generated the reconfiguration.
Reason 2 = An instance death was detected.
Reason 3 = Communications Failure
Reason 4 = Reconfiguration after suspend
```

从 reason =1 来看，数据库实例被强行终止重启也不是通信故障的问题。如果是通信的问题，那么 reason 值通常应该等于 3。reason=1 表明这是数据库节点自身监控时触发的 reconfig 操作。这也与我们前面分析节点 1 的日志的结论是符合的。

我们从 * kjfcln: DRM aborted due to CGS rcfg. 这段关键信息也可以确认，CGS reconfig 的原因也正是由于 DRM 操作失败导致。同时也可以看到，在 3:29 开始的 Begin DRM（82935）操作，一直到 3：44 出现故障时，这个 DRM 操作都没有结束（如果结束，会出现 End DRM（82935）类似关键字）。

由此也不难看出，实际上，该集群数据库可能在 3:29 之后就已经出现问题了。这里简单补充一下 Oracle DRM 的原理：在 Oracle RAC 环境中，当某个节点对某个资源访问频率较高时，而该资源的 master 节点不是 local 节点时，那么可能会触发 DRM（Dynamic Resource Management）操作。在 Oracle 10gR1 引入该特性之前，如果数据库需要更改某个资源的 master 节点，那么必须将数据库实例重启来完成。很显然，这一特性的引入无疑改变了一切。同时，从 Oracle 10gR2 开始，又引入了基于 object/undo 级别的 affinity。这里所谓的 affinity，本质上是引入操作系统的概念，即用来表示某个对象的亲和力程度。在数据库来看，即为对某个对象的访问频率程度。

在 Oracle 10gR2 版本中，默认情况下，当某个对象的被访问频率超过 50 时，而同时该对象的 master 又是其他节点时，那么 Oracle 则会触发 DRM 操作来修改 master 节点，这样的好处是可以大幅降低 gc grant 之类的等待事件。在进程 DRM 操作的过程中，Oracle 会将该资源的相关信息进行临时 frozen，然后将该资源在其他节点进行 unfrozen，然后更改资源的 master 节点。这里我们需要注意的是，这里 frozen 的资源其实是 GRD（Global Resource Directory）中的资源。在整个 DRM 的过程之中，访问该资源的进程都将被临时挂起。正因为如此，当系统出现 DRM 操作时，很可能导致系统或进程出现异常的。

实际上关于 Oracle DRM 的 Bug 也非常多，尤其是 Oracle 10gR2 版本中。针对本次故障，我们基本上可以认定是如下的 Bug 导致。

```
Bug 6960699 - "latch: cache buffers chains" contention/ORA-481/kjfcdrmrfg: SYNC TIMEOUT/
OERI[kjbldrmrpst:!master] (ID 6960699.8)

Dynamic ReMastering (DRM) can be too aggressive at times causing any combination
of the following symptoms :

- "latch: cache buffers chains" contention.
- "latch: object queue header operation" contention
- a RAC node can crash with and ora-481 / kjfcdrmrfg: SYNC TIMEOUT ... step 31
- a RAC node can crash with OERI[kjbldrmrpst:!master]
```

最后建议客户屏蔽 Oracle DRM 功能之后，经过监控发现运行了相当长一段时间后都没有出现类似问题。

通过这个 RAC 的案例分析，大家可能注意到，相关的集群日志非常多。在分析过程中，需要保持非常清晰的思路，才能基于问题将这些信息串联起来，最终成为定位和解决问题的重要信息。而事后基于问题的总结和研究，更加是巩固知识、带动我们成长的关键。希望大家能够从中学到故障分析、学习提升的思路和方法。

13.3 始于垒土：应用无法连接数据库问题分析

前不久某运营商客户反映某套业务系统在 2016 年 8 月 4 日凌晨出现过无法访问数据库的

情况。当接到客户请求之后我才通过 VPN 登录进行日志分析。首先我分析数据库告警日志发现，8 月 4 日凌晨 54 分开始出现 unable to spawn jobq slave process 相关错误，如下所示。

```
Thu Aug 04 00:54:10 CST 2016
Process J006 died, see its trace file
Thu Aug 04 00:54:10 CST 2016
kkjcre1p: unable to spawn jobq slave process
Thu Aug 04 00:54:10 CST 2016
Errors in file /oracle/app/oracle/admin/fwktdb/bdump/fwktdb_cjq0_5487.trc:
Process J006 died, see its trace file
Thu Aug 04 00:54:11 CST 2016
kkjcre1p: unable to spawn jobq slave process
Thu Aug 04 00:54:11 CST 2016
Errors in file /oracle/app/oracle/admin/fwktdb/bdump/fwktdb_cjq0_5487.trc:
……
Thread 1 advanced to log sequence 111405 (LGWR switch)
  Current log# 3 seq# 111405 mem# 0: /dev/vx/rdsk/datadg02/fwkt_redo1_3_raw_512m
Thu Aug 04 03:09:55 CST 2016
Process J007 died, see its trace file
Thu Aug 04 03:09:55 CST 2016
kkjcre1p: unable to spawn jobq slave process
Thu Aug 04 03:09:55 CST 2016
Errors in file /oracle/app/oracle/admin/fwktdb/bdump/fwktdb_cjq0_5487.trc:
……
Thu Aug 04 06:07:50 CST 2016
Process J002 died, see its trace file
Thu Aug 04 06:07:50 CST 2016
kkjcre1p: unable to spawn jobq slave process
Thu Aug 04 06:07:50 CST 2016
Errors in file /oracle/app/oracle/admin/fwktdb/bdump/fwktdb_cjq0_5487.trc:
```

从上述告警日志来看，没有明显的 ORA 相关错误。但是我们细心一点就会发现上述日志中存在一些告警信息，比如 kkjcrelp:unable to spawn jobq slave process，该告警也值得我们关注。那么这个告警信息的产生一般有哪些原因呢？根据 kkjcrelp:unable to spawn jobq slave process 错误来看，一般有以下几种原因。

（1）数据库进程参数 processes 参数设置偏小。

（2）job 参数 job_queue_processes 设置过小。

（3）系统资源（CPU/IO/Memory）不足，例如内存不足，导致新产生的进程无法获取资源。

根据经验我们知道，这极有可能是资源的问题。因为如果是数据库实例 processes 参数设置偏小，那么 alert log 通常会伴随 ORA-00200 类似的错误，然而我们却并没有发现。因此我们可以很容易的排除第一种可能性。至于第 2 种和第 3 种可能原因，这里我们暂时还无法排除，还需要进一步分析相关日志才能下结论。

既然是应用程序无法访问，那么数据库监听日志应该会有一些相关记录。我继续检查数据库监听日志发现，4 日凌晨确实出现了大量的 TNS 相关错误，如下所示。

```
04-AUG-2016 01:33:58 *
(CONNECT_DATA=(SERVICE_NAME=fwktdb)(CID=(PROGRAM=SigService) (HOST=wljhinas35)(USER=inas))) *
(ADDRESS=(PROTOCOL=tcp)(HOST=135.160.9.35)(PORT=48172)) * establish * fwktdb * 12518
TNS-12518: TNS:listener could not hand off client connection
 TNS-12547: TNS:lost contact
  TNS-12560: TNS:protocol adapter error
   TNS-00517: Lost contact
    Solaris Error: 32: Broken pipe
04-AUG-2016 01:33:58 *
……
04-AUG-2016 01:34:06 *
(CONNECT_DATA=(SERVICE_NAME=fwktdb)(CID=(PROGRAM=WlanService) (HOST=nxwljh1.localdomain)(USER=inas))) *
(ADDRESS=(PROTOCOL=tcp)(HOST=135.160.9.42)(PORT=50713)) * establish * fwktdb * 12518
TNS-12518: TNS:listener could not hand off client connection
 TNS-12547: TNS:lost contact
  TNS-12560: TNS:protocol adapter error
   TNS-00517: Lost contact
    Solaris Error: 32: Broken pipe
```

上面这部分内容，我相信大家并不陌生，这是非常常见的一些错误。我相信很多人第一感觉是搜索 Oracle MOS，确认 TNS-12518 是什么意思，什么原因。想必大家应该会看到这样的文章，ORA-12518 / TNS-12518 Troubleshooting（ID 556428.1）。该文档描述的解决方案有很多种，这里我大致描述一下，有如下几种。

- 操作系统时钟问题，建议设置如下参数。

```
stener.ora: INBOUND_CONNECT_TIMEOUT_listenername
SQLnet.ora: SQLNET.INBOUND_CONNECT_TIMEOUT。
```

- 操作系统可用内存不足，建议增加内存。

- 数据库参数 processes 设置过小，建议调大一些。

- 启用大内存支持，突破内核的单进程只能消耗 1GB 内存的限制，这实际上通常是指的 windows 环境。

- PGA 内存不足，而 SGA 设置过大，导致进程内存不足。建议缩小 SGA，给 PGA 预留更多内存。

- 配置 MTS 共享模式。

- 如果监听报错 TNS-12518，并进程 crash，那么很可能是命中一些因为系统压力过大或者内存溢出等相关 oracle Bug，例如 6139856。

实际上当我们遇到上述类似错误时，不应该直接往下判断。首先在分析这个问题时，我们要确认一点客户所说是否是真实的。也就是说这个问题之前是否出现过，还是仅仅是 8 月 4 日凌晨出现过 1 次。带着这样的疑问，我继续检查分析监听日志，发现实际上 8 月 3 日也出现了无法连接数据库的情况。

```
03-AUG-2016 18:24:48 * (CONNECT_DATA=(SID=fwktdb)(CID=(PROGRAM=JDBC Thin
Client) (HOST=__jdbc__) (USER=bea))) * (ADDRESS=(PROTOCOL=tcp)(HOST=135.161.23.93)
(PORT=35059)) * establish * fwktdb * 12518
 TNS-12518: TNS:listener could not hand off client connection
  TNS-12547: TNS:lost contact
   TNS-12560: TNS:protocol adapter error
    TNS-00517: Lost contact
     Solaris Error: 32: Broken pipe
……
03-AUG-2016 18:24:52 *
(CONNECT_DATA=(SERVICE_NAME=fwktdb)(CID=(PROGRAM=SigService)
(HOST=wljhinas33)(USER=inas))) * (ADDRESS=(PROTOCOL=tcp)(HOST=135.160.9.33)(PORT=59837)) *
establish * fwktdb * 12518
 TNS-12518: TNS:listener could not hand off client connection
  TNS-12547: TNS:lost contact
   TNS-12560: TNS:protocol adapter error
    TNS-00517: Lost contact
Solaris Error: 32: Broken pipe
```

对于 Oracle 的错误分析，我给大家的建议都是应该从下往上看，比如下面所列的错误。

```
TNS-12518: TNS:listener could not hand off client connection
 TNS-12547: TNS:lost contact
  TNS-12560: TNS:protocol adapter error
   TNS-00517: Lost contact
    Solaris Error: 32: Broken pipe
```

Oracle 监听日志在这里其实抛出了一连串的告警信息，我们需要有所判断，确认其中的哪行信息是关键信息，是产生问题的根源。很明显最后一行的 Solaris Error:32：Broken pipe 是此问题的关键所在。

针对 Solaris error:32 broken pipe 错误，这里我不自行进行解释，我选择引用 Oracle 官方的解释。

```
The error 32 indicates the communication has been broken while the listener is trying
to hand off the client connection to the server process or dispatcher process.
```

这里我简单解释一下上述文档内容的描述，简单地讲就是：Oracle 监听程序尝试去处理客户端到服务器端进程或者调度器（dispatcher processes）进程之前的连接时，将客户端进程通信强行中断了。那么监听程序为什么要终止连接呢？文档解释说有如下几种可能性原因。

```
  1. One of reason would be processes parameter being low, and can be verified by the
v$resource_limit view.
  2. In Shared Server mode, check the 'lsnrctl services' output and see if the dispatcher
has refused any connections, if so, then consider increasing the number of dispatchers.
  3. Check the alert log for any possible errors.
  4. Memory resource is also another cause for this issue. Check the swap, memory usage
of the OS.
  5. If RAC/SCAN or listener is running in separate home, check the following note。
```

从 Oracle 文档描述来看，产生该问题原因主要有如下几种。

（1）processes 参数设置过小。

（2）共享服务器模式下，dispatcher 不足。

（3）内存资源不足。

由于客户这里实际上是 Oracle 10.2.0.4 环境，因此就不需要考虑上述第 5 条描述了。

从上述问题描述的内容来看，与我们之前讲述 TNS-12518 错误几种可能性的原因有些类似。那么我们如何去定位这个问题呢？针对类似的问题，我通常的建议是使用最简单的方法：排除法。

首先我们来从数据库层面判断是否可以直接排除第 1 种可能性原因。我分析了数据库在 8 月 4 日凌晨 1-2 点的 AWR 数据库，发现数据库进程并没有达到 processes 参数设置限制，如图 13-1 所示。

Resource Name	Current Utilization	Maximum Utilization	Initial Allocation	Limit
processes	1,628	2,048	2048	2048
sessions	1,635	2,140	2257	2257

图 13-1

所以可以排除第 1 种可能性原因，同时由于数据库是专用模式，因此也可以排除第 2 种可能性原因。因此，这里似乎可以大致断定导致此次问题的原因是第 3 种可能性了，即系统资源使用方面。

我们知道操作系统资源主要分为 CPU/IO 以及内存资源，接下来从数据库监控角度来判断一下数据库资源在故障期间内的使用情况如何。通过器监控发现，CPU 资源消耗正常，swap 内存消耗存在一定的波动，但是整体来看也是正常的，如图 13-2、图 13-3 所示。

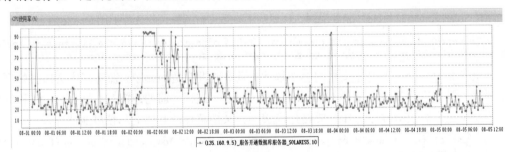

图 13-2

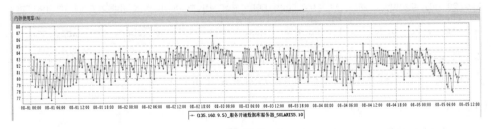

图 13-3

13.3 始于垒土：应用无法连接数据库问题分析

基于前面的分析不难看出，系统资源在使用上没有任何问题，没有出现资源过度消耗或资源不足的情况。到这里整个问题的分析似乎陷入了僵局。

Oracle 数据库是基于操作系统，因此实际上，当数据库出现异常之后，我们在进行问题分析时，首先应该确认操作系统本身是否正常，比如内核参数设置是否正确等。因此自然而然应该进一步检查数据库服务器操作系统日志是否存在相关蛛丝马迹。

正如心中所想，再检查 Solaris 操作系统的日志发现，确实存在相关错误。

```
Aug  4 00:54:24 nxfwktdb SC[SUNWscor.oracle_server.monitor]:nxfwktdb_rg:ora-fwktdb_rs:
[ID 564643 local7.error] Fault monitor detected error DBMS_ERROR: 20 DEFAULT Action=NONE : Max.
number of DBMS processes exceeded
Aug  4 00:54:54 nxfwktdb SC[SUNWscor.oracle_server.monitor]:nxfwktdb_rg:ora-fwktdb_rs:
[ID 564643 local7.error] Fault monitor detected error DBMS_ERROR: 1045 DEFAULT Action=NONE :
Fault monitor user lacks CREATE SESSION privilege; logon denied.
……
Aug  4 02:26:57 nxfwktdb SC[SUNWscor.oracle_server.monitor]:nxfwktdb_rg:ora-fwktdb_rs:
[ID 564643 local7.error] Fault monitor detected error DBMS_ERROR: 1045 DEFAULT Action=NONE :
Fault monitor user lacks CREATE SESSION privilege; logon denied.
Aug  4 02:28:26 nxfwktdb last message repeated 3 times
Aug  4 02:28:56 nxfwktdb SC[SUNWscor.oracle_server.monitor]:nxfwktdb_rg:ora-fwktdb_rs:
[ID 564643 local7.error] Fault monitor detected error DBMS_ERROR: 20 DEFAULT Action=NONE : Max.
number of DBMS processes exceeded
```

从上述日志来看，确实存在 processes 超过限制的情况。同时还能发现有监控用户由于缺乏权限，仍然在不断尝试登陆数据库，这也是一个安全隐患。针对操作系统日志中的 Max number of DBMS processes execeeded 错误信息，Oracle 官方文档有如下类似解释。

```
There can be several contributors to this failure. Solaris kernel may have resource
(memory/cpu) shortage, performance problem or the database itself could be having some issues
for example:
```
detected error DBMS_ERROR: 20 DEFAULT Action=NONE : Max. number of DBMS processes exceeded

> 说明：参考文档 Solaris Cluster 3.x HA-Oracle Fault Monitor Dumps Core With "Aborting fault monitor child process" or "probe response time exceeded timeout" (ID 1381317.1)

根据 Oracle Solaris 的文档解释来看，上述错误 ERROR 20 表示 Solaris 操作系统本身内核资源使用可能存在异常，或者存在性能问题，或者数据库本身可能存在某些问题。这个分析与我们前面的种种分析似乎比较接近。这里我需要说明的是，我期间分析了故障前后的 AWR 和 ASH 相关数据，没有发现明显异常，因此可以断定数据库本身是正常的。

分析到这个层面，我相信大家心中已经有了答案。有没有可能是操作系统本身有问题呢？这里需要注意的是，操作系统本身有问题，并不代表是指的操作系统资源使用有问题，也有可能是 Solaris 相关内核参数设置问题。实际上我们通过前面的操作系统监控信息就可以排除系统资源消耗异常的可能性。所以这里我们很有必要对 Solaris 系统本身的相关内核参数进行全面检查一次，通过检查发现确实存在参数设置不合理的情况，如下所示。

```
---操作系统内核参数
root@nxfwktdb # cat /etc/system|grep 'sys'
.
set semsys:seminfo_semmni=100
set semsys:seminfo_semmns=2048
set semsys:seminfo_semmsl=256
set semsys:seminfo_semvmx=32767
set shmsys:shminfo_shmmax=17179869184
set shmsys:shminfo_shmmin=1
set shmsys:shminfo_shmmni=100
set shmsys:shminfo_shmseg=10
root@nxfwktdb # cat /etc/project
system:0::::
user.root:1::::
noproject:2::::
default:3::::
group.staff:10::::

---数据库会话总数
SQL> select count(1) from v$session;

  COUNT(1)
----------
      1866
```

对于上述的内核参数，这里我不做过多解释，大家可以参考 Oracle 官方安装文档，里面有详细的描述。从 Soalris 10 开始，已经废弃了如下几个参数：

- semsys:seminfo_semmns——操作系统信号灯最大个数；
- shmsys:shminfo_shmmin——共享内存段最小值；
- semsys:seminfo_semvmx——操作系统信号灯的最大值。

同时，从 Solaris 10 开始，已经不再建议通过修改/etc/system 的方式来调整内核参数了。建议通过 project 的方式来进行控制，而客户这里没有为 Oracle 用户单独创建 project，而使用了默认的 default project，如下：

```
root@nxfwktdb # id -p oracle
uid=101(oracle) gid=100(dba) projid=3(default)
```

那么我们来进一步检查 default project 设置是否恰当。其中 Solaris 10 版本中使用 process.max-sem-nsems 来替代 semsys:seminfo_semmsl 参数的控制，该参数表示一个信号组中信号灯等最大数量，由于 Oracle 用户默认使用了 default project，因此这里我们来检查 default project 的设置情况，如下所示。

```
root@nxfwktdb # prctl -i project default
project: 3: default
NAME    PRIVILEGE       VALUE    FLAG   ACTION                      RECIPIENT
project.max-contracts
        privileged      10.0K       -   deny                                -
        system          2.15G     max   deny                                -
```

```
project.max-device-locked-memory
        privileged      1.96GB    -     deny                    -
        system          16.0EB    max   deny                    -
project.max-port-ids
        privileged      8.19K     -     deny                    -
        system          65.5K     max   deny                    -
project.max-shm-memory
        privileged      1.56TB    -     deny                    -
        system          16.0EB    max   deny                    -
project.max-shm-ids
        privileged      128       -     deny                    -
        system          16.8M     max   deny                    -
project.max-msg-ids
        privileged      128       -     deny                    -
        system          16.8M     max   deny                    -
project.max-sem-ids
        privileged      128       -     deny                    -
        system          16.8M     max   deny                    -
project.max-crypto-memory
        privileged      7.83GB    -     deny                    -
        system          16.0EB    max   deny                    -
project.max-tasks
        system          2.15G     max   deny                    -
project.max-lwps
        system          2.15G     max   deny                    -
project.cpu-shares
        privileged      1         -     none                    -
        system          65.5K     max   none                    -
zone.max-lwps
        system          2.15G     max   deny                    -
zone.cpu-shares
        privileged      1         -     none          -         -
```

我们可以看到 default project 中定义了 project.max-sem-ids，它表示系统中信号组的最大值，即 128，表示目前该系统最大 default project 中所定义的用户创建的最大信号组之和不能超过 128。同时/etc/system 中的参数 semsys:seminfo_semmsl 为 256，表示该 default project 的每一个信号组最大只能创建 256 个信号灯。可以通过 ipcs 命令直接查看这一点。

```
root@nxfwktdb # ipcs -sb
IPC status from <running system> as of Tue Aug 16 16:00:01 CST 2016
T         ID      KEY        MODE        OWNER    GROUP  NSEMS
Semaphores:
s         12      0x4b623218 --ra-r-----  oracle   dba    256
s         11      0x4b623217 --ra-r-----  oracle   dba    256
s         10      0x4b623216 --ra-r-----  oracle   dba    256
s         9       0x4b623215 --ra-r-----  oracle   dba    256
s         8       0x4b623214 --ra-r-----  oracle   dba    256
s         7       0x4b623213 --ra-r-----  oracle   dba    256
s         6       0x4b623212 --ra-r-----  oracle   dba    256
s         5       0x4b623211 --ra-r-----  oracle   dba    256
s         4       0x4b623210 --ra-r-----  oracle   dba    256
s         0       0x71050a00 --ra-ra-ra-  root     root   1
```

```
root@nxfwktdb # ipcs -ma
……
T         ID    KEY            MODE        OWNER       GROUP   CREATOR      CGROUP  NATTCH        SEGSZ
CPID  LPID  ATIME    DTIME  CTIME
Shared Memory:
m          1    0x333184c8  --rw-r-----  oracle      dba     oracle       dba     1870  12884934656
5253  24096 17:07:40 17:07:40 0:36:07
```

上述查询结果的 NSEMS 列即表示每个信号组中最大的信号灯数量。同时从 NATTCH 的值可以看出，目前当前 Oracle 用户已经创建了 1 870 个进程。实际上 seminfo_semmsl 参数的设置应该是大于 processes 参数设置的。目前数据库 processes 参数设置为 2 048，且数据库进程已经超过 1 850，因此建议将该参数调整的更大一些。

从前面的数据来看，目前 default project 中所有用户能够消耗的最大信号灯约等于 128×256=32 768。尽管这个数值看上去已经够大，其实不然。在操作系统命令来看，一个进程和信号灯等对应关系通常是一对多。而且从查询结果来看 Oracle 实际上只分配了 9 个信号组，每组最大的信号灯数量是 256。如果根据这个计算也就是 9×256=2 304。而目前数据库实例的进程数量已经接近 2 000。

因此最后我们建议将 seminfo_semmsl 参数调整的更大一些，建议设置为 512 以上。

以上三则案例基本上跟数据库的内存管理和系统参数设置相关。每一个故障都不是单一方面的问题产生的，也涉及各方面的知识。对于 DBA 来说，具备完整的技术观非常重要。

13.4 变与不变：应用 SQL 突然变慢优化分析

去年底，某运营商客户反馈某个核心业务模块突然出现严重性能问题，产生了极为不好的用户体验，客服接到了大量投诉电话。这样的问题，处理起来相对容易一些。首先我们根据 v$session.program 很容易就可以定位到业务程序，并结合 event、SQL_ID 和 SQL_TEXT 很快就确认是运行较慢的业务 SQL 语句，经过分析发现是 SQL 选择了错误的 Index，导致执行计划效率不高。这里我通过在客户生产库创建相同的对象，然后执行类似 SQL，完全可以重现问题。

首先我们来分析一下出问题的 SQL 语句，如下所示。

```
SELECT
 SERIAL_NBR
  FROM BILLUSER.I_B2O_OWE_C_BAK
 WHERE (PROC_TYPE = '13' or PROC_TYPE = '14' or PROC_TYPE = 'S13' or
       PROC_TYPE = 'S14')
   AND BEGIN_OWE_DATE >=
       TO_DATE('2015-12-02 12:00:00', 'YYYY/MM/DD HH24:MI:SS')
   AND ACC_NBR = '18115258610'
```

需要说明一点，这里的 I_B2O_OWE_C_BAK 表是复制的 I_B2O_OWE_C 表。

13.4 变与不变：应用 SQL 突然变慢优化分析

该 SQL 语句从结构上看很简单，初步怀疑是这个 SQL 的执行计划可能存在异常，这里我们首先来分析一下执行计划，如下所示。

```
Plan hash value: 239766388
-------------------------------------------------------------------------------
| Id  | Operation                   | Name                | Starts | E-Rows | A-Rows |   A-Time   | Buffers |
-------------------------------------------------------------------------------
|   0 | SELECT STATEMENT            |                     |      1 |        |      0 |00:00:00.59 |    3049 |
|*  1 |  TABLE ACCESS BY INDEX ROWID| I_B2O_OWE_C_BAK     |      1 |      1 |      0 |00:00:00.59 |    3049 |
|*  2 |   INDEX RANGE SCAN          | I_B2O_OWE_C_BAK_BDATE|     1 |      3 |  19871 |00:00:00.01 |     236 |
-------------------------------------------------------------------------------

Predicate Information (identified by operation id):
---------------------------------------------------
  1 - filter((TO_NUMBER("ACC_NBR")=:SYS_B_6 AND INTERNAL_FUNCTION("PROC_TYPE")))
  2 - access("BEGIN_OWE_DATE">=TO_DATE(:SYS_B_4,:SYS_B_5))
```

我相信，有 SQL 优化经验的同学应该都能看出来，上述的执行计划确实有些异常，SQL 通过走 Index range scan 的方式，最后返回了仅仅 1 条数据，然而逻辑读的消耗却高达 3000 之多，这很明显是有问题的。同时我们还可以看出 Oracle 优化器评估出来的 R-rows 和实际的 A-rows 差距非常大。也正是由于这一点，导致该执行计划选择了 I_B2O_OWE_C_BAK_BDATE 索引后进行回表操作，产生了大量的逻辑读（逻辑读消耗几乎都在回表操作上）。

由于是客户反馈，那么在处理时我们首先想到的是，这个问题 SQL 是不是真如客户所讲那样，在当天才出现问题，而之前都是正常呢？还是说有可能之前也出现过异常，只是业务没有反馈而已。

要验证这一点，需要查询该 SQL_ID 的历史执行计划信息。通过查询确认该 SQL 实际上在 12 月 1 日就出现了异常。但是由于当天业务量较小，没有对业务产生实质性的影响，如下所示。

```
   SNAP_ID SQL_ID          PLAN_HASH_VALUE OPTIMIZER_ FORCE_MATCHING_SIGNATURE
---------- -------------- ---------------- ---------- ------------------------
     80546 1pzv7x3nhszva       1247210496 ALL_ROWS         8159205349038030075
……
     80569 1pzv7x3nhszva       1247210496 ALL_ROWS         8159205349038030075
     80570 1pzv7x3nhszva       1247210496 ALL_ROWS         8159205349038030075
     80571 1pzv7x3nhszva        239766388 ALL_ROWS         8159205349038030075
     80571 1pzv7x3nhszva       1247210496 ALL_ROWS         8159205349038030075
     80573 1pzv7x3nhszva       1247210496 ALL_ROWS         8159205349038030075
     80573 1pzv7x3nhszva        239766388 ALL_ROWS         8159205349038030075
     80596 1pzv7x3nhszva        239766388 ALL_ROWS         8159205349038030075
…….
     80609 1pzv7x3nhszva        239766388 ALL_ROWS         8159205349038030075
     80609 1pzv7x3nhszva       1247210496 ALL_ROWS         8159205349038030075
     80610 1pzv7x3nhszva        239766388 ALL_ROWS         8159205349038030075
```

```
80610  1pzv7x3nhszva      1247210496 ALL_ROWS      8159205349038030075
80623  1pzv7x3nhszva      1247210496 ALL_ROWS      8159205349038030075
80624  1pzv7x3nhszva      1247210496 ALL_ROWS      8159205349038030075
```

其中快照 ID 的对应的时间段信息如下所示。

```
   SNAP_ID BEGIN_INTERVAL_TI
---------- -----------------
     80546 20151129 23:00:34
     80547 20151130 00:00:48
     ……
     80569 20151130 22:00:02
     80570 20151130 23:00:20
     80571 20151201 00:00:40
     80572 20151201 01:00:59
     80573 20151201 02:00:16
     80574 20151201 03:00:38
     ……
     80596 20151202 01:01:00
     ……
     80609 20151202 14:00:49
     80610 20151202 15:00:16
     80611 20151202 16:00:50
     80612 20151202 17:00:19
```

其中 plan_hash_value 为 239766388 的执行计划为错误的执行计划。当产生错误的执行计划之后，由于 Oracle bind peeking 的问题，就可能导致数据库在之后的一段时间内一直沿用错误的执行计划。

我开始怀疑可能是统计信息的问题，由于客户后续已经重新收集了统计信息。因此现在我想，是否可以将统计信息还原到出问题之前的某天，然后执行 SQL 语句，看其执行计划是怎么样的。

实际上 Oracle 本身就提供了这样一个功能，利用 DBMS_STATS 可以将表的统计信息还原到过去某个时间点，操作步骤如下所示。

```
SQL> exec
dbms_stats.restore_table_stats(ownname=>'BILLUSER',tabname=>'I_B2O_OWE_ C_BAK',AS_OF_
TIMESTAMP=>'01-DEC-15 12.09.22.020077 AM +08:00');

PL/SQL procedure successfully completed.

SQL> select owner,table_name,COLUMN_NAME,NUM_DISTINCT,DENSITY,NUM_BUCKETS,LAST_ ANALYZED,
HISTOGRAM from dba_tab_col_statistics where table_name='I_B2O_OWE_C_BAK';

OWNER         TABLE_NAME         COLUMN_NAME      NUM_DISTINCT    DENSITY NUM_BUCKETS
LAST_ANALYZED         HISTOGRAM
------------- ------------------ --------------- ------------- ---------- -----------
BILLUSER      I_B2O_OWE_C_BAK    SOURCE              0          0          0 2015-11-17 00:10:25 NONE
BILLUSER      I_B2O_OWE_C_BAK    TDBS                3 3.4099E-08          3 2015-11-17 00:10:25 FREQUENCY
BILLUSER      I_B2O_OWE_C_BAK    SERIAL_NBR   14663108 6.8198E-08          1 2015-11-17 00:10:25 NONE
BILLUSER      I_B2O_OWE_C_BAK    PROC_TYPE           5 3.4099E-08          5 2015-11-17 00:10:25 FREQUENCY
```

13.4 变与不变：应用 SQL 突然变慢优化分析

```
    BILLUSER        I_B2O_OWE_C_BAK    PROD_ID        2326397 4.2985E-07    1 2015-11-17 00:10:25 NONE
    BILLUSER        I_B2O_OWE_C_BAK    CREATE_DATE    5660648 1.3521E-06    254 2015-11-17
00:10:25 HEIGHT BALANCED
    BILLUSER        I_B2O_OWE_C_BAK    BEGIN_OWE_DATE 5833484 7.9253E-07    254 2015-11-17
00:10:25 HEIGHT BALANCED
    BILLUSER        I_B2O_OWE_C_BAK    STATE          3 3.4099E-08    3 2015-11-17 00:10:25 FREQUENCY
    BILLUSER        I_B2O_OWE_C_BAK    STATE_DATE     5927070 6.3963E-07    254 2015-11-17 00:10:25
HEIGHT BALANCED
    BILLUSER        I_B2O_OWE_C_BAK    ACC_NBR        2306453 1.2826E-06    254 2015-11-17 00:10:25
HEIGHT BALANCED
    BILLUSER        I_B2O_OWE_C_BAK    PRODUCT_TYPE   3 3.4099E-08    3 2015-11-17 00:10:25
FREQUENCY
    BILLUSER        I_B2O_OWE_C_BAK    BILLING_MODE_ID 3 .333333333    1 2015-11-17 00:10:25 NONE
    BILLUSER        I_B2O_OWE_C_BAK    AREA_CODE      5      .2    1 2015-11-17 00:10:25 NONE
    BILLUSER        I_B2O_OWE_C_BAK    JRFS           0      0    0 2015-11-17 00:10:25 NONE

    ---基于列的详细统计信息
    COLUMN_NAME        DATA_TYPE   M   NUM_VALS   NUM_NULLS  DNSTY   LOW_V              HI_V
    ---------------    ---------   -   --------   ---------  ------  -----------------  -----------------
    SERIAL_NBR         NUMBER      Y   14663,108          0  0.0000  118812796          142619179
    PROC_TYPE          VARCHAR2    Y          5           0  0.0000  11                 15
    PROD_ID            NUMBER      Y   2326,397           0  0.0000  10282208           561259458
    CREATE_DATE        DATE        Y   5660,648           0  0.0000  2015-2-10 21:50:52 2015-11-16 23:53:30
    BEGIN_OWE_DATE     DATE        N   5833,484           0  0.0000  2015-2-10 21:50:51 2015-11-16 23:55:26
    STATE              VARCHAR2    Y          3           0  0.0000  101                3
    STATE_DATE         DATE        Y   5927,070           0  0.0000  2015-2-10 23:0:47  2015-11-16 23:59:41
    ACC_NBR            VARCHAR2    Y   2306,453           0  0.0000  13309500000        18995499998
    PRODUCT_TYPE       NUMBER      Y          3           0  0.0000  401731             403094
    BILLING_MODE_ID    NUMBER      N          3           0  0.3333  1                  4
    AREA_CODE          VARCHAR2    N          5           0  0.2000  0951               0955
    JRFS               VARCHAR2    N          0   14765,004  0.0000
    TDBS               NUMBER      Y          3           0  0.0000  0                  2
    SOURCE             VARCHAR2    N          0   14765,004  0.0000
```

通过还原统计信息之后，发现该表的统计信息最新值实际上是 11 月 17 日凌晨，也就是说从 11 月 17 日到 12 月 1 日，该表的统计信息都没有发现发现过变动（实际上到 12 月 1 日客户感觉到出问题之后，手工收集过统计信息）。由于最近半月数据变化比较大，但统计信息却并没有更新，说明统计信息过旧也可能导致 Oracle 仍然沿用旧的统计信息进行执行计划评估，进而产生错误的执行计划。至少目前为止我们还不能排除这个可能性。

这里首先来看一下 Oracle card 的计算公式，如下所示。

```
= 基数计算公式：1/num_distinct*(num_rows-num_nulls)
>= 基数计算公式：((high_value-limit)/(high_value-low_value)+1/num_distinct)*(num_rows-num_nulls)
```

对于该问题 SQL 的索引，如下所示。

```
OWNER     TABLE_NAME       INDEX_NAME             BLEVEL LEAF_BLOCKS NUM_ROWS CLUSTERING_FACTOR DISTINCT_KEYS
--------- ---------------- ---------------------- ------ ----------- -------- ----------------- -------------
BILLUSER  I_B2O_OWE_C_BAK  I_B2O_OWE_C_BDATE_BAK       2       42001 15833626           7108231       6150468
BILLUSER  I_B2O_OWE_C_BAK  I_B2O_OWE_C_ACCNBR_BAK      2       50751 15833626          15723479       2421731
```

那么根据 Oracle Card 计算公式，对于两个索引的估算情况如下所示。

- 对于其中的索引 I_B2O_OWE_C_ACCNBR_BAK，由于 SQL 的过滤条件为 = 符号。当统计信息不变的情况之下，根据公式所计算的 Card 值都是不变的。
- 对于索引 I_B2O_OWE_C_BDATE_BAK，由于其过滤条件是 >=符号。根据 Card 的计算公式，是需要通过 high value 等值来进行计算的。

这里我们怀疑极有可能跟这个 BDATE 时间列的 high value 等统计信息有关系。为了验证这个观点，这里我进行了如下的测试验证过程。

情况一：当时间条件取值大于 high value 时。

```
SQL> SELECT /*+ gather_plan_statistics */ SERIAL_NBR FROM BILLUSER.I_B2O_OWE_C_BAK WHERE
(PROC_TYPE='13' or PROC_TYPE='14' or
  2   PROC_TYPE='S13' or PROC_TYPE='S14')  AND BEGIN_OWE_DATE >=
  3   TO_DATE('2015-11-16 23:56:00','YYYY/MM/DD HH24:MI:SS') AND ACC_NBR ='18909502122'
  4  /

no rows selected
SQL> select * from table(dbms_xplan.display_cursor(NULL,NULL,'ALLSTATS'));
PLAN_TABLE_OUTPUT
--------------------------------------------------------------------------------
SQL_ID  4fjrp8d5gqz1z, child number 0
--------------------------------------
SELECT /*+ gather_plan_statistics */ SERIAL_NBR FROM BILLUSER.I_B2O_OWE_C_BAK WHERE
(PROC_TYPE='13' orPROC_TYPE='14' or PROC_TYPE='S13' or PROC_TYPE='S14')  AND BEGIN_OWE_DATE >=
TO_DATE('2015-11-16 23:56:00','YYYY/MM/DD HH24:MI:SS') AND ACC_NBR ='18909502122'

Plan hash value: 3356388844
--------------------------------------------------------------------------------
| Id | Operation                     | Name                   | Starts | E-Rows | A-Rows |   A-Time   | Buffers | Reads |
--------------------------------------------------------------------------------
|  0 | SELECT STATEMENT              |                        |   1    |        |    0   |00:00:06.29 |   450K  |  8861 |
|* 1 |  TABLE ACCESS BY INDEX ROWID  | I_B2O_OWE_C_BAK        |   1    |    1   |    0   |00:00:06.29 |   450K  |  8861 |
|* 2 |   INDEX RANGE SCAN            | I_B2O_OWE_C_BDATE_BAK  |   1    |    3   | 1068K  |00:00:00.01 |  2838   |   0   |
--------------------------------------------------------------------------------
Predicate Information (identified by operation id):
---------------------------------------------------
   1 - filter(("ACC_NBR"='18909502122' AND INTERNAL_FUNCTION("PROC_TYPE")))
   2 - access("BEGIN_OWE_DATE">=TO_DATE('  2015-11-16 23:56:00', 'syyyy-mm-dd hh24:mi:ss'))
```

情况二：当时间条件取值不超过 high value 时。

```
SQL> SELECT /*+ gather_plan_statistics */ SERIAL_NBR FROM BILLUSER.I_B2O_OWE_C_BAK WHERE
(PROC_TYPE='13' or PROC_TYPE='14' or
  2   PROC_TYPE='S13' or PROC_TYPE='S14')  AND BEGIN_OWE_DATE >=
  3   TO_DATE('2015-11-16 23:52:00','YYYY/MM/DD HH24:MI:SS') AND ACC_NBR ='18909502122'
  4  /
no rows selected
```

13.4 变与不变：应用 SQL 突然变慢优化分析

```
SQL> select * from table(dbms_xplan.display_cursor(NULL,NULL,'ALLSTATS'));
PLAN_TABLE_OUTPUT
-------------------------------------------------------------------------------
SQL_ID  6vrh3x0qqnxnh, child number 0
-------------------------------------
SELECT /*+ gather_plan_statistics */ SERIAL_NBR FROM BILLUSER.I_B2O_OWE_C_BAK WHERE
(PROC_TYPE='13' or
    PROC_TYPE='14'  or  PROC_TYPE='S13'  or  PROC_TYPE='S14')     AND  BEGIN_OWE_DATE  >=
TO_DATE('2015-11-16
    23:52:00','YYYY/MM/DD HH24:MI:SS') AND ACC_NBR ='18909502122'

Plan hash value: 1431363508
-------------------------------------------------------------------------------
| Id | Operation                    | Name                | Starts | E-Rows | A-Rows |   A-Time    | Buffers | Reads |

|  0 | SELECT STATEMENT             |                     |    1   |        |     0  |00:00:00.40 |    36   |   35  |
|* 1 |  TABLE ACCESS BY INDEX ROWID | I_B2O_OWE_C_BAK     |        |     1  |     1  |      0
|00:00:00.40 |     36 |     35 |
|* 2 |   INDEX RANGE SCAN           | I_B2O_OWE_C_ACCNBR_BAK |     |     1  |     7  |     33
|00:00:00.03 |      3 |      2 |
-------------------------------------------------------------------------------
Predicate Information (identified by operation id):
-------------------------------------------------------
  1 - filter(("BEGIN_OWE_DATE">=TO_DATE(' 2015-11-16 23:52:00', 'syyyy-mm-dd hh24:mi:ss') AND
              INTERNAL_FUNCTION("PROC_TYPE")))
  2 - access("ACC_NBR"='18909502122')
```

从上面的测试验证可以看出，当 where 条件的时间列取值超过该列的 high value 时，Oracle 选择了错误的 Index。而当时间列取值小于该列的 high value 时，Oracle 则会选择正确的执行计划。

当取值小于 high value 时，此时 BEGIN_OWE_DATE 的选择性为 （(high_value-limit)/(high_value-low_value)+1/num_distinct）即为（2015-11-16 23:55:26 - 2015-11-16 23:50:00）/（2015-11-16 23:55:26-2015-2-10 21:50:51）+（7.9253E-07）= 0.000013691。这样计算出来的结果表明 begin_owe_date 列的选择性要比 accnbr 列的选择性低。这种情况下 Oracle 会选择正确的索引。当取值大于 high value 时，begin_owe_date 列的选择性所计算出来的只要比 accnbr 要高，这样就会导致 Oracle 选择错误的索引。

为什么会出现这样奇怪的情况呢？ 实际上是因为当谓词条件取值超过 high value 时，此时 Oracle 无法判断，只能进行猜测。而这种情况之下，通常都会导致所计算出来的 Card 不合理，导致 SQL 选择错误的执行计划。尤其是针对 date 列、sequence 等情况，相对要常见得多，这是 Oracle CBO 的一个老问题。

实际上，为了避免这种情况出现，我们要尽可能地保证统计信息的准确性，这样可以最大程度降低地执行计划改变的风险。**同时也要改变一种错误的认识，即统计信息不改变，SQL 的执行计划不一定一成不变，而事实往往恰恰相反，很可能改变。**

当然，对于前面的问题，要解决是相对简单的。由于我们确认 ACCNBR 列的选择性永远都是非常高的，因此可以通过 SQL Profile 来固定这个执行计划，可以确保后续的稳定性。

最后针对前面这个 SQL 执行计划改变的案例进简单总结。由于 11 月 16 日至 11 月 30 日该业务表数据变动不大，未触发 Oracle 对表进行统计信息收集的阈值（数据变更超过 10%）时，导致统计信息一直未进行更新。直到 12 月 1 日凌晨，随着大量数据的加载和更新，超过统计信息收集的阈值后，Oracle 进行了统计信息收集。然而由于 Oracle 默认的统计信息收集 Job 是每周一到周六的晚上 22 点才进行收集，因此这就会出现一天时间的间隔。在该间隔时间段内，由于统计信息不准确，SQL 语句索引列的 high value 值过旧的问题，导致 Oracle 优化器评估出的 Card 值较小，最后选择了错误的执行计划。

13.5 实践真知：INSERT 入库慢的案例分析

某次，运营商客户的计费库反应其入库程序很慢，应用方通过监控程序发现主要慢在对于几个表的插入操作上。按照我们的通常理解，insert 应该是极快的，为什么会很慢呢？难道是大量 redo 日志产生或者 undo 不足？

据客户反馈，在反应之前应用程序都是正常的，这个问题是突然出现的，而且是每个月中下旬开始出现慢的情况。这让我百思不得其解。通过检查 event 也并没有发现什么奇怪的地方，于是我通过 10046 跟踪了应用的入库程序，如下即是应用方反应比较慢的表的插入操作，经过验证确实非常慢。

```
INSERT INTO XXXX_EVENT_201605C (ROAMING_NBR,.....,OFFER_INSTANCE_ID4)
VALUES (:ROAMING_NBR,......:OFFER_INSTANCE_ID4)

call     count       cpu    elapsed       disk      query    current       rows
-------  ------  --------  ---------  ---------  ---------  ---------  ---------
Parse        17      0.00       0.00          0          0          0          0
Execute      18      1.06      27.41       4534        518      33976       4579
Fetch         0      0.00       0.00          0          0          0          0
-------  ------  --------  ---------  ---------  ---------  ---------  ---------
total        35      1.06      27.41       4534        518      33976       4579
……
Elapsed times include waiting on following events:
  Event waited on                             Times   Max. Wait  Total Waited
  ----------------------------------------   Waited  ---------  ------------
  db file sequential read                      4495       0.03         24.02
  gc current grant 2-way                       2301       0.00          0.77
  SQL*Net more data from client                 795       0.00          0.02
  ……
  latch: gcs resource hash                        1       0.00          0.00
```

以上分析可以发现，插入了 4579 条数据，一共花了 27.41 秒。其中有 24.02 秒消耗在 db file sequential read 上。很明显这是索引的读取操作，实际上检查 10046 trace 裸文件，发现等待的

13.5 实践真知：INSERT 入库慢的案例分析

对象确实是该表上的 2 个索引。同时从上面 10046 trace 可以看出，该 SQL 执行之所以很慢，主要是因为存在了大量的物理读，其中 4579 条数据的插入，物理读为 4534。这说明什么问题呢？ 这说明，每插入一条数据大概产生一个物理读，而且都是 index block 的读取。 如下是 10046 trace 跟踪产生的裸文件内容。

```
Received ORADEBUG command 'event 10046 trace name context forever,level 12' from process Unix proc
WAIT #2: nam='db file sequential read' ela= 14797 file#=1056 block#=1209394 blocks=1 obj#=19787394
WAIT #2: nam='gc current grant 2-way' ela= 187 p1=893 p2=1528427 p3=3355 4433 obj#=19787394
WAIT #2: nam='db file sequential read' ela= 6045 file#=893 block#=1528427 blocks=1 obj#=19787394
WAIT #2: nam='db file sequential read' ela= 5332 file#=1030 block#=1467374 blocks=1 obj#=19787394
WAIT #2: nam='gc current grant 2-way' ela= 191 p1=1030 p2=1469787 p3=3355 4433 obj#=19787394
WAIT #2: nam='db file sequential read' ela= 4185 file#=1030 block#=1469787 blocks=1 obj#=19787394
WAIT #2: nam='gc current grant 2-way' ela= 486 p1=713 p2=1404968 p3=3355 4433 obj#=19787500
WAIT #2: nam='db file sequential read' ela= 4795 file#=713 block#=1404968 blocks=1 obj#=19787500
WAIT #2: nam='gc current grant 2-way' ela= 755 p1=712 p2=1678915 p3=3355 4433 obj#=19787500
WAIT #2: nam='db file sequential read' ela= 5805 file#=712 block#=1678915 blocks=1 obj#=19787500
WAIT #2: nam='db file sequential read' ela= 6483 file#=891 block#=1481712 blocks=1 obj#=19787500
WAIT #2: nam='gc current grant 2-way' ela= 227 p1=1030 p2=1576455 p3=3355 4433 obj#=19787500
WAIT #2: nam='db file sequential read' ela= 3993 file#=1030 block#=1576455 blocks=1 obj#=19787500
WAIT #2: nam='db file sequential read' ela= 11604 file#=1056 block#=1499298 blocks=1 obj#=19787500
WAIT #2: nam='db file sequential read' ela= 3948 file#=1068 block#=1487764 blocks=1 obj#=19787500
WAIT #2: nam='db file sequential read' ela= 6384 file#=891 block#=1572520 blocks=1 obj#=19787500
WAIT #2: nam='db file sequential read' ela= 194 file#=1068 block#=1444814 blocks=1 obj#=19787500
WAIT #2: nam='db file sequential read' ela= 3435 file#=1068 block#=1471632 blocks=1 obj#=19787500
......省略部分内容
```

```
OWNER      OBJECT_NAME                        OBJECT_ID OBJECT_TYPE
---------- -------------------------------- ---------- -----------
BILL       I2_ILE_DATA_EVENT_201605C          19787500 INDEX
BILL       I1_ILE_DATA_EVENT_201605C          19787394 INDEX
BILL       PK_A_EVENT_201605C                 18245536 INDEX
```

很明显，通过将该 index cache 移到 keep 池可以解决该问题。 实际上也确实如此，通过 cache 后，应用反馈程序快了很多。 那么对该问题，这里其实有几个疑问，为什么这里的 SQL insert 时物理读如此之高？这里我进一步来检查者 2 个 Index 的情况，如下所示。

```
[oracle@x tmp]$ cat bill_xx6.trc |grep 'db file sequential read'|grep 19787394|wc -l
2228
[oracle@x tmp]$ cat bill_xx6.trc |grep 'db file sequential read'|grep 19787500|wc -l
2267
[oracle@x tmp]$ cat bill_xx6.trc |grep 'db file sequential read'|grep 18245536|wc -l
0
```

```
OWNER INDEX_NAME                   BLEVEL LEAF_BLOCKS CLUSTERING_FACTOR PCT_FREE   NUM_ROWS
----- -------------------------- ------- ----------- ----------------- -------- ----------
BILL  I1_ILE_DATA_EVENT_201605C        3      371668         105055486       10  113812207
BILL  PK_A_EVENT_201605C               3      475596          22421073       10  118686445
BILL  I2_ILE_DATA_EVENT_201605C        3      470824         101005004       10  109236718
```

通过以上数据可以发现，顺序读等待最高的 2 个恰好其 clustering_factor 比较高。有没有可能跟这个有关系呢？

要解释这个问题，其实并不容易。首先这里我们需要理解 Oracle B-tree Index 的结构，如图 13-4 所示。

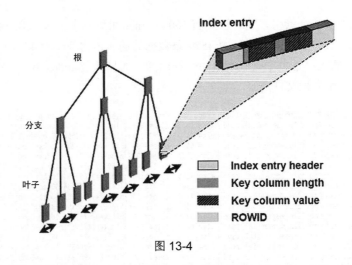

图 13-4

当 Oracle 向 index branch block 或 leaf block 中插入数据时,其实都是等价的,Oracle 这里其实并不需要进行代价的估算。随机选择 branch block 或者 leaf block,让其插入键值即可。所以这里其实也就跟我们前面提到的 Index 聚簇因子没有任何关系了。实际上我通过创建测试表,分别模拟聚簇因子大和小的情况,进行多次插入以及 10046 trace 跟踪分析,发现时间几乎一致。从测试来看也可以排除其影响。

不过 Index 聚簇因子也是一直表现形式。当聚簇因子与表的行数非常接近时,说明数据分散程度比较高。假设现在现在有一个表 Test,表上创建了一个 Index IDX1。当向表中插入数据时,Oracle 会进行 index 的同步维护,也就是需要同时向 index block 中插入键值。如果此时 IDX1 索引的离散读非常高,那么也就是说每次插入数据时 Oracle 在 cache 中命中 index block 的概率可能就非常低,这样就不得不进行物理读操作,进而影响插入的效率。

想到这一点,我相信大多数人已经知道如何处理该问题了。最简单的方式就是将整个 Index 缓存到内存中。这样就可以避免在插入数据时进行大量物理读操作了。

那么如何解释客户说的每个月中下旬这个应用模块就开始慢的问题呢? 结合前面的分析,其实这一点我们很容易解释。每个月初客户都会创建类似的空表,当数据量很小的情况下,Index 也是非常小的。即使 index 键值非常离散,那么 index block 在 cache 中缓存的可能性都比较大。因为索引比较小,而 SGA 的 buffer cache 够大。然而随着时间的推移,数据量越来越大,index 也会越拉越大,这也就降低了 index block 被 cache 的几率。

根据客户的业务特点,每个月都会创建类似的表和 Index,因此我部署了一个脚本具体如下所示。

```
#!/usr/bin/sh

export ORACLE_BASE=/oracle/app/oracle
```

```
export ORACLE_HOME=/oracle/app/oracle/product/102
export ORACLE_SID=xx
export ORACLE_TERM=xterm
export NLS_LANG=American_america.zhs16gbk

export PATH=$ARCH_HOME:$ORACLE_HOME/bin:$ORACLE_HOME/OPatch:/urs/ccs/bin:/bin:/usr/bin:$PATH
C_DATE=`date +%Y%m`
SQLplus "/as sysdba" << EOF
alter index BILL.I1_ILE_DATA_EVENT_${C_DATE}A storage (buffer_pool keep);
alter index BILL.I1_ILE_DATA_EVENT_${C_DATE}B storage (buffer_pool keep);
alter index BILL.I1_ILE_DATA_EVENT_${C_DATE}C storage (buffer_pool keep);
alter index BILL.I1_ILE_DATA_EVENT_${C_DATE}D storage (buffer_pool keep);
alter index BILL.I1_ILE_DATA_EVENT_${C_DATE}E storage (buffer_pool keep);
alter index BILL.I2_ILE_DATA_EVENT_${C_DATE}A storage (buffer_pool keep);
alter index BILL.I2_ILE_DATA_EVENT_${C_DATE}B storage (buffer_pool keep);
alter index BILL.I2_ILE_DATA_EVENT_${C_DATE}C storage (buffer_pool keep);
alter index BILL.I2_ILE_DATA_EVENT_${C_DATE}D storage (buffer_pool keep);
alter index BILL.I2_ILE_DATA_EVENT_${C_DATE}E storage (buffer_pool keep);
EXIT;
EOF
```

通过这样的方式临时解决了该问题。而且从 2016 年 5 月解决之后至今客户没有再反馈过该业务模块有问题，看来是彻底解决了这个问题。

其实这里还有一个问题。一般来讲，物理内存都是有限的，而且 SGA 的 keep 池不可能设置得过大。同时如果缓存的 Index 比较多，比较大，那么 Oracle 的 keep pool 对于缓存对象的处理机制是如何的？

通过以下测试来探索一下 Keep 池对于缓存的对象的处理机制。

（1）创建 2 个测试表，并创建好相应的 index，如下所示。

```
SQL> conn roger/roger
SQL> create table t_insert as select * from sys.dba_objects where 1=1;
SQL> create index idx_name_t on t_insert(object_name);
SQL> analyze table t_insert compute statistics for all indexed columns;
SQL> select INDEX_NAME,BLEVEL,LEAF_BLOCKS,DISTINCT_KEYS,CLUSTERING_FACTOR,NUM_ROWS from
dba_indexes where table_name='T_INSERT';

INDEX_NAME        BLEVEL LEAF_BLOCKS DISTINCT_KEYS CLUSTERING_FACTOR   NUM_ROWS
--------------- -------- ----------- ------------- ----------------- ----------
IDX_NAME_T             1         246         29808             24664      49859
SQL> alter system set db_keep_cache_size=4m;
SQL> create table t_insert2 as select * from sys.dba_objects where 1=1;
SQL> create index idx_name_t2 on t_insert2(object_name);
SQL> insert into t_insert select * from sys.dba_objects;
……多次 insert
SQL> commit;
```

从前面的信息我们可以看出，object_name 上的 index 聚簇因子比较高，说明其数据分布比较离散。

（2）我们现在将 index 都 cache 到 keep 池中，如下所示。

```
SQL> alter index idx_name_t storage (buffer_pool keep);
SQL> alter index idx_name_t2 storage (buffer_pool keep);
SQL> alter system flush buffer_cache;
```

这里需要注意的是，仅仅执行 alter 命令是不够的，还需要手工将 index block 读取到 keep 池中，如下所示。

```
SQL> conn /as sysdba
SQL> @get_keep_pool_obj.SQL
no rows selected
SQL> select /*+ index(idx_name_t,t_insert) */ count(object_name) from roger.t_insert;
COUNT(OBJECT_NAME)
------------------
             99721
SQL> @get_keep_pool_obj.SQL
SUBCACHE      OBJECT_NAME                      BLOCKS
------------  ------------------------------   ----------
KEEP          IDX_NAME_T                       499
DEFAULT       T_INSERT                         431

SQL> select /*+ index(idx_name_t2,t_insert2) */ count(object_name) from roger.t_insert2;
COUNT(OBJECT_NAME)
------------------
             99723
SQL> @get_keep_pool_obj.SQL
SUBCACHE      OBJECT_NAME                      BLOCKS
------------  ------------------------------   ----------
KEEP          IDX_NAME_T                       40
KEEP          IDX_NAME_T2                      459
DEFAULT       T_INSERT2                        522
DEFAULT       T_INSERT                         431
SQL> select /*+ index(idx_name_t,t_insert) */ count(object_name) from roger.t_insert;
COUNT(OBJECT_NAME)
------------------
             99721
SQL> @get_keep_pool_obj.SQL
SUBCACHE      OBJECT_NAME                      BLOCKS
------------  ------------------------------   ----------
KEEP          IDX_NAME_T                       467
KEEP          IDX_NAME_T2                      32
DEFAULT       T_INSERT2                        522
DEFAULT       T_INSERT                         431
```

通过上述的多次简单测试可以看出，keep pool 对于缓存的 block 的清除机制是先进先出原则。这相对于 Oracle Buffer Cache 的机制来讲，要简单许多。

最后针对这个问题，可以总结一下，产生类似问题的原因可能有哪些呢？实际上可能有很多种，如下所示。

（1）Index 碎片过于严重。

（2）Index next extent 分配问题。

（3）Buffer Cache 设置偏小。

（4）Index 分区方式不合理。

（5）Redo、undo 相关问题。

上述方面内容都可能产生本文中提到的问题，只不过表现形式可能不同而已。这是一个常规而又隐秘的问题，我们通常容易忽视它，但是又很容易出现产生一些问题。这些问题会对业务系统性能产生重大影响，希望能够引起大家的重视。

13.6　按图索骥：Expdp 遭遇 ORA-07445 的背后

据某客户反馈，其管理的某套数据库由于非正常关机重启之后，通过 expdp 进行数据导出发现报错，当报错之后，expdp 数据导出命令直接终止退出，报错信息如下所示。

```
处理对象类型 SCHEMA_EXPORT/JOB
. . 导出了 "STATS"."T_REPORT_MONTH_TEMPS"              988.2 MB 1292221 行
ORA-39014: 一个或多个 worker 进程已过早地退出。
ORA-39029: worker 进程 1 (进程名为 "DW01") 过早地终止
ORA-31672: Worker 进程 DW01 意外停止。

作业 "SYS"."SYS_EXPORT_SCHEMA_04" 因致命错误于 23:58:10 停止
```

此时我让客户检查数据库告警日志，发现有如下 ORA-07445 错误。

```
Errors in file /u01/app/oracle/admin/orcl/bdump/orcl_dw01_28608.trc:
ORA-07445: 出现异常错误: 核心转储 [klufprd()+321] [SIGSEGV] [Address not mapped to object]
[0x000000000] [] []
```

通过分析上面的信息，可以得到如下几个结论。

（1）expdp 的写进程报错，因为日志产生的进程是 dw01 进程。

（2）dw 进程报错的原因是遭遇了 ora-07445 [klufprd（）+321]错误。

（3）[klufprd（）+321] 这个函数非常少见。但是从前面 2 点可以知道这肯定与 buffer cache 有关系。

客户咨询，有没有什么好的解决方法。实际上当想到上述几点之后，脑海中马上就出现了一个临时解决方案。也就是说，通过 alter system flush buffer_cache 刷新缓存之后，然后再次进行 expdp 数据导出操作。

据客户反馈，在进行 flush buffer_cache 之后，再次运行 expdp 数据导出命令，发现 expdp 操作仍然会报类似错误，但是 expdp 不会异常终止，会继续完成后面其他对象的导出。通过临时解决方法，能实现这一点，基本上也算告一段落了，至少客户导出数据的目标是实现了。

然而对于我而言，作为一个好奇心很重的 DBA，势必要弄清楚，Oracle 这里为什么会发生 ORA-07445 错误。为什么 expdp 会异常终止退出，这些目前来看都是谜团。

进一步检查日志，发现有如下错误。

```
*** SESSION ID:(2760.1968) 2016-04-08 00:14:14.347
            row 01808438.0 continuation at
            file# 6 block# 33784 slot 14 not found
******************************************
KDSTABN_GET: 0 ..... ntab: 1
curSlot: 14 ..... nrows: 14
******************************************
*** 2016-04-08 00:14:14.348
ksedmp: internal or fatal error
ORA-00600: internal error code, arguments: [kdsgrp1], [], [], [], [], [], [], []
Current SQL statement for this session:
SELECT /*+NESTED_TABLE_GET_REFS+*/ "STATS"."T_REPORT_MONTH".* FROM "STATS"."T_REPORT_MONTH"
----- Call Stack Trace -----
```

擦亮眼睛发现，这里提到的这个表，恰好就是 expdp 报错所遇到的表，只不过我们在刷新 buffer cache 之后，expdp 可以跳过这个表继续完成其他对象的导出。从上述的信息来看，这里确实存在一些错误。当然，客户也意识到了这样一点，习惯性的通过 dbv 自带工具对该表所在的数据文件进行坏块检查，然而却并没有发现文件中存在数据坏块。

从这里的信息来看，Oracle 发现所需要的这行记录 row 01808438.0 应该在 file 6 block 33784 中找到，但是却并没有发现。（注意，这里的 file 6 block 33784 本身是完好的，因为 dbv 检测并没有发现坏块）。

那么这里的 row 01808438.0 表示什么含义呢？

其实这是表示的 nrid，这可以理解为一个指针。其中前面一部分是表现 rdba 地址，后面表现行编号。那么针对这个问题，我们如何进一步进行分析呢？其实很简单，分别将 block 33784 以及 rdba 01808438（16 进制）进行 dump，然后进行块级别分析。如下是 rdba 地址转换的脚本。

```
SQL> SELECT dbms_utility.data_block_address_block(25199672) "BLOCK",
  2         dbms_utility.data_block_address_file(25199672) "FILE"
  3  FROM dual;

     BLOCK       FILE
---------- ----------
     33848          6
```

由于前面的报错日志中提到的是 row 01808438.0，那么我们首先来分析 file 6 block 33848 的 dump。

```
Block header dump: 0x01808438
 Object id on Block? Y
 seg/obj: 0xc03d01  csc: 0xb37.78b5ae28  itc: 3  flg: E  typ: 1 - DATA
```

13.6 按图索骥：Expdp 遭遇 ORA-07445 的背后

```
         brn: 0  bdba: 0x1807d8a ver: 0x01 opc: 0
         inc: 0  exflg: 0

 Itl           Xid                  Uba         Flag  Lck        Scn/Fsc
0x01   0x000a.02d.000cdc5c  0x00809c91.6507.21  --U-   2   fsc 0x0001.78b6a4b1
0x02   0x000a.014.000cdd00  0x00806957.650d.15  --U-   2   fsc 0x0000.78b6ec5d
0x03   0x000a.025.000cdd5d  0x00801e50.650f.0a  --U-   2   fsc 0x0000.78b71584

data_block_dump,data header at 0x1fb2f87c
===============
tsiz: 0x1f80
hsiz: 0x34
pbl: 0x1fb2f87c
bdba: 0x01808438
     76543210
flag=--------
ntab=1
nrow=17
frre=-1
fsbo=0x34
fseo=0xd2
avsp=0x33b
tosp=0x33c
0xe:pti[0]    nrow=17  offs=0
0x12:pri[0]   offs=0x1e34
......
0x30:pri[15]     offs=0x6e2
0x32:pri[16]     offs=0x583
block_row_dump:
tab 0, row 0, @0x1e34
tl: 332 fb: --H-F--- lb: 0x0  cc: 79
nrid:  0x018083f8.e
col  0: [ 5]  c4 04 5a 27 1b
col  1: [ 7]  47 59 30 32 30 30 31
col  2: [ 4]  c3 15 11 04
col  3: [12]  31 38 37 33 34 34 32 30 30 30 30 36
col  4: [12]  31 34 30 34 34 34 32 30 30 30 30 31
col  5: [30]
......
```

上述的 dump 信息表示的是 rdba 地址 01808438 的第 0 行，也就是我们大家所理解的第一行数据。我们可以发现在这行记录中，行头存在一个 nrid 地址（0x018083f8.e）。既然这里提到 nrid，那么我们就有必要来解释一下。当出现行迁移或者行链接时，数据块内用 nrid 来标示下一个 rowid 地址。

首先我们说一下 Oracle 的行迁移。行迁移分为几种情况，最常见的一种其实是数据块内的。一个数据块中单条记录的最大列数是 255 列，当一行记录的列超过 255 时，其他的列数据库会被 Oracle 分成另外一个 row piece 存在同一个数据块中（当然也有可能存到其他数据块）。也就是说超过 255 列的行数据，会被分成多个 row piece。而当我们读取这个行数据时，怎么知道是一个完整的整体呢？答案就是 nrid。Oracle 通过 nrid 来将这多个 row piece 串在一起，组成一个完整的行数据。另外其中常见情况就是更新时，数据块内剩余空间不足容纳更新之后的列数据时，也会产生行迁移。

而行链接通常是一个数据块不足以容纳一条数据，需要申请多个数据块来容纳一条记录，而这种情况通常都是 lob 的使用场景下才会出现。正因为如此，大家所见的场景通常其实都是行迁移，而非行链接。

想到这一点，那么我们再回头去看下前面的错误。row 01808438.0 表示这个数据块的第 0 行，而该数据块的第 0 行所存在的 nrid 地址是 0x018083f8.e；那么我们进一步到 0x018083f8 数据块中去寻找第 e 行记录，却发现结果是这样的。

```
Object id on Block? Y
seg/obj: 0xc03d01  csc: 0xb37.78bb5e9f  itc: 3  flg: E  typ: 1 - DATA
    brn: 0  bdba: 0x1807d8a ver: 0x01 opc: 0
    inc: 0  exflg: 0

 Itl           xid                  Uba         Flag  Lck        Scn/Fsc
0x01   0x000a.013.000cdc01  0x01c02834.6573.33  --U-   2  fsc 0x0000.78cbf31d
0x02   0x000a.001.000cda7a  0x0080150a.64d3.21  C---   0  scn 0x0b37.78b584df
0x03   0x000a.01e.000cdade  0x00801510.64d3.13  C-U-   0  scn 0x0b37.78b99f21

data_block_dump,data header at 0x2b4fc709007c
===============
tsiz: 0x1f80
hsiz: 0x2e
pbl: 0x2b4fc709007c
bdba: 0x018083f80x018083f8
     76543210
flag=--------
ntab=1
nrow=14
frre=-1
fsbo=0x2e
fseo=0x568
avsp=0x53a
tosp=0x53a
0xe:pti[0]   nrow=14 offs=0
0x12:pri[0]  offs=0x1d78
0x14:pri[1]  offs=0x1c37
......
0x2a:pri[12]    offs=0x6c2
0x2c:pri[13]    offs=0x568
block_row_dump:
tab 0, row 0, @0x1d78
tl: 520 fb: -----L-- lb: 0x0  cc: 255
......
tab 0, row 13, @0x568
tl: 346 fb: --H-F--- lb: 0x1  cc: 79
nrid:  0x018083f8.c
col  0: [ 5]  c4 04 5a 3a 0a
col  1: [ 7]  47 59 30 32 30 30 31
```

```
col  2: [ 4] c3 15 11 04
......
col 76: [ 1] 80
col 77: [ 1] 80
col 78: [ 1] 80
end_of_block_dump
End dump data blocks tsn: 6 file#: 6 minblk 33784 maxblk 33784
```

可以发现，我们定位到 0x018083f8 数据块的第 e 行 (也就是第 14 行)，其实就是该数据块的最后一行数据。然而我们发现该行数据也存在一个 nrid。该 nrid 是 0x018083f8.c，这表示该 33784 数据块第 12 行记录。和第 13 行是组合成一条完整行记录的。

换句话说，我们前面报错的那条记录，应该有 2 个 row piece，其中一个 row piece 是存在的，其中一个 row piece 本应该存在在 33784 数据块中。但是存在 33784 中的这个 row piece 的 nrid 是指向另外一行记录。很明显这是不匹配的。

正是因为如此，由于对应的 row piece 不正确，Oracle 才报了上述的错误。

实际上遇到该错误之后，大多数人都会以为是索引的问题，通过 drop 重建可以解决，然而这里的问题比较特殊。本质上是表自身的某条数据有问题，所以这就是为什么客户重建索引会报错的原因。

```
SQL> CREATE INDEX "STATS"."MONTHINDEX_STATUS2" ON "STATS"."T_REPORT_MONTH" ("TARGET_298",
"UNIT_LEVEL", "TARGET_VAL", "MONTH_FLG")
  2    TABLESPACE "STATDATA" ;
第 1 行出现错误:
ORA-00600: 内部错误代码, 参数: [kdsgrp1], [], [], [], [], [], [], []
```

最后我们清楚原因之后就可以很好解决这个问题了。通过 rowid 的方式跳过这行有问题的记录，将其他数据取出，重建表即可。

13.7　城门失火：Goldengate 引发的数据库故障

2014 年 11 月 8 日 21 点左右某客户的数据库集群出现 swap 耗尽的情况，导致数据库无法正常使用。检查数据库日志，发现相关的错误如下所示。

```
Sat Nov 08 20:50:23 CST 2014
Process startup failed, error stack:
Sat Nov 08 20:50:41 CST 2014
Errors in file /oracle/product/10.2.0/admin/xxx/bdump/xxx1_psp0_1835540.trc:
ORA-27300: OS system dependent operation:fork failed with status: 12
ORA-27301: OS failure message: Not enough space
ORA-27302: failure occurred at: skgpspawn3
Sat Nov 08 20:50:41 CST 2014
Process m000 died, see its trace file
```

根据数据库 alert log 的报错信息，我们可以知道从 20:50 左右开始出现 ORA-27300、

ORA-27301 错误。根据 Oracle MOS 文档 Troubleshooting ORA-27300 ORA-27301 ORA-27302 errors [ID 579365.1]的描述可以知道,这个错误产生的原因就是内存不足导致。出现该错误的主机为 Oracle RAC 的 1 节点。该主机物理内存大小为 96GB,Oracle SGA 配置为 30GB,PGA 配置为 6GB,操作系统 Swap 配置为 16GB。

正常情况下,物理主机的内存是可以满足正常使用的。由于在 20:56 开始出现无法 fork 进程,即使无法分配内存资源,说明在该时间点之前物理主机的内存使用已经出现问题了。通过 Nmon 监控数据可以发现,实际上从当天下午 18 点左右,操作系统物理内存 free memory 就开始大幅下降,如图 13-5 所示。

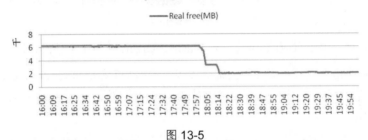

图 13-5

从图 13-5 可以看出,数据库主机节点 1 的物理内存大约从 18:01 开始突然下降得很厉害,到 18:14 左右时,物理内存 free Memory 已经不足 2GB 了。而该主机的物理内存中,大部分为 Process%所消耗,如图 13-6 所示。

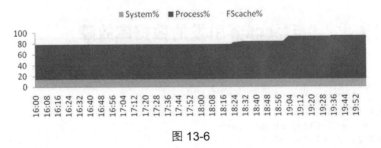

图 13-6

可以看到,该节点从 18:20 左右 Process% 消耗的内存开始突然增加,到 19:52 时,基本上消耗了所有的物理内存。这里需要注意的,这里的 Process% 内存消耗,是指该主机上的所有应用程序的进程消耗,包括 Oracle 的所有进程,以及其他应用程序的进程。

然而我根据 Oracle 的 AWR 历史数据进行查询,发现并没有明显的会话消耗内存很高的情况。

```
SQL> select INSTANCE_NUMBER,SNAP_ID,STAT_NAME,value from dba_hist_osstat
```

13.7 城门失火：Goldengate 引发的数据库故障

```
  2  where stat_name like 'VM%' and snap_id > 13550 and snap_id < 13559
  3  order by 1,2;

INSTANCE_NUMBER    SNAP_ID STAT_NAME                VALUE
--------------- ---------- ---------------- -------------
              1      13551 VM_IN_BYTES       4691725926408
              1      13551 VM_OUT_BYTES     14798577905664
              1      13552 VM_OUT_BYTES     14799491960832
              1      13552 VM_IN_BYTES       4691727376392
              1      13553 VM_OUT_BYTES     14800572719088
              1      13553 VM_IN_BYTES       4691727429624
              1      13554 VM_IN_BYTES       4691777949696
              1      13554 VM_OUT_BYTES     14820690083832
              1      13555 VM_OUT_BYTES     14857568350200
              1      13555 VM_IN_BYTES       4693160173560
              1      13556 VM_OUT_BYTES     14876324397048
              1      13556 VM_IN_BYTES       4695865995264
              1      13558 VM_OUT_BYTES     14882330329080
              1      13558 VM_IN_BYTES       4829460062208
              2      13551 VM_OUT_BYTES      2273165344776
              2      13551 VM_IN_BYTES        347420766216
              2      13552 VM_OUT_BYTES      2273229529104
              2      13552 VM_IN_BYTES        347420766216
              2      13553 VM_OUT_BYTES      2273286496272
              2      13553 VM_IN_BYTES        347420766216
              2      13554 VM_OUT_BYTES      2324453691408
              2      13554 VM_IN_BYTES        347433598968
              2      13555 VM_IN_BYTES        347559141384
              2      13555 VM_OUT_BYTES      2383075213320
              2      13556 VM_IN_BYTES        347674648584
              2      13556 VM_OUT_BYTES      2430000705552
              2      13557 VM_IN_BYTES        473531183112
              2      13557 VM_OUT_BYTES      2499316277256
              2      13558 VM_OUT_BYTES      2507250249744
              2      13558 VM_IN_BYTES        473575673856
```

看到上列数据不要惊讶，上列数据为累积数据，我们需要前后相减进行计算。

```
16:00 --17:00点:
SQL> select (4691727376392-4691725926408)/1024/1024 from dual;
(4691727376392-4691725926408)/1024/1024
---------------------------------------
                              1.3828125
SQL> select (14799491960832-14798577905664)/1024/1024 from dual;
(14799491960832-14798577905664)/1024/1024
-----------------------------------------
                               871.710938      ---换出的内存,单位为M.

17:00 --18:00点:
SQL> select (4691727429624-4691727376392)/1024/1024 from dual;
(4691727429624-4691727376392)/1024/1024
---------------------------------------
                               .050765991
SQL> select (14800572719088-14799491960832) /1024/1024 from dual;
```

```
(14800572719088-14799491960832)/1024/1024
----------------------------------------
 1030.69139    ---换出的内存,单位为 M.

18:00 --19:00 点:
SQL> select (4691777949696-4691727429624)/1024/1024 from dual;
(4691777949696-4691727429624)/1024/1024
----------------------------------------
 48.1796951
SQL> select (14820690083832-14800572719088)/1024/1024 from dual;
(14820690083832-14800572719088)/1024/1024
----------------------------------------
 19185.4141    ---换出的内存,单位为 M.

19:00 --20:00 点:
SQL> select (4693160173560-4691777949696)/1024/1024 from dual;
(4693160173560-4691777949696)/1024/1024
----------------------------------------
 1318.1914
SQL> select (14857568350200-14820690083832)/1024/1024 from dual;
(14857568350200-14820690083832)/1024/1024
----------------------------------------
 35169.8555

20:00 --21:00 点:
SQL> select (4695865995264-4693160173560)/1024/1024 from dual;
(4695865995264-4693160173560)/1024/1024
----------------------------------------
 2580.47266
SQL> select (14876324397048-14857568350200)/1024/1024 from dual;
(14876324397048-14857568350200)/1024/1024
----------------------------------------
 17887.1602
```

从查询结果来看,Oracle 确实检查到了大量的换页操作,从 18 点~19 点开始。但是我查询数据库 18~21 点的 AWR 数据却并没有发现有非常大的 SQL 操作。

```
SQL> select INSTANCE_NUMBER,snap_id,sum(SHARABLE_MEM)/1024/1024 from dba_hist_SQLstat
  2  where snap_id > 13550 and snap_id < 13558
  3  group by INSTANCE_NUMBER,snap_id order by 1,2;
INSTANCE_NUMBER     SNAP_ID  SUM(SHARABLE_MEM)/1024/1024
--------------- ---------- ---------------------------
              1       13551                  28.9083166
              1       13552                  30.0213976
              1       13553                  28.7059259
              1       13554                  29.1716347
              1       13555                  29.1961374
              1       13556                  35.6658726
              2       13551                  19.5267887
              2       13552                  20.9447975
              2       13553                  23.5789862
              2       13554                  21.0861912
              2       13555                  22.5129433
```

13.7 城门失火：Goldengate 引发的数据库故障

| 2 | 13556 | 23.0631037 |
| 2 | 13557 | 21.7776823 |

据业务了解，数据库节点 2 节点每周六会进行一次批量的操作，这可能会产生影响。基于这一点，我们分析了节点 2 节点的 nmon 数据，发现内存使用率也非常高，如图 13-7 所示。

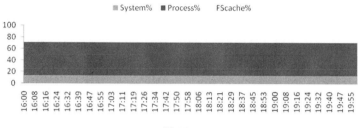

图 13-7

从节点 2 的数据来看，内存的变化也是在 18:00 左右。然而变化的却是 FScache%，Process% 的指标是没有变化的。根据这一点可以判断，在该时间段内数据库产生了大量的归档日志，进而导致文件系统 cache 所占的内存比例大幅上升。进一步检查数据库发现确实这段时间产生的日志相对较多。

由于在该时间段产生了大量的操作，因此就可能就会产生大量的 gc 等待。从节点 2 的 awr 数据来看，确实产生了大量的 gc 等待事件，如下所示。

```
Event Waits Time(s) Avg Wait(ms) % Total Call Time Wait Class
gc buffer busy 11,042,588 20,687 2 28.1 Cluster
gc current block 2-way 1,113,439 16,922 15 23.0 Cluster
gc current block congested 19,115 10,336 541 14.1 Cluster
CPU time  9,914  13.5
gc cr multi block request 6,430,854 3,965 1 5.4 Cluster
```

那么这里有没有可能是由于大量 gc 的产生，导致 Oracle RAC 的核心进程 LMS 等进程的负载增加，导致内存消耗的剧增呢？实际上，这一点是可以排除的。如果是 Oracle 的进程内存消耗较高，那么节点 2 的内存消耗变动，在 18:00 左右开始上升的应该是 Process%，而不是 FScache%。

到这里我们可以总结如下两点。

（1）节点 1 内存耗光，主要是 Process % 消耗。

（2）节点 2 的内存的变化主要是发生在 FScache%。

基于这两点，我们可以排除是数据库本身的进程消耗的大量内存导致 swap 被耗尽。在进一步分析时，发现于节点 1 部署了 Oracle Goldengate 同步软件，所以我怀疑极有可能是该软件导致。基于这个猜想，我进行了简单分析。如下是 ogg 的 report 信息。

```
2014-11-08 18:01:38  INFO     OGG-01026  Rolling over remote file ./dirdat/nt006444.
2014-11-08 18:03:43  INFO     OGG-01026  Rolling over remote file ./dirdat/nt006445.
......
2014-11-08 20:38:18  INFO     OGG-01026  Rolling over remote file ./dirdat/nt006551.
2014-11-08 20:52:02  INFO     OGG-01026  Rolling over remote file ./dirdat/nt006553.
```

我们可以发现，在故障时间段内该进程的操作明显要比其他时间段要频繁得多。从 18:00 到 20:38，Goldengate 抽取进程产生的 trail 文件超过 100 个。由于 ogg 本身也是部署在 Oracle 用户下，因此这也不难解释为什么节点 1 从该时间点开始内存的消耗会突然增加，而且 nmon 监控显示是 Process%消耗增加。通过 Nmon 的监控数据，我们也可以发现 paging 的操作主要是 goldengate 的 extract 进程导致，如图 13-8 所示。

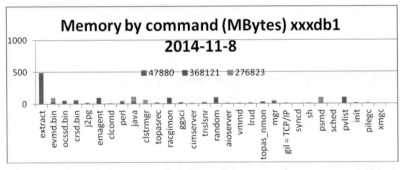

图 13-8

Goldengate 进程之所以会消耗大量的内存，是因为 extract 进程首先需要将数据读取到本地节点的内存中（即 Goldengate 的 cachesize）。然后将 cache 中的数据解析为 Goldengate 特有的日志格式，并写入到 trail 文件中。通常情况下，当遭遇大事务时，Goldengate extract 进程消耗的内存会大幅增加，因为 Goldengate 默认是以事务为单位（如果没有大事务拆分处理的话）。

上面数据是截取的部分内容。从时间点来看，和之前的 nmon 监控内存的变化是完全符合的。这可以完美解释为什么查询节点 1 的 AWR 数据发现从 18 点之后开始出现大量 swap，但是数据库本身的却查不到内存消耗高的进程。

因此最终确认导致此次故障的最本质原因是由于 Goldengate 抽取进程消耗大量内存，最后产生大量的 swap，进而最终影响了业务。

不过这里我并不推荐大家通过设置 cachesize 参数来控制 goldengate 进程的内存消耗，这样有可能降低 Goldengate 的处理能力。但是我们可以通过其他手段来处理这个问题，比如将大事务进程拆分等。

以上是在我工作中遇到的比较经典的案例，在此与大家分享，希望对大家的成长有帮助。也祝愿大家能够找到适合自己的兴趣点和学习方法。

第 14 章 勤奋与汗水

——我的思考与经验（李轶楠）

我最早在论坛上注册的 ID 是 ORA-600，后来这个网名成为了大家对我们的称呼，使用的人超过知道我名字的人，而且李轶楠三个字以各种错误方法被拼写。

14.1 我的职业生涯与思考

我的职业生涯可以简单概括为：5 年开发，5 年漂泊，5 年创业。因偶然的机缘了解到 Oracle，并全面转行进入 Oracle 领域。在成长过程中 ITPUB 给了我一个大舞台，在这里不仅认识了很多的朋友，共同探讨一些技术问题，交流一些心得体会。在技术能力提高的同时，也弥补了去现场机会较少的短板，通过交流与案例讨论，极大提高了自己技术服务的能力。

随着在论坛上经验和知名度的积累，我得到越来越多客户的认可，也开始为有需要的朋友和客户做一些打包的维护项目服务。在这些服务的过程中，我逐渐意识到随着运维需求的增加和客户服务意识的增强，客户对于专业服务团队在数据库服务领域的需求越来越迫切，因此有了创业的念头。我希望和朋友们一起合作，把服务做大作强，同时也希望能够通过这种方式，把我们的一些服务理念带给客户，为客户带来一些不一样的服务感受，真正从专业服务中得到一些获益。在社区的氛围里，在对行业的认识之下，我最终选择加入恩墨团队，走上创业的道路。

14.2 如何看待企业运维

随着企业信息化的发展，越来越多的大型行业、企业用户对信息化的要求变得更高。他们不再局限于能做什么，而更强调能做到什么、能得到什么。尤其是一些行业用户，经过多年的信息化发展，分散式管理已经不再适合他们的需求。他们开始强调集中管理、综合应用，所以越来越多的项目开始变得大型化、规模化、集中化。在这样的需求下，如何保障这种应用系统能稳定、高效、可靠地运行，就变得非常重要。

在大型应用系统中，由于业务量、数据量的巨大，很容易造成前期规划不当，中期维护复杂，后期调整难度大的情况。因此，必须在前期就要尽可能考虑全面。

我们一直希望把 **DBA** 的工作从简单的后台维护和救火，转换到前期的规划设计上。DBA 在一个项目的生命周期中应该是无处不在的，一个有丰富经验，尤其是有过项目开发经验的 DBA，对于项目生命周期的各个环节中可能出现的情况感受颇深。因此，不论项目的前期、中期还是后期，DBA 都应该积极参与，尽其能力为项目的性能提供理念、方法和思路。

- **项目的前期**。主要是需求分析、架构设计、应用设计等工作。在这个过程中，DBA 可以参与或者提供建议。在需求分析时，DBA 可以提供业务、技术实现手段的评判，对一些需求进行适度的限制以保障优化方法的实现；在架构规划时，可以根据用户对高可用、容灾、健壮性、性能等多方面的要求，提供 RAC、HA、DG、OGG、使用版本、硬件资源需求等多方面的数据库架构规划；在应用设计阶段，对应用的需求评估，确定使用前台语言实现还是使用 PL/SQL 实现，并设计 E-R 模型，构建应用的数据模型。项目前期的更多工作集中在规划设计上，虽然这时并不需要 DBA 去做什么，但 DBA 的经验对于前期设计的准确性与完整性有很大的帮助。

- **项目的中期**。主要是代码开发、数据库部署、性能测试等工作。在代码开发阶段，DBA 将会辅助开发人员制定代码规范和数据规范，并对高效 SQL 进行评估，确保在大数据需求下高质量的 SQL 代码，现在我们把这个环境的工作称为 SQL 审核。同时通过对需求的了解和代码的掌握，确定应用对数据库的资源需求特性和资源消耗量。在此基础上，更准确地规划存储、网络、主机、中间件等资源，并且合理配置数据库，为不同需求构建不同类型的表，在相应字段上构建索引以满足各种查询需求。而对于性能测试阶段，DBA 则需要做更细节的监控和精细分析工作。通过各种测试数据的分析，确定目前系统仍然存在的潜在风险，从稳定、可靠、高效等多方位确定调整方案，确保系统上线后的平稳运行。

- **项目的后期**。主要就是运维，这就是大家认为的传统意义上 DBA 应该做的事情。日常运维中最主要的就是监控、预警和巡检、优化。通过对系统的理解，DBA 可以设定一些监控点和阈值（也可以通过定期巡检来审核），通过这些方法提前发现和排除系统潜在的隐患，以保证系统的正常运行。

可以说，DBA 的参与对整个系统的性能和稳定性都有着非常重大的意义，所以别再把 DBA 当作大厦的救火器，其实它是大厦的脊梁。

14.3 对性能问题的认识

说到性能问题，大部分人可能都抱有性能就是数据库性能的一个错误观点。其实性能问题绝不能简单地说成是数据库性能，而应该考虑的是整体应用性能。因此，需要优化的地方涉及应用系统的各个环节，就像前面所讲，不论是前期的规划设计，还是中期的运维调整，还是后期的 SQL 分析，都是性能的一部分。所以在项目整个生命期内，从需求分析、架构设计、应用设计、数据库设计、存储规划、网络规划、中间件规划到代码开发、数据库部署等各个环节，都需要考虑性能的影响。

性能问题与数据量有关，但并没有绝对值。影响性能的因素很多，数据量、并发量、主机配置、网络环境、应用架构、代码质量等等都会对性能带来很大影响。我所见过的最小的一个数据库，最大的单表数据量仅有几 M，也就是 10 万条数据左右，整个数据库不到 200M 的数据，并发会话只有 100 左右，但最慢的查询一次就要执行十几分钟，所以很难简单地说数据量多大就要考虑性能优化。

既然影响性能的因素很多，而其中不少因素又是在系统上线后很难调整的，所以前期规划设计的优化肯定很重要。如果前期不做好，即使系统上线运行了，在一段时间之后，各种性能问题也会随后出现，而到了那个时候，就会发现很多优化方法都受到了种种限制，以至于无法顺利实施，这就是前期没有考虑性能问题所带来的隐患。因此，前期规划设计绝对不能忽视，必须提前考虑各种设计方案在性能上的影响，综合平衡多方面的需求。

很多人认为，现在 Oracle 的技术越来越成熟，功能也越来越完备，这样对于 Oracle 的开发会变得更加的简单，不需要多高的技术水平就可以用 Oracle 完成数据库的建设。我并不同意这种观点。要想系统的性能可控、预测和度量，就不要把希望寄托在别人身上。不需要优化的系统，只是在系统负载量不大、业务不繁忙的时候才可能存在，所有大型系统都必须摒弃这个观点。

现在很多数据库都在通过各种技术手段使开发简单化、实施傻瓜化，这是技术发展的一个必然阶段。任何用户都不希望使用的是一个自己搞不定的软件，但是傻瓜式的构建只适合傻瓜式（或者叫通用性）的需求。这种开发和实施的自动化、简单化，实际上是数据库软件智能化的一个发展过程，但这种智能化目前为止还非常弱，不论是自动诊断还是自动调整，都有很大的局限性和延迟性。如果大型系统完全依赖于自动管理与调整，将会有非常多的问题发生。同时，目前我们所用到的自动化与智能化仅限于数据库内，而一套应用系统需要考虑健壮性、可用性、高效性、安全性等方面。这些方面能否满足需求，并不仅仅是通过数据库的自动化就能解决的，需要考虑系统、架构、需求、代码、存储、网络等多个环节，因此，绝对不要认为数据库技术发展了，我们就可以放松系统规划设计、代码优化、存储调整等一系列技术需求。数据库技术发展，仅仅意味着 Oracle 希望能减轻我们的工作，并不能根本上改变系统优化的模式。

14.4 学习方法

在学习技术的过程中，我认为勤奋、自学是非常重要的，汗水永远是成功的基石。

为了工作，去 Google 寻找答案。但为了自己，看书和试验更重要。

不要说你没有环境可以测试，所以学不会。环境是可以自己创造的，你可以装虚拟机，你也可以在网上与别人讨论他们的案例，这都是产生学习环境的基本方法，说没有环境就学不会，那只是逃避的借口。与别人的交流和讨论是快速成长的必要手段。多看书，多练习，踏踏实实地学习，脚踏实地才能够逐步起飞。

DBA 是个好职业，但 DBA 不是终点，而是另一个起点。事实证明，很多经验丰富的 DBA，逐渐成为系统运维的管理者，或者企业级架构师，它们在另外的岗位上，很好的运用他们的 DBA 经验为企业的 IT 发展提供着必要的技术、经验支撑。

14.5 所有奇异的故障都有一个最简单的本质

在数据的世界里，任何奇异的事情都可能发生。也许司空见惯的一个现象会在某场景下导致严重的数据库故障，这就要求运维者要时刻保持警惕，要细致、耐心地分析现象，并通过表象认识事物的本质。

接下来我们通过两个真实案例认识运维中的诡异故障。并通过这两个案例的分析跟大家分享我在处理问题中的一般思路。供大家参考和学习。

在案例分享之前，先跟大家普及一下故障分析的常规思路，也便于初学者在遇到问题的时候知道从何处入手。

- 辨识问题。在遇到故障时，首先要辨识一下当前的场景主要是性能问题还是真的故障。
- 对待性能问题。如果是性能问题，那就需要收集当时的系统性能信息与数据库性能信息，如 awr、ash，或者系统的 nmon、top 之类采样信息。
- 对待故障问题。如果是故障，那就要检查数据库的告警日志与跟踪文件了。非常典型的就是 alert_<SID>.log，这里面往往给了我们进一步追踪痕迹的指引。除此之外，各个进程的 trace 文件，asm 的 trace 文件以及 RAC 的各种 log、trace 文件都会给出一些故障的原因或者印记。
- 官方信息检索。另外，当遭遇这些问题的时候，如果有 MOS 账号的话，建议首先去 MOS 中查看是否有故障相关的 Bug 或者技术文章，这既是快速诊断问题、解决问题的途径，也是 DBA 快速成长的重要手段。

所有奇异的故障的奇异之处都在于——你对深层的技术本质不够理解。

当我们了解了技术的本质之后，你会发现所有奇异的故障原本都很简单。

14.6 案例一：意料之外的 RAC 宕机祸首——子游标

这个案例是关于一次子游标过多导致的 RAC 节点宕机的故障。子游标与实例宕机会有什么关系？我们通过具体的案例看看如何从表象入手，抽丝剥茧，层层深入发现线索，并逐步还原犯罪场景，找出实例宕机的真实原因。

14.6.1 信息采集，准确定位问题

某天晚上，我们接到客户的电话，说某一核心系统有一个节点挂了，实例自动重启。虽然没有影响业务，但这种无缘无故的重启发生在核心系统上，直接威胁到业务运行，希望我们能协助尽快找到原因。

听到客户的描述，心里第一个反应是：还好，只是单节点重启，问题本身应该不严重，至少对业务影响并不大。

客户目前急需的是快速给出问题的原因，以便根据情况作出后续的规避措施，所以需要尽快拿到现场的一些数据来做分析。毕竟分析所需要的信息数据距离故障时间越近，精准度就越高，越有利分析。

从客户处得到的故障信息如下所示。

- 重启的是整个数据库架构的 2 节点，这个库是核心系统，晚上也有业务。
- 重启实例的情况其实以前也发生过，只不过发生的不多[1]
- 当前数据库版本为 11.2.0.1。[2]

当然只听客户描述是不够的。于是，我们远程登录了客户的服务器，开始做进一步检查。以下是收录的关键线索。最核心的是节点 2 的告警日志，部分内容如下所示。

```
Fri Jun 26 20:24:52 2015
    Errors    in    file    /u01/app/oracle/diag/rdbms/orcl/orcl2/trace/orcl2_p001_13581.trc
(incident=204565):
    ORA-00600: 内部错误代码, 参数: [kksfbc-wrong-kkscsflgs],[39913467504],[33],[],[],[],
[],[],[],[],[],[]
    Incident details in:
```

---

[1] 潜台词是这不是个案，以前的重启意味着这个问题一直存在。

[2] 看到每个大版本的第一个小版本，总是觉得这种系统会 Bug 缠身。虽然夸大了点，但我们确实遇到了不少这种小版本的 Bug。

```
    /u01/app/oracle/diag/rdbms/orcl/orcl2/incident/incdir_204565/
orcl2_p001_13581_i204565.trc
    ......
    Fri Jun 26 21:50:26 2015
    LCK0 (ospid: 29987) waits for latch 'object queue header operation' for 77 secs.
    Errors in file /u01/app/oracle/diag/rdbms/orcl/orcl2/trace/orcl2_lmhb_29939.trc
(incident=204141):
    ORA-29771: process MMON (OSID 29967) blocks LCK0 (OSID 29987) for more than 70 seconds
    Incident details in:
/u01/app/oracle/diag/rdbms/orcl/orcl2/incident/incdir_204141/ orcl2_lmhb_29939_i204141.trc
    MMON (ospid: 29967) is blocking LCK0 (ospid: 29987) in a wait
    LMHB (ospid: 29939) kills MMON (ospid: 29967).
    Please check LMHB trace file for more detail.
    Fri Jun 26 21:51:06 2015
    Restarting dead background process MMON
    Fri Jun 26 21:51:06 2015
    MMON started with pid=213, OS id=16612
    Fri Jun 26 21:54:10 2015
    LMS1 (ospid: 29929) waits for latch 'object queue header operation' for 81 secs.
    LCK0 (ospid: 29987) has not called a wait for 85 secs.
    Errors in file /u01/app/oracle/diag/rdbms/orcl/orcl2/trace/orcl2_lmhb_29939.trc
(incident=204142):
    ORA-29770: global enqueue process LMS1 (OSID 29929) is hung for more than 70 seconds
    Fri Jun 26 21:54:20 2015
    Errors in file /u01/app/oracle/diag/rdbms/orcl/orcl2/trace/orcl2_lmhb_29939.trc
(incident=204143):
    ORA-29770: global enqueue process LCK0 (OSID 29987) is hung for more than 70 seconds
    Incident details in: /u01/app/oracle/diag/rdbms/orcl/orcl2/incident/incdir_204142/
orcl2_lmhb_29939_i204142.trc
    ERROR: Some process(s) is not making progress.
    LMHB (ospid: 29939) is terminating the instance.
    Please check LMHB trace file for more details. Please also check the CPU load, I/O load
and other system properties for anomalous behavior
    ERROR: Some process(s) is not making progress.
    LMHB (ospid: 29939): terminating the instance due to error 29770
    Fri Jun 26 21:54:21 2015
    opiodr aborting process unknown ospid (26414) as a result of ORA-1092
    Fri Jun 26 21:54:21 2015
    ORA-1092 : opitsk aborting process
    Fri Jun 26 21:54:21 2015
    System state dump is made for local instance
    System State dumped to trace file /u01/app/oracle/diag/rdbms/orcl/orcl2/trace/orcl2_
diag_29889.trc
    Instance terminated by LMHB, pid = 29939
```

在告警日志中我们发现一个很明显的 ORA-600 错误，同时也发现一些其他的 ORA 报错，见上面着重标识部分。于是我们对这些错误分别进行了分析。

ORA-600 [kksfbc-wrong-kkscsflgs] (Doc ID 970885.1) 确实是一个 Bug，在 MOS 上有详细说明，如表 14-1 所示。

14.6 案例一：意料之外的 RAC 宕机祸首——子游标

表 14-1

Bug	Fixed	Description
9067282	11.2.0.1.2, 11.2.0.1.BP01, 11.2.0.3, 12.1.0.1	ORA-600 [kksfbc-wrong-kkscsflgs] can occur
9066130	10.2.0.5, 11.1.0.7.2, 11.2.0.2, 12.1.0.1	OERI [kksfbc-wrong-kkscsflgs] / spin with multiple children
8828328	11.2.0.1.1, 11.2.0.1.BP04, 11.2.0.2, 12.1.0.1	OERI [kksfbc-wrong-kkscsflgs]
8661168	11.2.0.1.1, 11.2.0.1.BP01, 11.2.0.2, 12.1.0.1	OERI[kksfbc-wrong-kkscsflgs] can occur

但 MOS 上并未说明该 Bug 会导致实例宕机，这个 600 错误看来应该与此次重启关系不大。好吧，作为一个问题记下来就是了。

在故障时间点，LMHB 进程 check 发现 MMON 进程阻塞了 LCK0 进程，超过 70 秒，因此尝试 kill MMON 进程，该进程被 kill 之后将会自动重启。

可以看到在 6 月 26 日 21:51:06 时间点，MMON 进程重启完成。

接下来，在 6 月 26 日 21:54:10，LMS1 进程报错无法获得 latch（object queue header operation）超过 85 秒。[1]

为了更清楚地理清线索，我们根据节点 2 的告警日志信息，整理出如下的时间流。

```
Jun 26 20:24:52   ORA-00600 [kksfbc-wrong-kkscsflgs]

Jun 26 21:50:26   LCK0 (ospid: 29987) waits for latch 'object queue header operation' for 77 secs
                  MMON (OSID 29967) blocks LCK0 (OSID 29987) for more than 70 seconds
                  MMON (ospid: 29967) is blocking LCK0 (ospid: 29987) in a wait
                  LMHB (ospid: 29939) kills MMON (ospid: 29967)

Jun 26 21:51:06   MMON started with pid=213, OS id=16612

Jun 26 21:54:10   LMS1 (ospid: 29929) waits for latch 'object queue header operation' for 81 secs
                  LCK0 (ospid: 29987) has not called a wait for 85 secs
                  ORA-29770: global enqueue process LMS1 (OSID 29929) is hung for more than 70 seconds

Jun 26 21:54:20   ORA-29770: global enqueue process LCK0 (OSID 29987) is hung for more than 70 seconds
                  ERROR: Some process(s) is not making progress.
                  LMHB (ospid: 29939) is terminating the instance
                  LMHB (ospid: 29939): terminating the instance due to error 29770
```

从最后的信息可以看出，在 21:54:20 时间点，LMHB 进程强行终止了数据库实例。而终止实例的原因是 LMHB 进程发现 LCK0 进行 hung 住了，而且超过了 70 秒。从前面的信息也可

---

1 LMHB 是 11gR2 中引入的后台进程，官方文档介绍是 Global Cache/Enqueue Service Heartbeat Monitor，Monitor the heartbeat of LMON，LMD，and LMSn processes, LMHB monitors LMON，LMD，and LMSn processes to ensure they are running normally without blocking or spinning。该进程负责监控 LMON、LMD、LMSn 等 RAC 关键的后台进程，保证这些 background process 不被阻塞或 spin。LMHB 是 Lock Manager Heartbeat 的缩写。

以看出，实际上在 21:54:10 时间点，LCK0 进程就已经没有活动了，而且在该时间点 LMS1 进程也一直在等待 latch。很明显，如果 LMS1 进程无法正常工作，Oracle 为了保证集群数据的一致性，为了避免脑裂，必然将问题节点强行驱逐重启。

14.6.2 层层分析，揪出罪魁祸首

在故障日志中，LMS1 进程一直处于等待状态，那么 LMS1 在等什么呢？LCK0 为什么被 Hung 住了？我们通过更详细的信息来分析。

首先让我们来看看 LMS1 到底在干什么？

检查 orcl2_lmhb_29939_i204142.trc，而该 trace 文件产生的时间点是 6 月 26 日 21:54:10。

```
SO: 0x9a1175160, type: 2, owner: (nil), flag: INIT/-/-/0x00 if: 0x3 c: 0x3
 proc=0x9a1175160, name=process, file=ksu.h LINE:11459, pg=0
(process) Oracle pid:14, ser:1, calls cur/top: 0x9b17e5330/0x9b17e0e60
          flags : (0x6) SYSTEM flags2: (0x100), flags3: (0x0)  intr error: 0, call error: 0,
sess error: 0, txn error 0 intr queue: empty
  ksudlp FALSE at location: 0
  (post info) last post received: 0 0 116
              last post sent-location: ksl2.h LINE:2160 ID:kslges
              last process posted by me: 9811032c8 1 14
    (latch info) wait_event=0 bits=a
          Location from where call was made: kcbo.h LINE:890 ID:kcbo_unlink_q_bg:
      waiting for 993cfec60 Child object queue header operation level=5 child#=117
          Location from where latch is held: kcl2.h LINE:3966 ID:kclbufs:
          Context saved from call: 0
          state=busy(shared) [value=0x4000000000000001] wlstate=free [value=0]
          waiters [orapid (seconds since: put on list, posted, alive check)]:
           14 (95, 1435326858, 4)        21 (94, 1435326858, 7)
           waiter count=2
          gotten 73961423 times wait, failed first 4752 sleeps 1927
          gotten 33986 times nowait, failed: 4
          possible holder pid = 36 ospid=29987
      on wait list for 993cfec60
      holding    (efd=5) 9b59be480 Child gc element level=3 child#=20
          Location from where latch is held: kcl2.h LINE:3535 ID:kclbla: Context saved from call: 0
          state=busy(exclusive) [value=0x200000000000000e, holder orapid=14] wlstate=free [value=0]
      holding    (efd=5) 9b45cac50 Child cache buffers chains level=1 child#=61221
          Location from where latch is held: kcl2.h LINE:3140 ID:kclbla:  Context saved from call: 0
          state=busy(exclusive) [value=0x200000000000000e, holder orapid=14] wlstate=free [value=0]
    Process Group: DEFAULT, pseudo proc: 0x9b11ca008
    O/S info: user: oracle, term: UNKNOWN, ospid: 29929
    OSD pid info: Unix process pid: 29929, image: oracle@ebtadbsvr2 (LMS1)
```

从 LMS1 进程的信息来看，LMS1 进程所等待的资源（object queue header operation）正被 ospid=29987 持有，那么 29987 又是什么呢？

进一步分下 ospid=29987 是什么？让我们接着往下看。

14.6 案例一：意料之外的 RAC 宕机祸首——子游标

```
SO: 0x9911283b0, type: 2, owner: (nil), flag: INIT/-/-/0x00 if: 0x3 c: 0x3
proc=0x9911283b0, name=process, file=ksu.h LINE:11459, pg=0
(process) Oracle pid:36, ser:2, calls cur/top: 0x9b17e58e0/0x9b17e58e0
          flags : (0x6) SYSTEM flags2: (0x0),   flags3: (0x0)   intr error: 0, call error:
0, sess error: 0, txn error 0 intr queue: empty
  ksudlp FALSE at location: 0
    (post info) last post received: 0 0 35
              last post received-location: ksr2.h LINE:603 ID:ksrpublish
              last process to post me: 981110608 118 0
              last post sent: 0 0 36
              last post sent-location: ksr2.h LINE:607 ID:ksrmdone
              last process posted by me: 9911283b0 2 6
    (latch info) wait_event=0 bits=20
      holding    (efd=3) 993cfec60 Child object queue header operation level=5 child#=117
          Location from where latch is held: kcl2.h LINE:3966 ID:kclbufs:
          Context saved from call: 0
          state=busy(shared) [value=0x4000000000000001] wlstate=free [value=0]
          waiters [orapid (seconds since: put on list, posted, alive check)]:
           14 (95, 1435326858, 4)
           21 (94, 1435326858, 7)
          waiter count=2
  Process Group: DEFAULT, pseudo proc: 0x9b11ca008
  O/S info: user: oracle, term: UNKNOWN, ospid: 29987
  OSD pid info: Unix process pid: 29987, image: oracle@ebtadbsvr2 (LCK0)
```

最后一句很明显地告诉我们，29987 原来就是 LCK0 进程。这意味着 LMS1 进程所等待的资源正被 LCK0 进程所持有。而同时还有另外一个进程 orapid=21 也在等待该进程。通过分析我们发现，orapid=21，是 DBW2 进程，即数据库写进程。[1]

从数据库告警日志来看，在 6 月 26 日 21:54:10 有提示 LCK0 进程已经超过 85 秒没有响应。

```
LCK0 (ospid: 29987) has not called a wait for 85 secs
```

根据时间点来计算，大概在 6 月 26 日 21:52:45 点左右开始，LCK0 进程没有响应了。那么很明显，我们只要知道 LCK0 进程为什么会 hung，就知道了此次故障的原因。

那么来看看 LCK0 的 trace 文件，能不能看到一些线索。LCK0 进程的 trace 信息如下所示。

```
*** 2015-06-26 21:50:29.329    Process diagnostic dump for oracle@ebtadbsvr2 (LCK0), OS id=29987,
pid: 36, proc_ser: 2, sid: 1729, sess_ser: 3
-------------------------------------------------
loadavg : 6.47 26.96 34.97
Memory (Avail / Total) = 10598.05M / 64421.55M
Swap (Avail / Total) = 20256.00M /  20479.99M
F S UID        PID  PPID  C PRI  NI ADDR SZ WCHAN  STIME TTY          TIME CMD
0 S oracle   29987     1  0  76   0 - 10541946 semtim Apr05 ?     01:25:21 ora_lck0_orcl2
Short stack dump:
```

---

[1] 这里解释一下，orapid 是 oracle 中的进程 id，而 pid 则是 os 上的进程 id。所以 orapid=21 从 v$process 中可以查到是 dbw2，而 orapid=14 是 LMS1。

```
       ksedsts()+461<-ksdxfstk()+32<-ksdxcb()+1782<-sspuser()+112<-__restore_rt()<-semop()+7
<-skgpwwait()+156<-kslgess()+1799<-ksl_get_shared_latch()+620<-kclbufs()+272<-kclanticheck
()+412<-kclahrt()+88<-ksbcti()+212<-ksbabs()+1732<-kclabs()+186<-ksbrdp()+923<-opirip()+62
3<-opidrv()+603<-sou2o()+103<-opimai_real()+266<-ssthrdmain()+214<-main()+201<-__libc_star
t_main()+244<-_start()+36

       *** 2015-06-26 21:54:18.438
       Process diagnostic dump for oracle@ebtadbsvr2 (LCK0), OS id=29987,
       pid: 36, proc_ser: 2, sid: 1729, sess_ser: 3
       -------------------------------------------------------------------------------
       loadavg : 2.04 13.34 27.63
       Memory (Avail / Total) = 9519.06M / 64421.55M
       Swap (Avail / Total) = 20256.00M / 20479.99M
       F S UID        PID  PPID  C PRI  NI ADDR SZ WCHAN  STIME TTY          TIME CMD
       0 R oracle   29987     1  0  85   0 - 10541965 ?   Apr05 ?         01:26:55 ora_lck0_orcl2
       Short stack dump:
       ksedsts()+461<-ksdxfstk()+32<-ksdxcb()+1782<-sspuser()+112<-__restore_rt()<-kcbo_get_
next_qheader()+255<-kclbufs()+321<-kcldirtycheck()+231<-kclahrt()+93<-ksbcti()+212<-ksbabs
()+1732<-kclabs()+186<-ksbrdp()+923<-opirip()+623<-opidrv()+603<-sou2o()+103<-opimai_real(
)+266<-ssthrdmain()+214<-main()+201<-__libc_start_main()+244<-_start()+36
```

从上述 LCK0 进程的几个时间点的信息来看，第一个时间点 21:50:29，wchan 为 semtim。wchan，表示进程 sleeping 的等待表现形式。semtim 表示在该时间点，LCK0 进程一直处于 sleep 状态。所谓的 sleep 状态，是进程持有的资源是不会释放的。

而在第 2 个时间点 21:54:18，LCK0 进程的 wchan 状态是 "?"，这表示未知，如果是为空，则表示该进程处理 running 状态。在 21：50 到 21：52 时间段内，LCK0 进程仍然没有恢复正常。那么 LCK0 到底被什么阻塞了（或者说它需要的资源被谁占用了）？

同时也可以看到内存和 swap 都空闲很多，CPU 也并不很忙。

继续检查 trace，在 21：50 时间点我们发现，LCK0 进程是被 MMON 进程锁阻塞了。

```
       SO: 0x9d10f97c0, type: 2, owner: (nil), flag: INIT/-/-/0x00 if: 0x3 c: 0x3
        proc=0x9d10f97c0, name=process, file=ksu.h LINE:11459, pg=0
       (process) Oracle pid:31, ser:1, calls cur/top: 0x965657408/0x9b17e3f18
                 flags : (0x2) SYSTEM flags2: (0x20),  flags3: (0x0)  intr error: 0, call error:
0, sess error: 0, txn error 0 intr queue: empty
         ksudlp FALSE at location: 0
           (post info) last post received: 0 0 35
                       last post received-location: ksr2.h LINE:603 ID:ksrpublish
                       last process to post me: 9911283b0 2 6
                       last post sent: 0 0 26
                       last post sent-location: ksa2.h LINE:282 ID:ksasnd
                       last process posted by me: 9911283b0 2 6
           (latch info) wait_event=0 bits=26
             holding    (efd=7) 993cfec60 Child object queue header operation level=5 child#=117
                 Location from where latch is held: kcbo.h LINE:884 ID:kcbo_link_q:
                 Context saved from call: 0
                 state=busy(exclusive) [value=0x200000000000001f, holder orapid=31] wlstate=free
[value=0]
                 waiters [orapid (seconds since: put on list, posted, alive check)]:
```

14.6 案例一:意料之外的 RAC 宕机祸首——子游标

```
              36 (82, 1435326627, 1)           21 (81, 1435326627, 1)
           waiter count=2
      holding    (efd=7) 9b5a5d630 Child cache buffers lru chain level=2 child#=233
           Location from where latch is held: kcb2.h LINE:3601 ID:kcbzgws:
           Context saved from call: 0
           state=busy [holder orapid=31] wlstate=free [value=0]
      holding    (efd=7) 9c2a99938 Child cache buffers chains level=1 child#=27857
           Location from where latch is held: kcb2.h LINE:3214 ID:kcbgtcr: fast path (cr pin):
           Context saved from call: 12583184
           state=busy(exclusive) [value=0x200000000000001f, holder orapid=31] wlstate=free
[value=0]
      Process Group: DEFAULT, pseudo proc: 0x9b11ca008
      O/S info: user: oracle, term: UNKNOWN, ospid: 29967
      OSD pid info: Unix process pid: 29967, image: oracle@ebtadbsvr2 (MMON)
```

从上面的 trace 可以看到,之前一直被等待的 993cfec60 资源(Child object queue header operation)正被 MMON 进程持有。

21:50:29 的时候 LCK0 在等待 MMON 释放资源,而此时 MMON 出现异常,因此在 21:51:06 MMON 进程被 LMHB 强制重启。然后在重启后,由于 MMON 的异常,导致 21:54:18 该资源仍无法被 LCK0 进程正常持有,最终导致 21:54:20 LMHB 进程强制重启了整个实例。

因此,最终的罪魁祸首是 MMON 进程。

```
Fri Jun 26 21:50:26 2015
 LCK0 (ospid: 29987) waits for latch 'object queue header operation' for 77 secs.
 Errors in file /u01/app/oracle/diag/rdbms/orcl/orcl2/trace/orcl2_lmhb_29939.trc
(incident=204141):
ORA-29771: process MMON (OSID 29967) blocks LCK0 (OSID 29987) for more than 70 seconds
 Incident details in:
/u01/app/oracle/diag/rdbms/orcl/orcl2/incident/incdir_204141/ orcl2_lmhb_29939_i204141.trc
 MMON (ospid: 29967) is blocking LCK0 (ospid: 29987) in a wait
 LMHB (ospid: 29939) kills MMON (ospid: 29967).
 Please check LMHB trace file for more detail.
 Fri Jun 26 21:51:06 2015
 Restarting dead background process MMON
 Fri Jun 26 21:51:06 2015
 MMON started with pid=213, OS id=16612
```

MMON 进程 Oracle 是用于进行 AWR 信息收集的。既然案情发生的原因与它有关,那么接下来的分析自然是查看它的相关 trace 了。

检查 MMON 的相关 trace 可以看到,MMON 进程负责处理对象的统计信息。从 trace 中可以看到大量的游标包含了太多的子游标。

```
User=b1456358 Session=c146d760 ReferenceCount=1 Flags=CNB/[0001] SavepointNum=558d5760
 LibraryHandle:   Address=2f79eb08  Hash=3dec6f4a LockMode=N PinMode=0 LoadLockMode=0
Status=VALD
   ObjectName: Name=  select time_mp, scn, num_mappings, tim_scn_map from smon_scn_time
where scn =   (select max(scn) from smon_scn_time where scn <= :1)
   FullHashValue=c36d5a579fdc3e19733192893dec6f4a Namespace=SQL AREA(00) Type=CURSOR(00)
Identifier=1038905162 OwnerIdn=0
```

```
        Statistics:      InvalidationCount=0  ExecutionCount=23741   LoadCount=107  ActiveLocks=1
TotalLockCount=6093 TotalPinCount=1
        Counters:        BrokenCount=1   RevocablePointer=1    KeepDependency=106   KeepHandle=106
BucketInUse=6092 HandleInUse=6092
        Concurrency:     DependencyMutex=2f79ebb8(0, 0, 0, 0) Mutex=2f79ec30(0, 87578, 0, 0)
        Flags=RON/PIN/TIM/PN0/DBN/[10012841]
        WaitersLists:
        Lock=2f79eb98[2f79eb98,2f79eb98]
        Pin=2f79eba8[2f79eb78,2f79eb78]
        Timestamp:  Current=04-05-2015 09:48:42
        LibraryObject:  Address=dff3bc60 HeapMask=0000-0001-0001 Flags=EXS[0000] Flags2=[0000]
PublicFlags=[0000]
        ChildTable:  size='112'
            Child:  id='0'  Table=dff3cb60 Reference=dff3c5b0 Handle=2f79e908
            Child:  id='1'  Table=dff3cb60 Reference=dff3c8e0 Handle=2f4e2d90
            Child:  id='2'  Table=dff3cb60 Reference=df3e7400 Handle=2f8c9e90
            Child:  id='3'  Table=dff3cb60 Reference=df3e76e8 Handle=2f8abce8
            ......
            Child:  id='101' Table=dc86f748 Reference=df02d368 Handle=288e6460
            Child:  id='102' Table=dc86f748 Reference=dd65c3e0 Handle=274d0b40
            Child:  id='103' Table=dc86f748 Reference=dd65c6c8 Handle=29aa92f8
            Child:  id='104' Table=dc86f748 Reference=dd65c9b8 Handle=26f3a460
            Child:  id='105' Table=dc86f748 Reference=dd65ccd0 Handle=25c02dd8
        NamespaceDump:
        Parent Cursor:   SQL_id=76cckj4yysvua parent=0x8dff3bd48 maxchild=106 plk=y ppn=n
        Current Cursor Sharing Diagnostics Nodes:
        ......
        ......
        Child Node: 100  ID=34 reason=Rolling Invalidate Window Exceeded(2) size=0x0
        already processed:
        Child Node: 101  ID=34 reason=Rolling Invalidate Window Exceeded(2) size=0x0
        already processed:
```

类似上面的信息非常多。很明显，上述父游标（parent cursor）包含了大量的子游标，这是为什么在 20 点～21 点（节点 2 还未重启前的时段）的 awr 报告中出现大量 cursor: mutex S 的原因，表 14-2 是这个时段的等待事件。

表 14-2

Event	Waits	Time(s)	Avg wait (ms)	% DB time	Wait Class
DB CPU		47,072		93.05	
cursor: mutex S	31,751,317	18,253	1	36.08	Concurrency
gc cr multi block request	359,897	1,281	4	2.53	Cluster
gc buffer busy acquire	10,465	686	66	1.36	Cluster
library cache lock	9,285	550	59	1.09	Concurrency

在 MMON 正常通过内部 SQL 收集系统信息时，根本不应该出现这种情况，而此时 MMON

进程却出现异常，这个异常看来应该是与 cursor 子游标过多有关了。

最后数据库实例被强行终止的原因是 LCK0 进程出现异常导致 LMHB 进程强行终止 instance，在最后 instance 终止之前的 diag dump 中，LCK0 进程的状态仍然是 hang 的状态，同时也阻塞了 3 个其他 session，如下所示。

```
SO: 0x9914dbce8, type: 4, owner: 0x9911283b0, flag: INIT-/-/0x00 if: 0x3 c: 0x3
proc=0x9911283b0, name=session, file=ksu.h LINE:11467, pg=0
 (session) sid: 1729 ser: 3 trans: (nil), creator: 0x9911283b0
flags: (0x51) USR/- flags_idl: (0x1) BSY/-/-/-/-/-
flags2: (0x408) -/- DID: , short-term DID: txn branch: (nil)
oct: 0, prv: 0, SQL: (nil), pSQL: (nil), user: 0/SYS
ksuxds FALSE at location: 0
service name: SYS$BACKGROUND
Current Wait Stack:
Not in wait; last wait ended 1 min 39 sec ago
There are 3 sessions blocked by this session.
Dumping one waiter:
inst: 2, sid: 1009, ser: 1
wait event: 'latch: object queue header operation'
p1: 'address'=0x993cfec60      p2: 'number'=0xa2      p3: 'tries'=0x0
row_wait_obj#: 4294967295, block#: 0, row#: 0, file# 0
min_blocked_time: 81 secs, waiter_cache_ver: 14285
Wait State:
fixed_waits=0 flags=0x20 boundary=(nil)/-1
```

对于数据库进程，如果状态不是 dead，而是 busy，而且持有 latch 不释放，那么这就意味着该进程已被挂起，LCK0 持有的 latch 是 object queue header operation。oracle mos 文档关于该 event 的描述如下：Scans of the object queue in the buffer cache can hold the "object queue header operation"。

14.6.3 对症下药，排除数据故障

基于上述的分析，我们最终判断，LCK0 进程出现问题的原因与游标子游标过多有关。同时，这又与在 11.2.0.1 版本上的 child cursor 总数阈值限制过高有关。实际在版本 10g 中就引入了该 Cursor Obsolescence 游标废弃特性，10g 的 child cursor 的总数阈值是 1024，即子游标超过 1024 即被过期，但是这个阈值在 11g 的初期版本中被移除了。这就导致出现一个父游标下大量子游标即 high version count 的发生，由此引发了一系列的版本 11.2.0.3 之前的 cursor sharing 性能问题。这意味着版本 11.2.0.1 和 11.2.0.2 的数据库，将可能出现大量的 Cursor: Mutex S 和 library cache lock 等待事件这种症状，进而诱发其他故障的发生。

因此，通常我们建议 11.2.0.4 以下的版本应**尽快将数据库版本升级到** 11.2.0.4（11.2.0.3 中默认就有_cursor_obsolete_threshold 了，而且默认值为 100），或者通过_cursor_features_enabled 和 106001 event 来强制控制子游标过多的情况。

14.6.4 深入总结，一次故障长久经验

该案例的分析还是有些曲折，因为常见导致单节点故障最常见的情况主要是心跳、节点资源之类，而该案例的诱发原因相对有些诡异。先是出现了 ora-600 错误，然后又报了 kill mmon 的信息，这都让案情分析有些扑朔迷离，当然，最终我们还是找出了问题的主要根源。

通过该起案件我们可以得到如下几点体会。

- 数据库版本的选择绝对是影响系统稳定性的关键要点。
- 不要以为性能问题只是影响用户体验，有些性能问题是会诱发系统一系列问题的。
- 问题的分析不要想当然，通过 trace 逐步递进，结合原理与经验，才能更为准确的确定问题。
- 子游标过多千万不能小视，最好能找出根源并解决它。如果确实不好解决，那么可通过设置隐含参数 _cursor_features_enabled 和 106001 event 强制失效子游标的方式来防止子游标过多的情况，操作如下所示。

```
SQL> alter system set "_cursor_features_enabled"=300 scope=spfile;
SQL> alter system set event='106001 trace name context forever,level 1024' scope=spfile;
```

正常重启数据库即可。

14.7 案例二：异常诡异的 SQL 性能分析

2015 年 9 月末的一天，客户告知其核心数据库突然发生了一个诡异现象，甚至导致业务系统该功能无法正常处理。经过简单询问，发现仅仅是一条 SQL 导致的，而很诡异的是，这条 SQL 在第二次执行时，执行计划会发生了变化，导致执行效率极低，影响业务运行。根据客户的陈述，该问题可随时重现，无论换个会话还是换个客户端工具都会受到影响。即使把共享池 flush 掉，再次执行 SQL 仍然会发生同样的现象。

14.7.1 信息收集

下面我们就来看看案情现场重现。

一条 SQL 在同一个会话中执行两次，第一次执行时间为 10 秒。但第二次执行时效率很低，执行时间超过 1 分钟。下面是 SQL 文本。

```
SELECT /*bbbbb*/A.C_DOC_NO AS C_PLY_APP_NO, A.C_PLY_NO AS C_PLY_NO, B.N_PRM AS N_PRM,
  NVL(TO_CHAR(A.T_APP_TM, 'YYYY-MM-DD HH24:MI:SS'), CHR(0)) AS T_APP_TM,
       A.C_BLG_DPT_CDE AS C_DPT_CDE, A.C_PROD_NO AS C_PROD_NO,
  NVL(B.C_APP_NME, CHR(0)) AS C_APP_NME, NVL(B.C_APP_TEL, CHR(0)) AS C_APP_TEL
    FROM T_PLY_UNDRMSG A, T_PLY_BASE B, T_FIN_PLYEDR_COLDUE C
```

14.7 案例二：异常诡异的 SQL 性能分析

```
WHERE 1 = 1
  AND ROWNUM < 1000
  AND A.C_DOC_NO = B.C_PLY_APP_NO AND A.C_DOC_NO = C.C_PLY_APP_NO(+)
  AND A.C_SOURCE = '1'
  AND A.C_SEND_MRK NOT IN ('2')
  AND DECODE(TRIM(C.C_OPT_NO), CHR(0), NULL, TRIM(C.C_OPT_NO)) IS NULL
  AND (NVL(C.C_PRM_TYP, CHR(0)) IN (CHR(0), 'R1'))
  AND (NVL(C.C_ARP_FLAG, CHR(0)) IN (CHR(0), '0', '3', '4')) AND (NVL(C.N_TMS, 0) IN (0, 1))
  AND B.C_HAND_PER = '1012337'
  AND A.T_APP_TM BETWEEN
      TO_DATE('2015-09-29 00:00:00', 'YYYY-MM-DD HH24:MI:SS') AND
      TO_DATE('2015-09-30 23:59:59', 'YYYY-MM-DD HH24:MI:SS')
  AND A.T_INPUT_TM BETWEEN
      TO_DATE('2015-09-29 00:00:00', 'YYYY-MM-DD HH24:MI:SS') AND
      TO_DATE('2015-09-30 23:59:59', 'YYYY-MM-DD HH24:MI:SS')
```

观察该 SQL 执行计划信息，发现第二次执行计划发生了变化。其中 T_PLY_BASE 表的索引扫描变成了分区表扫描，而且驱动表和被驱动表也发生了改变，第二次执行计划中的 COST 也是在这里出现了明显增高。

第一次执行计划如图 14-1 所示。

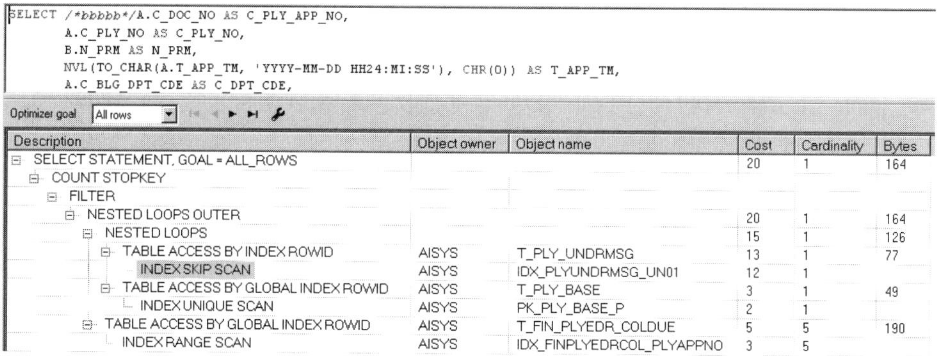

图 14-1

第二次执行计划如下所示。

```
call     count       cpu    elapsed       disk      query    current        rows
------- ------  -------- ---------- ---------- ---------- ----------  ----------
Parse        1      0.02       0.04          0          0          0           0
Execute      1      0.00       0.00          0          0          0           0
Fetch        1     57.00     626.79    1043187    1816768          0           0
------- ------  -------- ---------- ---------- ---------- ----------  ----------
total        3     57.02     626.83    1043187    1816768          0           0

Misses in library cache during parse: 1
Optimizer mode: ALL_ROWS
Parsing user id: 413
Number of plan statistics captured: 1
```

```
Rows (1st) Rows (avg) Rows (max)  Row Source Operation
---------- ---------- ----------  ---------------------------------------------------
         0          0          0  COUNT STOPKEY (cr=0 pr=0 pw=0 time=12 us)
         0          0          0   FILTER  (cr=0 pr=0 pw=0 time=9 us)
         0          0          0    NESTED LOOPS OUTER (cr=0 pr=0 pw=0 time=9 us cost=78029
size=164328 card=1002)
         0          0          0     NESTED LOOPS  (cr=0 pr=0 pw=0 time=8 us cost=77047
size=27468 card=218)
         0          0          0      PARTITION RANGE ALL PARTITION: 1 20 (cr=0 pr=0 pw=0 time=7
us cost=76262 size=19747 card=403)
         0          0          0       PARTITION LIST ALL PARTITION: 1 7 (cr=1476837 pr=852917
pw=0 time=486948524 us cost=76262 size=19747 card=403)
         0          0          0        TABLE ACCESS FULL T_PLY_BASE PARTITION: 1 140
(cr=1786130 pr=1015408 pw=0 time=614908158 us cost=76262 size=19747 card=403)
         0          0          0      TABLE ACCESS BY INDEX ROWID T_PLY_UNDRMSG (cr=0 pr=0 pw=0
time=0 us cost=4 size=77 card=1)
         0          0          0       INDEX RANGE SCAN PK_PLY_UNDRMSG_HIST_20131203 (cr=0
pr=0 pw=0 time=0 us cost=3 size=0 card=1)(object id 406120)
         0          0          0     TABLE ACCESS BY GLOBAL INDEX ROWID T_FIN_PLYEDR_COLDUE
PARTITION: ROW LOCATION ROW LOCATION (cr=0 pr=0 pw=0 time=0 us cost=5 size=190 card=5)
         0          0          0      INDEX RANGE SCAN IDX_FINPLYEDRCOL_PLYAPPNO (cr=0
pw=0 time=0 us cost=3 size=0 card=5)(object id 180111)
```

很明显，的确是第二个执行计划出现了问题，导致了性能的严重下降。但是问题是，为什么同一个 SQL 第二次执行时执行计划会变呢？甚至同一个 SQL 连续两次执行也是如此？

理论上同一个会话上执行的同一个 SQL，第二次执行为软解析（或者软软解析），此时数据库应该重用执行计划，而不是产生新的执行计划。

14.7.2　新特性分析

在 11g 上出现了一些新特性，而其中一个典型会导致 SQL 执行计划发生改变的场景就是 ACS（自适应游标 adaptive_cursor_sharing）。但 ACS 典型出现的场景应该是使用了绑定变量的 SQL，但该 SQL 并未使用绑定，数据库中也并未通过 corsor_sharing 参数强制绑定，看起来应该不是 ACS，那么是什么原因呢？

在 11g 不但出现了 ACS 这样的自动优化新特性，还出现了另一个自动优化特性**基数反馈**（Cardinality Feedback）。而通过执行计划中的信息与基数反馈特性的对比，基本可以推断该问题是由 11g 新特性统计信息 Feedback 导致的 Bug，只需要关闭该特性再做验证即可确认。

参考 MOS Statistics（Cardinality）Feedback - Frequently Asked Questions（文档 ID 1344937.1），文档有对 11GR2 Statistics Feedback 新功能引起执行计划变化的描述、如何确认及解决方法。

本次问题就是典型 11g 新特性——统计信息 Feedback 导致的 Bug，这样的问题相对比较常见，我们一般推荐关闭自适应游标共享和统计信息回馈（事实上我们已经总结了不少应该关闭或者调整的新特性），通过两个参数就可以动态关闭它们。在我们很多其他客户的核心库中均已进行过设置，不会对系统造成损害，建议大多数 11g 核心系统最好关闭。

在执行了以下处理后，SQL 执行不但恢复正常，而且运行效率进一步得到了提高。

1. 优化参数

关闭自适应游标共享和统计信息回馈 11g 新特性。

```
alter system set "_optimizer_use_feedback" = false scope = both;
alter system set "_optimizer_adaptive_cursor_sharing" = false scope = both;
```

2. 优化索引

同时也对这条 SQL 的执行计划相关索引进行了优化。建议在表 T_UND_RMSG 的 T_INPUT_TM 列上的创建单列索引，这样就避免了出现跳扫的执行计划（同时还可将该 SQL 执行计划强制失效），或者通过 comment 命令使相关 SQL 强制重新解析（注意，这两种方法都会将该表的所有 SQL 执行计划全部过期失效，代价较高）。

```
Create index xxxx on T_UND_RMSG(T_INPUT_TM);
Comment on table T_UND_RMSG is 'xxxx';
```

当然，在 11g 上有一个更为推荐的方法——DBMS_SHARED_POOL.PURGE，这种方法将只失效特定执行计划异常的子游标，下面给出参考样例。

```
select address, hash_value from v$SQLarea where SQL_id = 'a6aqkm30u7p90';
ADDRESS            HASH_VALUE
----------------   ----------
C000000EB7ED3420   3248739616
exec dbms_shared_pool.purge('C000000EB7ED3420,3248739616','C');
```

14.8　总结

以上两则案例从表面上看都是有些诡异的，但作为一个专业的 DBA，一定要耐心细致，通过表象的层层分析发现故障的本质。除此而外，视而不见是运维的癌症，运维无小事，任何异常的现象，哪怕再小，都是值得深入分析和探究的。希望以上分享的案例对大家有所帮助。

参考文献

[1] 盖国强.《深入浅出——Oracle DBA 入门、进阶与诊断案例》[M].1 版.北京：人民邮电出版社，2006.

[2] IDC．刀片服务器推动企业基础架构走向新 IT 时代[R/OL].[2016.06]. http://server.it168.com/a2016/0629/2750/000002750902.shtml.

[3] Oracle.Oracle Database SQL language reference[EB/OL]. Oracle Database Online Documentation 12c Release 1 （12.1）. http://docs.oracle.com/database/121/SQLRF/toc.htm.

[4] Error Messages. [EB/OL]. Oracle Database Online Documentation 12c Release 1 （12.1）. http://docs.oracle.com/database/121/ERRMG/toc.htm.

[5] Tom Kyte. print_table.SQL. [EB/OL]. Ask Tom.com.

[6] Bug:9439759 文档 ID 9439759.8. [EB/OL]. MOS 文档. https://support.oracle.com/epmos/faces/DocumentDisplay?_afrLoop=361759285254672&id=9439759.8&_afrWindowMode=0&_adf.ctrl-state=7ldzouks1_78.

[7] Bug:18799993 文档 ID 18799993.8. [EB/OL]. MOS 文档. https://support.oracle.com/epmos/faces/DocumentDisplay?_afrLoop=361796458496535&id=18799993.8&_adf.ctrl-state=7ldzouks1_127.

[8] Bug: 6859515 文档 ID 8929701.8. [EB/OL]. MOS 文档. https://support.oracle.com/epmos/faces/DocumentDisplay?_afrLoop=361836056201378&id=8929701.8&_adf.ctrl-state=7ldzouks1_184.

[9] Bug: 6442431 文档 ID 6442431.8. [EB/OL]. MOS 文档. https://support.oracle.com/epmos/faces/DocumentDisplay?_afrLoop=361862338213762&id=6442431.8&_adf.ctrl-state=7ldzouks1_241.

[10] 文档 ID 1389592.1. [EB/OL]. MOS 文档. https://support.oracle.com/epmos/faces/DocumentDisplay?_afrLoop=361886503490150&id=1389592.1&_adf.ctrl-state=7ldzouks1_298.

欢迎来到异步社区！

异步社区的来历

异步社区（www.epubit.com.cn）是人民邮电出版社旗下IT专业图书旗舰社区，于2015年8月上线运营。

异步社区依托于人民邮电出版社20余年的IT专业优质出版资源和编辑策划团队，打造传统出版与电子出版和自出版结合、纸质书与电子书结合、传统印刷与POD按需印刷结合的出版平台，提供最新技术资讯，为作者和读者打造交流互动的平台。

社区里都有什么？

购买图书

我们出版的图书涵盖主流IT技术，在编程语言、Web技术、数据科学等领域有众多经典畅销图书。社区现已上线图书1000余种，电子书400多种，部分新书实现纸书、电子书同步出版。我们还会定期发布新书书讯。

下载资源

社区内提供随书附赠的资源，如书中的案例或程序源代码。
另外，社区还提供了大量的免费电子书，只要注册成为社区用户就可以免费下载。

与作译者互动

很多图书的作译者已经入驻社区，您可以关注他们、咨询技术问题；可以阅读不断更新的技术文章，听作译者和编辑畅聊好书背后有趣的故事；还可以参与社区的作者访谈栏目，向您关注的作者提出采访题目。

灵活优惠的购书

您可以方便地下单购买纸质图书或电子图书，纸质图书直接从人民邮电出版社书库发货，电子书提供多种阅读格式。

对于重磅新书，社区提供预售和新书首发服务，用户可以第一时间买到心仪的新书。

用户帐户中的积分可以用于购书优惠。100积分=1元，购买图书时，在 里填入可使用的积分数值，即可扣减相应金额。

特 别 优 惠

购买本书的读者专享**异步社区购书优惠券**。

使用方法：注册成为社区用户，在下单购书时输入 S4XC5 使用优惠码 ，然后点击"使用优惠码"，即可在原折扣基础上享受全单9折优惠。（订单满39元即可使用，本优惠券只可使用一次）

纸电图书组合购买

社区独家提供纸质图书和电子书组合购买方式，价格优惠，一次购买，多种阅读选择。

社区里还可以做什么？

提交勘误

您可以在图书页面下方提交勘误，每条勘误被确认后可以获得100积分。热心勘误的读者还有机会参与书稿的审校和翻译工作。

写作

社区提供基于 Markdown 的写作环境，喜欢写作的您可以在此一试身手，在社区里分享您的技术心得和读书体会，更可以体验自出版的乐趣，轻松实现出版的梦想。

如果成为社区认证作译者，还可以享受异步社区提供的作者专享特色服务。

会议活动早知道

您可以掌握 IT 圈的技术会议资讯，更有机会免费获赠大会门票。

加入异步

扫描任意二维码都能找到我们：

异步社区	微信服务号	微信订阅号	官方微博	QQ 群：368449889

社区网址：www.epubit.com.cn

投稿 & 咨询：contact@epubit.com.cn